ABHANDLUNGEN DER DEUTSCHEN AKADEMIE
DER WISSENSCHAFTEN ZU BERLIN

Klasse für Medizin

Jahrgang 1962 Nr. 1

STRAHLENINDUZIERTE MUTAGENESE (RADIATION INDUCED MUTAGENESIS)

ERWIN-BAUR-GEDÄCHTNISVORLESUNGEN II 1961

Veranstaltet vom Institut für Kulturpflanzenforschung Gatersleben der Deutschen Akademie der Wissenschaften zu Berlin

6. bis 10. Juni 1961

Herausgegeben von

H. STUBBE

Redaktion

H. BÖHME

Mit 91 Abbildungen und 29 Tabellen

AKADEMIE-VERLAG · BERLIN

1962

Vorgelegt von Hrn. STUBBE in der Gesamtsitzung vom 14. September 1961
Zum Druck genehmigt am gleichen Tage, ausgegeben am 14. 7. 62

Erschienen im Akademie-Verlag GmbH, Berlin W 8, Leipziger Straße 3-4

Lizenznummer: 202 · 100/427/62
Gesamtherstellung: IV/2/14 · VEB Werkdruck Gräfenhainichen · 1737
Bestellnummer: 2001/62/IV/1 · ES 18 G 1 · Preis: DM 45,–

INHALTSVERZEICHNIS

Einleitende Worte

Vor zwei Jahren haben wir die ersten Erwin-Baur-Gedächtnisvorlesungen mit dem Thema „Chemische Mutagenese" durchgeführt. Die Problematik der zweiten Erwin-Baur-Gedächtnisvorlesungen kreist um das Thema „Strahleninduzierte Mutagenese".

Es sind nun schon 34 Jahre vergangen, seit H. J. Muller auf dem V. Internationalen Kongreß für Genetik in Berlin im Jahre 1927 über seine ersten *Drosophila*-Versuche, durch Einwirkung kurzwelliger Strahlen Mutationen auszulösen, berichtete. Damals setzte eine Flut von Untersuchungen in vielen Genetischen Instituten der Welt ein, um die Ergebnisse Mullers zu bestätigen. Aber bald darauf wurden ernsthafte Ansätze gemacht mit Hilfe der „experimentellen Untersuchungen des Mutationsprozesses und einer physikalischen Analyse der Versuchsergebnisse, eine allgemeine Vorstellung über die Natur des Gens und der Mutation zu bilden". Sie alle kennen die grundlegende Arbeit von Timoféeff-Ressovsky, Zimmer und Delbrück (1935), in der versucht wurde, aus Ereignissen, die sich unmittelbar am Gen abspielen, Schlüsse auf die Struktur der Gene und auf die Natur des Mutationsprozesses zu ziehen. Das Ergebnis dieser Gemeinschaftsarbeit waren auf der Grundlage der Treffertheorie entwickelte, ganz allgemeine Vorstellungen, daß die Gene bestimmte stabile, wohl definierte Moleküle seien und daß eine bestimmte Mutation eine bestimmte Umlagerung in einem bestimmten Molekül bedeute. Diese Umlagerung erfolgt in einem Elementarprozeß ohne weitere Sekundärprozesse, also als Folge einer Ionisation oder Anregung, die als „Treffer" anzusehen ist. Die Mutation wird durch Energiezufuhr von außen oder durch Schwankungen der Temperaturenergie bewirkt.

Diese Vorstellungen, die sicherlich immer als ein bedeutender Schritt in der Geschichte der modernen Genetik angesehen werden müssen, haben also eine rein physikalische Basis und schienen lange Zeit richtungweisend bei allen Überlegungen zum Problem der strahleninduzierten Mutagenese zu sein.

Die intensive Weiterarbeit auf diesem Gebiet hat nun jedoch Schritt für Schritt gezeigt, daß neben der rein physikalischen, ionisierenden Wirkung kurzwelliger Strahlen, die die Veränderung des genetischen Materials bewirken, noch eine ganze Anzahl chemischer Prozesse und Reaktionsprodukte in den Ablauf des Mutationsprozesses eingreifen und diesen in mannigfacher Weise beeinflussen. Dabei sind auch das Alter und der physiologische Zustand der bestrahlten Zelle für den Effekt von besonderer Bedeutung.

Diese Tatsache hat bestimmte Konsequenzen. Während sich das Thema unserer ersten Erwin-Baur-Gedächtnisvorlesung „Chemische Mutagenese" zunächst von dem Thema unserer zweiten Gedächtnisvorlesung „Strahleninduzierte Mutagenese" formal deutlich abzugrenzen schien und sicher auch klare Unterschiede bei der Betrachtung des Ablaufs des Mutationsprozesses bestehen, ergibt sich nun doch eine Anzahl bedeutender Gemeinsamkeiten.

Wenn wir versuchen, heute ein Bild vom Ablauf des strahleninduzierten Mutationsprozesses zu entwerfen, entdecken wir Lücken, die geschlossen werden müssen. Auf der einen Seite kennen wir die physikalische Natur des auslösenden Agens recht genau und wissen andererseits, daß seine Wirkung zu einer zunächst dosisabhängigen Anzahl mutierter Individuen führt. Unbekannt aber ist die Kette von chemisch-physikalischen Ereignissen und Reaktionen, die im Augenblick der Bestrahlung oder kurz danach am Ort der Entstehung der molekularen Veränderungen vor sich gehen und den biologischen Endeffekt bedingen und beeinflussen. Wir wünschen also Klarheit über die Art und Form der ersten im biologischen Material ablaufenden Reaktionen und über die durch diese eingeleiteten chemischen Prozesse. Eine Kenntnis der Wirkungsweise der den strahleninduzierten Mutationsprozeß modifizierenden Faktoren würde für die Erkenntnis des Gesamtprozesses und seine Zergliederung in einzelne Phasen von großer Bedeutung sein. Ich nehme an, daß diese Probleme im Mittelpunkt unserer Konferenz stehen werden.

Unsere Bemühungen, die feineren Vorgänge bei der strahleninduzierten Mutagenese zu verstehen, werden voraussichtlich zu immer besseren Vorstellungen über die Natur des genetischen Materials führen. Das ist eines der großen theoretischen Probleme der Genetik, dessen Lösung all denen, die sich mit den Fragen der molekularen Genetik befassen, ein ganz besonderes Anliegen ist.

In zweierlei Hinsicht wird unser Verlangen, die Prozesse der strahleninduzierten Mutagenese genau zu kennen, auch durch Fragen angewandter Wissenschaftszweige gefördert. Die Kenntnis dieser Prozesse ist einmal die notwendige Voraussetzung, um den stetig wachsenden Gefahren für das Erbgut des Menschen zu begegnen, die durch die immer steigende Anwendung energiereicher Strahlen in Industrie und Technik gegeben sind. Nur auf der Grundlage genauer Kenntnisse der molekularen Vorgänge werden sich Strahlenschutzstoffe entwickeln und rationell anwenden lassen. Zum anderen ist die ständige Vertiefung unserer Erkenntnisse auf diesem Gebiet unerläßlich für den Züchtungsforscher, der bemüht ist, das Spektrum der ausgelösten Effekte zu beeinflussen und seinen Absichten nutzbar zu machen.

Wie weit die Forschung in vielen Ländern vorangeschritten ist und wo die entscheidenden Lücken bestehen, soll unsere Konferenz zeigen. Wir sind dankbar dafür, daß wir in einem Internationalen Gremium diese brennenden Fragen hier in Gatersleben besprechen können. Vertreter von 13 Nationen haben sich bei uns versammelt, darunter manche auf dem Gebiet der strahleninduzierten Mutagenese führenden Forscher, darunter ausgezeichnete Vertreter benachbarter Disziplinen, ohne deren intensive Mitarbeit der Genetiker allein bald verloren wäre, weil in der modernen Forschung immer wieder die Grenzen seines engeren Fachgebietes überschritten werden müssen, wenn er zu neuen Erkenntnissen gelangen will.

Über die Anwesenheit jedes einzelnen von Ihnen sind wir glücklich und hoffen, daß die vor uns liegenden Tage reichen Anlaß bieten werden zu fruchtbaren wissenschaftlichen Gesprächen, ebenso aber zur Erneuerung oder zur erstmaligen Anknüpfung persönlicher kollegialer und freundschaftlicher Beziehungen, die uns im Bereich der Wissenschaft über alle Grenzen der Länder hinaus miteinander verbinden. Im Auftrage und im Namen des Herrn Präsidenten der Deutschen Akademie der Wissenschaften zu Berlin heiße ich Sie alle auf das herzlichste willkommen.

H. Stubbe

Radiation Induced Mutagenesis – A Biophysicist's Viewpoint

L. H. Gray, F. R. S.

British Empire Cancer Campain Research Unit in Radiobiology,
Mount Vernon Hospital, Northwood, Middlesex

Damage to the genome can result from the exposure of cells to all forms of ionizing radiation to ultraviolet light, or even to visible light, provided the cell has been sensitised by pretreatment with a suitable dye.

When a cell is exposed to ionizing radiation, energy in the form of ionisation and excitation is transferred to biological structures at random along the tracks of ionizing particles. The number of tracks will depend on the dose and on the type of radiation employed. In the case of α-radiation the density of ionisation along the track is particularly heavy and the number of particles is correspondingly small. 3 rad to a nucleus 10μ in diameter will generally be delivered as 300000 eV of energy capable of forming some 10000 ions along the track of a single α-particle.

Very soft X-rays generate electrons having initially an energy of about 1500 eV, so that about 1000 particles are required to contribute a dose of 25 rad to a 10 μ diameter nucleus.

The lower limit to the rate of deposition of energy along the track, viz. that for very fast electrons, is at the rate of only about 200 eV/μ and the radiation is better thought of not in terms of "tracks" but of discreet centres of energy deposition. In some of these centres of energy deposition 100 eV or more may be deposited in a region whose dimensions are measured in terms of tens of Angstroms, and in others only a single molecular excitation may take place.

Exposure to high voltage X-or γ-radiation unfortunately gives rise to a very complex type of energy distribution within the cell, since all rates of dissipation of energy along the tracks, from that of the fastest electron to that of the very slow delta rays, will take place in different regions of the cell simultaneously. In all cases of exposure to an external source of ionising radiation with which we shall be concerned, the deposition of energy may be thought of as essentially independent of the chemical composition of the cellular components. Some degree of selectivity occurs, however, when cells are made to incorporate radioactive atoms, and a high degree of selective deposition of energy in DNA may be achieved under certain conditions with ^{32}P or ^{3}H thymidine.

Ultraviolet light is always selectively absorbed in accordance with the wavelength employed, and visible light of wavelength 480 mμ absorbed by acridine orange contributes excitation energy in amount $\sim$ 2.6 eV direct to the DNA of the cell.

The most pronounced effect of all these forms of energy absorption is to lead to loss of proliferating capacity – or reproductive integrity – of the irradiated cells. This is a major obstacle to the study of radiation-mutagenesis. It also complicates our task because, in many cases, a major factor contributing to loss of reproductive integrity is damage to the genome – damage moreover which may only differ in degree from that which may be expressed as a mutation in a cell which survives. I have in mind structural damage to the chromosomes.

At the first division after irradiation, chromosome structural damage is abundant. At subsequent divisions it becomes increasingly rare, showing that most forms are cell-lethal, but it does not disappear completely either in plant meristems (DAVIDSON, 1960) or in clones of mammalian cells (MITZUTANI, OHNUKI, NAKANISHI, and POMERAT, 1961). Patterns of abnomality persist through many cell generations. In the leucocytes of irradiated persons, chromosome structural damage has been seen two and a half years after exposure (BENDER and GOUCH, 1961). Moreover, we now know from a recent report of BAIKIE, COURT BROWN, BUCKTON, HARNDEN, JACOBS, and TOUGH (1960) that a particular chromosomal defect of a kind which might readily be produced by irradiation can be associated with a particular disease. I refer to the report of these authors on twelve cases of chronic myeloid leukaemia.

A proportion of the leucocytes seen in division after short term culture of the peripheral blood from each patient showed what appeared to be an identical chromosomal defect affecting either chromosome 21 or 22 of the Denver classification. This defect was not present in all dividing leucocytes in anyone patient, and was never seen in cultures of skin cells from the same patients. Some, but not all, of these patients had previously received a therapeutic dose of radiation for some condition other than leukaemia.

Clearly this chromsome abormality, whether radiation induced or not, has the characteristics of a somatic mutation.

Many mutations in *Drosophila* have long been known to be associated with structural changes in the chromosomes which can be identified as changes in the band pattern of the salivary gland chromosomes. In the last analysis, if genetic information is coded as nucleotide sequences, all mutations are either deletions or changes in the structure, and not in the substance, of the genetic material. The size of the re-arrangement, however, spans a very wide range, from a few nucleotides to something visible in the light microscope.

I propose to divide my remarks into two parts, the first concerned with forms of radiation induced chromosome structural damage, most of which are probably non-viable, and the second with radiation induced mutations not as yet associated with visible structural damage of the genome.

The Induction of Chromosome Structural Damage

1. Alpha Radiation

In *Tradescantia* microspores and *Vicia* meristem cells, all forms of structural damage induced by α-radiation, whether affecting one chromosome only or involving exchange of material between two chromosomes, are induced with a frequency which varies linearly with the dose (KOTVAL and GRAY, 1947, THODAY, 1951).[1] The formal inference is that each

[1] THODAY found that in *Vicia* "minutes" varied more rapidly than the first power of the dose (see also READ, 1959). This needs further investigation since there are other unusual features about these observations, as for example, the fact that the yield of "minutes" only increases rather slowly with time to reach a maximum at about 24 hours after the irradiation, by contrast with, say, isochromatid aberrations which are maximum at 9 hours. Micronuclei, seen in interphase cells, are at a maximum at about 36 hours, and it seems possible that there may have been some confusion of these with "minutes" seen in cells entering their second division after irradiation.

aberration is a result of energy deposited by a single α-particle – an inference which is manifestly valid in this case since at the lowest dose at which observations were made, the sum of isolocus and exchange aberrations alone is more than twice as numerous as the number of nuclei traversed by more than one α-particle. The target area presented for the initiation of chromosome aberrations by α-particles must be very great, since an aberration results on nearly 40% of the occasions on which a nucleus is traversed by an α-particle.

The aberrations produced by α-particles are very insensitive to changes in the chemical environment of the cell at the time of irradiation, and in particular, to changes in the concentration of dissolved oxygen, as was first demonstrated in *Vicia* by THODAY and READ (1949).

The same is, in general, true of chemical reactions induced by α-radiation. It has, however, recently been shown (TRUMBORE and HART, 1959) that chemical reactions induced by α-radiation are strongly influenced by oxygen at sufficiently high pressures (940 atm.), and it would be interesting to see whether the same is true of chromosome structural damage.

2. X- and gamma radiation

Chromosome structural damage induced by X- and γ-radiation differs outstandingly from that induced by α-radiation in two ways:

1. The relation between aberration frequency and dose may, in general, for any type of aberration, be expressed as the sum of two terms, one of which is linear with dose and the other proportional to the square of the dose

$$F = \alpha D + \beta D^2 \tag{1}$$

The constant β depends on the duration of irradiation.

Characteristically, the constant α does not depend on the duration of irradiation, which is consistent with the interpretation of the first power term as representing damage initiated by a single ionising particle, even when, as in exchange formation, two separate structures are involved.

2. The aberration frequency is quite sensitive to the chemical environment of the cell at the time of irradiation, and in particular to the presence or absence of dissolved oxygen.

Chemical reactions induced by X- and γ-radiation are also generally influenced by the presence of dissolved oxygen. The chemical yield may be enhanced or depressed, or the nature of the end products may be changed. In the case of polymers, including DNA, oxygen tends to enhance chain-breaking at the expense of cross-linking.

The effect of oxygen on chromosome structural damage is invariably to increase the effectiveness of the irradiation.

In the cases most carefully examined, the effect of oxygen during irradiation has been found to increase the effectiveness of a given dose in a fixed ratio f_{O_2}, irrespective of whether the effect under consideration is being initiated by one particle or by more than one particle. The factor f_{O_2} is related to the concentration of oxygen in the environment of the cell approximately in accordance with the ALPER and HOWARD-FLANDERS (1956) equation

$$f_{O_2} = 1 + (m - 1)\frac{[O_2]}{[O_2] + K} \tag{2}$$

where m is the value of f_{O_2} at oxygen concentrations large compared with K, and K is the concentration at which f_{O_2} is midway between its anoxic and its maximum aerobic value.

Thus, when allowance is made for the influence of oxygen, equation (1) becomes

$$F = \alpha\,(f_{O_2} D) + \beta\,(f_{O_2} D)^2\,. \tag{3}$$

This implies that the dose response curves at different oxygen concentrations are superimposable. We have data (HAWES, HOWARD, and GRAY; in course of publication) which provide an experimental justification for this statement in the case of the induction of anaphase bridges and fragments in Ehrlich ascites tumour cells.

In a number of ways (GRAY, 1959 and 1961) it is possible to distinguish this effect of oxygen, which follows instantly upon changes in oxygen concentration, from the effect of a period of anoxic metabolism (see below), and we can be reasonably sure that the former reflects some interaction of a physico-chemical nature between oxygen and unstable intermediates of short life formed along the tracks of the ionising particles. As Miss ALPER will tell us, from her work with micro-organisms, some forms of irradiation damage initiated in the presence of oxygen are differently affected by post-irradiation metabolism from those produced in the absence of oxygen, and we have to keep this in mind in relation to the induction of chromosome structural damage.

We must also bear in mind a rather obvious, and in a sense trivial, indirect effect of respiratory activity on the level of damage sustained by an irradiated cell, but one which has often been overlooked with resultant confusion and misinterpretation.

Tissue respiration inevitably sets up a gradient of oxygen tension such that the further a cell is away from its oxygen source, e.g. capillary network in a mammal, the lower will be its oxygen tension. For example, a cell which is separated from its blood supply by a 100μ of respiring tissue ($Q_{O_2} = 5 \mu l \; O_2$/mgm dry wt./hour), may be completely anoxic although the oxygen tension of the blood is maintained at 40 mm Hg (THOMLINSON and GRAY, 1955). A rise in temperature, which increases the Q_{O_2}, will make the gradient steeper, and a fall in temperature or treatment with a respiratory poison will decrease Q_{O_2} and decrease the gradient, and so render the cells aerobic and more sensitive to radiation, by a factor of ~ 3. If the altered respiration is of long duration, then the metabolic effects discussed later might also be effective. Such situations are of very frequent occurence. The results of OSTER (1959) seem to show that oxygen gradients within the *Drosophila* testis are important and influence the yield of mutations resulting from a given radiation exposure differently for the different stages of spermiogenesis, on account of the different positions which the cells occupy in relation to the main oxygen supply to the testis.

I am indebted to Dr. BEATTY for permission to report some recent observations, which very beautifully illustrate the effect of oxygen gradients of respiratory origin within the *Tradescantia* anthers. Aberrations were scored in sections instead of squashes, so as to preserve the structure of the tissue and were scored separately for the cells which were nearest to the surface, furthest from the surface, and in the intermediate zone. The gas atmosphere was changed only for a short while before and during irradiation. In the complete absence of oxygen, aberration frequencies were low, but the same in all three zones. In an oxygen atmosphere, the frequencies were much higher, but again were the same in all three zones, because the respiration was insufficient to reduce oxygen tension below that corresponding to maximum sensitivity. When the anthers were in 5% oxygen, however, or even in air, the existence of oxygen gradients is unequivocably shown by the observed gradient in aberration frequencies. It is important to note that the gradients occur, even in an atmosphere of air, since they will be influenced by temperature and by a change in Q_{O_2}. The observations to which I refer were made on aberrations induced by irradiation of early interphase anthers. The Q_{O_2} of an anther decreases very markedly during the four days of interphase, but the anther is also increasing in size during this time (HOWARD, GRAY and DOUGLAS; unpublished observations). These changes affect the oxygen gradient in opposite senses. Early observations by GILES (1957) on the average sensitivity of microspores (squash preparation) as a function of the oxygen content of the surrounding gas, suggest that oxygen gradients are in fact different at late interphase when chromatid aberrations are scored, from those at early interphase when chromosome aberrations are scored.

A further point turns out to be of some importance in a number of situations. Should the oxygen be temporarily consumed in any locality, as it is by the radiochemical reactions initiated by irradiation, it may take seconds or even minutes, for the oxygen to be replenished by diffusion from the main oxygen source. Consequently, radiation delivered at high dose rates may be received by cells which during part of the time were at a lower average oxygen tension than they would have been if irradiated at low dose rates. The effect will be most marked when high doses as well as high dose rates are used. The experiments of DEWEY and BOAG (1960), with bacteria, designed to investigate this phenomenon by means of a pulsed radiation source, show that 2000–3000 rad is sufficient to consume all the free oxygen in a bacillus initially at an oxygen tension of about 8 mm Hg (1% in the gas phase). There is evidence of the operation of this high dose rate radiochemically induced anoxia in microspores (KIRBY-SMITH and ROTBLAT, personal communication), in the skin (WRIGHT, B. E. C. C. Annual Report, 1960), and in the bone marrow (ROTBLAT and LINDOP, B. E. C. C. Annual Report, 1960), as well as in bacteria.

3. Neutron radiation

Neutron radiation is intermediate in physical properties between X- and γ-radiation on the one hand and α-radiation on the other. On account of the relatively high rate of loss of energy along the track of individual recoil protons it necessarily follows, of course, that there will be a shorter track length per cubic micron of tissue per unit dose, i. e. fewer particles contribute to a given dose. Correspondingly, the term in D^2, in equation (1), is relatively much less important, and in fact is rarely detectable at the doses usually employed. Within the limits of experimental error, aberrations of all types vary linearly with dose and are independent of dose rate (THODAY, 1942; GILES, 1954; CONGER, RANDOLPH, SHEPPARD, and LUIPPOLD, 1958). As regards the yield of aberrations per unit dose and a dependence of the yield on dissolved oxygen, neutrons occupy a position intermediate between X-radiation and α-radiation entirely in keeping with their physical properties. No new principle emerges from the small number of studies which have been made so far with pure neutrons, though a problem does arise out of studies with mixed neutron and X-ray beams (SHELDON WOLFF, ATWOOD, RANDOLPH, and LUIPPOLD, 1958).

4. The physical analysis of the induction of chromosome aberrations by ionizing radiation

LEA (1946) presented in its complete form the classical analysis which he, CATCHESIDE, and others had developed during the preceeding five years. A very large body of experimental data, mostly relating to the production of chromosome and chromatid aberrations in *Tradescantia* microspores, were shown to be capable of quantitative explanation in terms of a very simple model which had already been proposed by Professor KARL SAX, to whom we owe a great debt of gratitude for his pioneer quantitative studies.

The model postulated that the primary event is the production of a chromosome or chromatid break.

Subsidiary essential postulates were:

1. That broken ends are of two kinds – those which are fusable with another broken end, and those which are not. By various arguments but most directly from the proportion of

isochromatid breaks which did not undergo sister re-union, and from the number of incomplete interchanges, it was estimated that about 9% of broken ends are non-fusable.

2. That the remaining 91% of broken ends which are capable of fusion join with another broken end during the interval between irradiation and metaphase. In the majority of cases these reconstitute the original chromosomes, and no evidence of the breakage event is obtained. When, however, four broken ends from two broken chromosomes exist within about 1μ of one another, re-joining is esentially at random, and therefore a proportion of the breaks can give rise to an exchange type of aberration. All exchanges are assumed to arise in this way. When the dose rate is low, or the dose is fractionated a pair of breaks formed early in the irradiation may have reconstituted before the second pair is formed at a sufficiently close distance for exchange of partners to take place. Thus exchange formation (translocation) takes place in competition with restitution. Dose rate and fractionation studies with *Tradescantia* microspores could be accounted for if it were assumed that the majority of the breaks restitutes with a half-lifetime of about 4 minutes at room temperature, but the data were not altogether consistent as regards the magnitude of this parameter.
3. The total number of breaks primarily produced is computed as the sum of those which take part in the formation of isochromatid breaks and exchanges (and their counterparts which escape detection), together with 11 (i.e. 1/0.09) times the number of observed breaks. In practice, for both X-rays and neutrons, the latter term is at least 5 times as great as the former (Lea, 1946 – Table 64), so that the estimated primary breakage frequency is essentially determined by the estimated number of simple breaks.
4. In order to account for the relative yield of aberrations of the various types observed after exposure to gamma, X-, neutron, and alpha radiation, it was postulated that the chance that the transit of a fast electron through a chromatid thread leads to a break is very small indeed, while the chance that an α-particle will do so is essentially unity. All types of radiation could be comprehended with sufficient accuracy by the deliberately oversimplified postulate that in order to produce a break the ionizing particle must:
(a) be sufficiently densely ionizing to deposit a total of about $17 \times 35 = 600$ eV of energy in the thread, and (b) transverse the chromatid thread (i.e. have a total path length of at least 0.1μ).

The second requirement was introduced in order to meet the following difficulty which arises in connection with the aberrations induced by very soft X-rays. Catcheside and Lea (1943) were unable to induce either chromatid or isochromatid breaks by exposing the chromosomes in *Tradescantia* pollen tubes to X-rays of 1.5 keV photon energy, and I understand that a repetition of this experiment by Neary and Evans (personal communication) which is still in progress, is tending to confirm the original observations. Photo-electrons liberated by these soft X-rays will have about the same LET as recoil protons generated by neutrons, and therefore, on the classical theory, would have been expected to be very efficient in producing breaks. It is understandable that they should produce no exchanges or isochromatid breaks, because the extreme range of such particles is only about 0.05μ. On account of this short range, they would be unable to span two chromatid threads and they might not be numerous enough to initiate an exchange by two independent breaks in regions sufficiently close together. A deposition of 1000 eV of energy within, or in the immediate vicinity of, a chromatid thread would, however, have seemed amply sufficient to produce a simple break. It was to avoid this dilemma that Lea introduced the second condition for break production, viz. that the particles must span the chromatid thread, assumed to be about 0.1μ in diameter. This always appeared a little arbitrary. Read (1958) removed this arbitrariness by considering each *Tradescantia* chromatid to consist of two threads, individually about 20 ÅU in diameter, but separated by $\sim 0.1\mu$, and introducing the requirement that at least one ion cluster must be formed in each thread at the same locus. He obtained good agreement with the estimated primary breakage frequencies for 2.5 keV, 7.5 keV, and 12.5 keV electrons, (Li-D) neutrons, and α-particles.

5. The influence of metabolism on the yield of exchange type aberrations

An important development in our understanding of the underlying mechanisms came with the realisation that the yield of exchange type aberrations is, in certain circumstances, a function of cellular metabolism. The aberrations studied by Sheldon Wolff, Atwood, Randolph, and Luippold (1958) were induced in soaked *Vicia* seed and observed at the first root tip division four days later. The aberration frequencies were rather accurately proportional to the square of the dose (Sheldon Wolff, 1959) – indicating that the aberrations were the outcome of two events initiated independently by two ionizing particles – provided the dose were not spread out over too long a time. When given as two fractional doses, however, separated by more than a certain time interval, interaction between the effects of the first and second doses becomes progressively weaker, until finally the aberration yield was reduced to half the original value, being only equal to the sum of the yields from the two separate half doses. The quantity affected by metabolism was found to be the time interval during which interaction remained possible. This interval was prolonged by blocking oxidative phosphorylation during the interval, and by treatment with chloramphenicol and aureomycin (Sheldon Wolff 1959). It was also a function of the magnitude of the first dose – or more precisely a function of the damaging effect of this dose on some unidentified system – since a large first dose prolonged the time interval during which interaction could occur, whereas a small first dose, especially when administered to anoxic cells, reduced the interval. These observations could be fitted quite naturally into the classical theory of Lea (1946) by postulating that under the conditions of restricted metabolism, broken chromosome ends formed by the first irradiation could not rejoin, and therefore remained open to participate in exchange formation with those formed in the second irradiation. It was later shown (Beatty and Beatty, 1960a) that in the case of chromosome aberrations induced in *Tradescantia* microspores, metabolism immediately after a single short exposure as well as between fractionated exposures, markedly effects the yield of exchange type aberrations. They came upon these effects by observing an unexpectedly high aberration frequency when *Tradescantia* influorescences were exposed anaerobically at very low dose rates (1 r/min) for periods of 4–8 hours (Beatty, Beatty, and Collins, 1956). It soon became apparent that the high aberration yield was due to the prolonged anoxia and not to the low dose rate. Metabolism was clearly involved since, the necessary period of anaerobiosis was shorter at higher temperatures. Energy of oxidative phosphorylation was implicated when it was found that the effect of a period of pre-irradiation anoxia could be offset by placing the influorescences in a solution of ATP either before, or immediately after irradiation, or by supplying an alternative energy source. In the extreme case the range of sensitivity observed by Beatty and Beatty (1960 b) amounted to a factor of 5, and was greater than that observed by Sheldon Wolff (1955). Since the influorescences were irradiated at the beginning of microspore interphase, which lasts a total of nearly five days, the changes in sensitivity could not possibly be ascribed to synchrony resulting from the metabolic disturbances.

Beatty and Beatty (1960a) also observed an increase in aberration yield when influorescences which had been irradiated aerobically were transferred to an atmosphere of carbon monoxide (in the dark) immediately after irradiation. The carbon monoxide treatment was fully effective 30 sec after irradiation but ineffective 30 min after irradiation. The time constant involved appeared to be about 4 min at 30°C. It is evident from their results (Table I, loc. cit.)

1. that the influence of various chemical treatments is negligible at low temperatures when the level of metabolism is extremely low, and
2. that whereas the upper limit of aberration yield increases by a factor of 5 over the temperature range 0.2°–40°C, the lower level scarcely varies at all.

Dr. DESCHNER and I looked carefully for an influence of temperature on aberration frequencies in Ehrlich ascites tumour cells, and found only a slight rise between 2°C and 37°C, which was the same for cells irradiated aerobically and anaerobically – so that the oxygen/nitrogen ratio was the same at all temperatures examined (DESCHNER and GRAY, 1959). Temperature changes in this case were confined to a pre-treatment period of about 30 minutes, during which the cells received their normal nutriments from the peritoneal fluid in which they were suspended. Experiments performed recently, in collaboration with Dr. HOWARD and Miss HAWES, however, showed that if the ascites tumour fluid is allowed to stand for many hours outside the body at 3°C, the yield of aberrations which result from subsequent irradiation, whether aerobic or anaerobic, rises steadily. The presence or absence of oxygen during the storage period seems to have little influence in the sensitivity.

The dependence of the aberration frequency on metabolism, so clearly established in plant material, therefore probably has its parallel in mammalian cells – and I think Dr. SOBELS may tell us that is has its parallel also in *Drosophila*. The observations of BEATTY and BEATTY are not so naturally fitted into the classical theory as those of SHELDON WOLFF, since we should have to assume that while restricted metabolism enhances restitution, it permits illegitimate fusion between broken ends.

A much more serious difficulty has been brought to light by REVELL (1959) on the basis of a careful scrutiny of the aberrations produced by low doses of X-radiation in *Vicia* meristem cells, which has led him to call in question the basic assumption of the classical theory, namely that the simple chromatid break is the primary event out of which all other aberrations arise. He proposed instead that the initiation of an exchange is the primary event.

REVELL's case rests in the first instance on his observation that of all the visible discontinuities in the stained metaphase chromosomes of an irradiated cell, the overwhelming majority – about 95% – are not true breaks, but non-staining gaps or achromatic lesions in a structure which retains continuity. He notes that many of the gaps can still be seen in obviously intact chromosomes at anaphase. Numerically, REVELL's "gaps" are about equal in frequency to the aberrations classified as simple breaks by THODAY (1951), and, as reported by THODAY, they vary linearly with dose.

The true breaks scored by REVELL were clearly associated with chromatid exchange aberrations in three ways:

1. They are first seen at about 2 hours after low doses of radiation, simultaneously with all other chromatid aberrations, "gaps" having been seen at somewhat earlier times (30 minutes).
2. Their frequency bears the expected numerical ratio of about 2.5: 1 to the imcomplete isochromatid breaks.
3. Their frequency is not linearly related to dose, but very close to that for isochromatids and exchanges, namely about $D^{1.6}$.

The third point is, of course, particularly telling.

I believe that EVANS (personal communication) is in substantial agreement with REVELL's re-classification of the majority of breaks induced in *Vicia* by X-rays as achromatic lesions.

It seems clear that the production of a simple break in the interphase chromosomes of *Vicia* by electrons of medium LET is a much rarer event than was formerly supposed.

Is this also true of *Tradescantia*? BENDER and SHELDON WOLFF have scored chromatid breaks (being careful to exclude achromatic gaps), isochromatid breaks, and exchanges in human and monkey epithelioid cells, and in Chinese hamster fibroblast cells, and compared their data with LEA's figures of chromosome aberrations in *Tradescantia* microspores (1961). The aberration frequencies are comparable in the three types of mammalian cell. The ratio of true chromatid to isochromatid breaks are 0.27, 0.18 and 0.12, averaging 0.19. These figures are comparable with the value of this ratio derived from REVELL's data for *Vicia* which is about 0.25, but in very marked contrast with the value 2.7 for the same ratio derived from LEA's figures for *Tradescantia* chromatid aberrations. This rather indirect argument suggests either that the frequency of true breaks in *Tradescantia* has been over-estimated by a factor of 10 or more, or that *Tradescantia* is not representative of chromosomes in general, and emphasises the urgent need for re-examination of the *Tradescantia* chromatid aberration frequencies. It is, in any case, important to note that the frequencies observed in mammalian cells appear to agree closely with those reported by REVELL for *Vicia*.

In REVELL's view, exchanges only take place between portions of chromatid which happen to lie adjacent to one another, and close together at the time of irradiation. He postulates three stages in the formation of an exchange, (1) some unspecified form of damage to each of the two portions of chromatid thread attributable directly to the action of the ionizing particles, (2) interaction between the two injured portions of chromatid thread to initiate an exchange, and (3) subsequently, the formation of the actual exchange. Numerically such a view is tenable, since, if the length of the chromatid threads in *Vicia* is taken to be $\sim 2300\mu$ (READ, 1960), there would be an average of about 350 eV of energy dissipated per micron of thread exposed to 100 rad, and hence a good chance that each of the two portions of thread which happens to lie together will be damaged by the irradiation. Only a few such juxtapositions of chromatid thread may be required in any one cell to account for the observed aberration frequencies. On the other hand, grave numerical difficulties would arise in the case of α-radiation, unless it is assumed that threads are lying in close proximity over nearly their whole length – but we have as yet no experimental grounds for calling in question the ability of an α-particle to break a chromatid thread in the manner postulated by the classical theory.

REVELL's observations do not prove that the breaks are not produced even by X- or γ-radiation, but only remove the grounds for postulating the formation of a break by means other than in association with exchange production.

We have to look elsewhere for positive evidence on this point. The yield of true breaks and of achromatic lesions induced in *Vicia* by α-radiation is not yet known.

We are in any case left with the "gaps" or achromatic lesions as a prominent feature of chromosome damage in cells exposed to X- or γ-radiation. Since the frequency of these aberrations is linearly related to dose, we may infer that they are initiated individually by a single ionizing particle. Their biological significance has yet to be evaluated.

No trace of achromatic lesion is seen in metaphase chromosomes of *Vicia* fixed 30 seconds after doses as high as 10^6 rad (HOWARD and GRAY; unpublished). REVELL first sees them 30 minutes after low doses of X-irradiation. They are probably, therefore, not simply a physicochemical change induced outright at the time of irradiation.

5. Ultraviolet and visible photons

LEE and PUCK (1960) reported survival curves for HELA cells exposed to 2357 ÅU ultraviolet radiation, for which the D_0 was about 100 erg/mm^2 incident on the cells. LOVELACE (1954) found no detectable chromosome aberrations at pollen tube metaphase when the same amount of ultraviolet energy at 2650 ÅU was delivered to dry *Tradescantia* pollen. He recorded only 0.02 chromatid and 0.03 isochromatid breaks per cell after exposure to ten times this dose, and SWANSON (1942) observed that the same dose produced about 0.1 chromatid breaks per cell and no isochromatids when pollen tube chromosomes were irradiated. KIRBY-SMITH and CRAIG (1957) used still higher doses of 2650 ÅU, viz. 10000 ergs/mm², and observed chromatid and isochromatid frequencies of about 0.4 per cell. The action spectrum was that of nucleic acid. No oxygen dependence and no photo-reactivation were observed. In general, the drier the pollen, the more sensitive it was to UV. Unshielded nucleoprotein exposed to 10000 erg/mm² of 2537 ÅU, would absorb 3.10^{10} erg/g of energy, equivalent to 300 million rad. Although the energy input to the chromosomes cannot be computed accurately because it is not known what allowance to make for shielding, it is evidently very large compared with the energy of ionizing radiation which would produce the same amount of chromosome structural damage.

The discovery of KIHLMAN (1959) that the dye sensitised excitation of genetic material leads to the induction of chromosome structural damage, promises to be very fruitful, since on the one hand there is a growing body of information concerning the interactions between molecules such as acridine orange and the components of chromosomes, and on the other hand about the mutagenic effects of these molecules at the rII locus of bacteriophage T_4.

On the basis of viscosity measurements and X-ray diffraction studies, LERMAN (1961) has suggested that the molecules of acridine orange become intercalated between the nucleotides in the DNA helix, causing an extension of the backbone and an increase in stiffness. The adsorption is completely reversible. Light energy of the appropriate wavelength may thus be transferred very directly to the DNA. When *Vicia* roots are pre-treated with acridine orange and exposed to visible light of appropriate wavelength, isochromatid breaks and chromatid exchanges are produced in considerable numbers, but only if the exposure to light is in the presence of oxygen or nitric oxide. The fact that aberrations can by this means be brought about by energy which, in the first instance, is donated to DNA, is an important piece of evidence, and the requirement for oxygen or nitric oxide links this mode of inducing chromosome structural damage interestingly with X- and γ-radiation.

The Induction of Mutations not as yet Associated with Visible Structural Damage to the Genome

Although frequency-dose relations yield only a limited amount of information, it is convenient to take the linear frequency-dose relation characteristic of mutations induced by ionizing radiation as firm ground from which to explore a field which has already grown quite complex.

The linear relation was first shown to hold for both the visible and sex-linked lethal mutations which from the irradiation of *Drosophila* sperm by TIMOFÉEFF-RESSOVSKY and DELBRÜCK (1936) some twentyfour years ago. This relation has frequently been re-examined and found to be valid (MULLER, 1954; OSTER, 1958; EDDINGTON and RANDOLPH, 1958). Apparent saturation effects at very high doses (BELGOVSKY, 1958) may be explained by radiochemically induced anoxia, cf. p. 3, or in other ways. The yield of mutations is dose rate independent when the mutations are induced by irradiation of the mature sperm as would be expected for an effect initiated by a single ionizing particle. These mutations, as also certain "point mutations" (BENTLEY GLASS and METTLER, 1958), are oxygen dependent, and although anoxia only reduces the yield to a small extent when mature sperm are irradiated, this is probably due chiefly to the low oxygen tension at the location of these cells under normal conditions in the testis (OSTER, 1959). When induced at an earlier stage of spermiogenesis, a very strong dependence on oxygen tension is observed (OSTER, *loc cit*, and MOSSIGE, private communication). Since, however, the sex-linked lethal mutations are undoubtedly very heterogenous in nature, and include a large but inaccurately known proportion of structural mutants, I shall not consider them further.

The frequency of mutants among micro-organism which survive exposure to ionising radiation has been found to be proportional to the dose by a number of workers (DEMEREC and LATARJET, 1946; GILES, 1951; DEMEREC and SAMS, 1960). Most of the mutants investigated so far are also oxygen dependent (ANDERSON, 1951 and GÜNTHER personal communication) In respect of dose dependence, mutations induced by ionizing radiation contrast with mutations induced by ultraviolet radiation which, as we shall see, frequently increase as the square of the dose.

A particularly careful and rigorous investigation of the dose-frequency relation for X-rays was carried out by DEMEREC and SAMS (1960) for reversion of auxotrophs to prototrophs at the *met-2*, *thr-1*, and *arg-3* loci in *E. coli*. At the *met-2* locus the relation was strictly linear from 17–4000 r. At the other two loci it was linear at low doses, but somewhat fewer mutants than expected from the linear relation were observed at high doses. The authors point out that this may be due to the simultaneous induction at high doses of changes (compatible with viability) which compensate for the reversions. Alternatively, since the cells were irradiated without gas equilibration at up to 2.10^{10} bacteria/ml, it is possible that they were normally almost, but not quite, anoxic when irradiated, so that the high dose could be less effective on account of a radiochemically induced anoxia (see p. 5). From the biophysical standpoint, the important point is that in the case of these mutations – which the authors believe to be true reversions and not suppressor mutations – the linear relation has been found to hold accurately down to doses of 17 r, or in one case 8.5 r, at which only a small proportion of the bacteria will be traversed by more than one electron.

The linear law may, therefore, confidently be assumed to be valid, and the frequency of mutation completely independent of dose rate at all doses below about 10 r. Frequencies per rad, at the *thr-1*, *met-2*, and *arg-3* loci were, respectively, $0.6 \cdot 10^{11}$, $0.9 \cdot 10^{-11}$, and $3.2 \cdot 10^{-11}$.

The question as to whether the radiation has been instrumental in restoring aberrant genetic material to its original form – as against the inactivation of a suppressor gene – is important. The authors believe that the three mutations in question were true reversals. This question was examined very carefully by DE SERRES (1958) for forward mutations induced by ionizing radiation. He induced 21 adenine requiring mutants in *Neurospora* at the *ad-3B* locus by exposure of wild type cells to X-radiation, and studied the spontaneous and radiation induced reversions. Pre-existing mutants at the loci were not included in the tests for reversion. Genetic analysis indicated that about half the radiation induced auxotrophs probably arose from extragenic influences such as chromosomal re-arrangements and deficiencies. All these were stable against spontaneous or radiation induced reversion to prototrophy. The remaining 11 mutants, which showed normal linkage relations in the *hist-2-ad-3-nic-2* regions, appeared to involve only intragenic changes. Of these 6 were stable, but 5 could be induced by X-irradiation to revert to prototrophy with frequencies 12, 3, 1, 4, and $17 \cdot 10^{-11}$/rad, mean $7 \cdot 10^{-11}$/rad. Two out of five were observed to revert spontaneously to prototrophy. It is also interesting that an X-ray induced partial-revertant subsequently underwent further spontaneous reversion to full adenine independence, and could be induced to do so at higher frequency by exposure to X-radiation.

We therefore appear to have very good grounds for believing that radiation can bring about a change in genetic material which is precisely reversible by further exposure to radiation.

In terms of current thought, irradiation may have led to a change in the base sequence in DNA, which can be precisely reversed by further irradiation. From the linear dose-frequency relation we know that this change can be initiated by a single ionizing particle.

The frequencies per rad which I quoted above, which are typical for reversion to prototrophy, indicate, however, that this is a remarkably rare event compared with the number of occasions on which the genetic material is ionized or excited. When the only quantitative data available for visible mutation frequencies were for eye colour in *Drosophila* (Timoféeff-Ressovsky and Dellbrück, 1936), for which the frequencies per rad are about a thousand times greater than for many reversions to prototrophy in micro-organism, it was possible to make the assumption that one ion pair (or cluster) anywhere in the "gene" (as it was conceived 24 years ago) initiated mutation, and thus to arrive at a not implausible figure – equivalent in mass to 10–20 base pairs – for the size of the gene. If we were to apply the same calculation to the data for reversion to prototrophy in *E. coli* at the loci examined by Demerec and Sams (1960), or those examined by de Serres in *Neurospora* (1958), we should arrive at a value for the mass of the "gene" equal to about 1/90th of a base pair, or 1/3rd of an atom. Alternatively, and more realistically, we may say that the requirement for the production of a mutation which restores the capacity of a portion of genetic material to the condition in which it can again regulate enzyme synthesis, is very much more restrictive even than that the substance of the gene should suffer a single ionization or excitation.

In actual fact, we have no right to infer that only a single deposition of energy is needed to initiate a mutation. The linear frequency relation tells us only that a single ionizing particle is involved, and when we turn to the initiation of mutations by UV, we do, indeed, find that in reversions to prototrophy – some of which Giles (1951) had previously shown to be true reversions and not suppressor mutations – the frequency at low doses does not increase linearly, but as the square of the dose, indicating that two separate acts of energy absorption are involved.

In this connection the recent work of Witkin (1961) is extremely interesting. Her data show that when UV irradiated auxotrophs of *E. coli* B/r deficient for tyrosine synthesis are plated on a rich medium, the frequency of mutants per survivor increased as $(\text{dose})^2$, as was found by Giles in *Neurospora* (1951). When, however, the organisms are plated under conditions unfavourable to protein synthesis, not only does the yield of mutations decline, but the frequency becomes more and more nearly proportional to dose. We might say that the second event was of no avail to enhance the mutation frequency under these metabolic conditions. On the other hand, when the UV irradiated organisms are plated on a rich medium to which certain basic dyes, particularly acridine orange, have been added, the mutation frequencies are not only very much higher, but again approach proportionality to dose. We might say in this case that the second event was unnecessary. I think we may have here an indication that the two UV photons which co-operate in the induction of a mutation under conditions in which the $(\text{dose})^2$ law applies may achieve two entirely different effects on the genome. It would be interesting to carry over this speculation to two other cases in which a linear dose dependence is observed with X-rays and a $(\text{dose})^2$ dependence for UV irradiation, viz. the remarkable induction of mutation in kappa phage of *Serratia* by extracellular irradiation observed by Kaplan, Winkler, and Wolf-Ellmauer (1960), and the induction of lysogeny in *E. coli* K12 (λ) observed by Marcovich and Latarjet (1958).

We have, of course, no justification for thinking that the mechanisms involved in UV mutagenesis are also involved in X-ray mutagenesis. Unlike the situation which we considered in regard to the induction of chromosome structural damage, for which UV was found to be enormously less effective than ionizing radiation, there is not a very great disparity between these two radiations in regard to the induction of reversion to prototrophy. Such a comparison is not meaningful under normal conditions of plating because of the difference between the dose-response relations, but if comparison were made under WITKIN's low level conditions, the UV energy absorbed per gram of nucleoprotein per mutation, is found to be 200 times as great as in Demerec experiments with X-rays, and if the comparison is made under conditions which gave WITKIN the highest yield of mutations the disparity is reduced to a factor of 10, X-ray energy being still the more efficient.

The phenomenon of photorestoration (JAGGER, 1958), whereby the yield of UV induced mutations as well as the extent of loss of colony forming ability, is reduced by subsequent exposure to visible light, provides very clear evidence that some injuries induced by UV irradiation proceed through steps which may be modified by metabolic processes, since GOODAL, RUPERT, and HERRIOTT (1957) have been able to extract from photoreactivable cells a component which can mediate photorestoration of UV damage in transforming principle when this component and the damaged DNA are exposed to visible light in vitro. Subsequent work (RUPERT, 1961) has shown that the essential ingredient in the extract is an enzyme which complexes with UV-damaged DNA of many different kinds, and is able to restore a proportion of the UV induced lesions when the complex is exposed to visible light. This opens up very exciting prospects to the biochemists for the investigation of the nature of the DNA lesions – some of which we know would have been recorded as mutations in the absence of photorestoration. RUPERT has pointed out that this enzyme may have an important role in the evolution of primitive organisms in the presence of sunlight, and one is led to consider the possibility that the enzyme may be a constituent of the DNA protein of photoreactivable organisms.

Mutagenesis induced by ionizing radiation is also subject to metabolic control, but in different ways.

A very pronounced effect of post-irradiation metabolism on reversion to prototrophy induced in phenyl-alanineless strain of *E. coli* by X-radiation has been investigated by STAPLETON, SBARRA, and HOLLAENDER (1955). These authors observed that metabolic conditions which give the greatest number of survivors give the smallest yield of mutations per survivor, and conversely.

The experiments of KIMBALL, GAITHER, and WILSON (1959) are of special interest, since they have observed a post-irradiation metabolic control of deleterious mutations induced in the micronucleus of *Paramecium* by a high ion density radiation (polonium alpha particles), as well as by X-radiation and UV radiation. These experiments provide a starting point for the investigation of the nature of metabolic control over the yield of mutations induced by ionizing radiation which, as with loss of colony forming ability in micro-organisms (ALPER and GILLIES, 1958), is in general rather less pronounced than metabolic control over damage induced by UV. It is already apparent, that a proportion of lesions induced by ionizing radiation in the genetic material of the cell may be repaired or eliminated by metabolic processes during the period which is terminated by DNA synthesis.

RUSSELL, RUSSELL, and KELLY's discovery (1960) of a fourfold difference between the yield of coat colour mutations resulting from acute, as compared with chronic exposure of mouse spermatogonia to X- and γ-radiation took most of us by surprise. We had believed that the induction of mutations by ionizing radiation was dose-rate independent, consistent with the linear relation between mutation frequency and dose. This belief was supported by extensive data from experiments in which mature *Drosophila* sperm were irradiated. Since RUSSELL, RUSSELL, and KELLY's data for mature sperm in mice also, so far as they go, show no difference between the effectiveness of acute and chronic exposures, there is no essential conflict between the mouse data and data from other sources. There is also no theoretical difficulty involved in accepting RUSSELL, RUSSELL, and KELLY's data for spermatogonia because the acute exposure points have large enough standard errors to accommodate a term rising more rapidly than the first power of the dose.

Being wise after the event, we can see how this result might be related to the experiments of STAPLETON, SBARRA, and HOLLAENDER (1955), and of KIMBALL, GAITHER, and WILSON (1959) just referred to, because the lowest dose used by RUSSELL in the acute irradiations (300r) is quite large enough to disturb metabolism in proliferating mammalian cells. It is, for example, sufficient to delay the onset of DNA synthesis and mitosis in regenerating liver, in ascites tumour cells, and in plant meristem cells.

A very marked influence of doses smaller than this on the induction of somatic mutations in clover has been reported recently by DAVIES (personal communication and 1961). These mutations at the V^{by} locus are induced by irradiating embryo clover leaves at the *2–16* cell stage. The mutation frequency relation includes terms higher than the first power of the dose, so the mutations very probably include, and may even consist entirely of structural aberrations, but there is no independent evidence on this point. After 2000r the frequency is about 2.4%, but this is nearly halved if two hours beforehand the plant has been exposed to a dose as small as 100r. Indeed, the effect of this pre-treatment is just detectable if the pre-treatment dose is as small as 12.5r, and it is not enhanced by increasing the pre-treatment dose to 1000r. A further remarkable feature reported by DAVIES is that the pre-treatment is ineffective if given under anoxic conditions.

The optimum separation in time between the pre-treatment and the test dose was two hours at 25°C, but longer for lower temperatures. DAVIES seeks to explain these very interesting results in terms of arrest of DNA synthesis by the pre-treatment irradiation.

Conclusion

In view of these manifold instances of metabolic control over mutagenesis by both UV and ionizing radiation, I feel that it would be premature – for me at any rate – to offer you any model of the initiation processes. If I did feel bold enough to attempt this I should have tried to relate the differences between UV and ionizing radiation – photoreactivation in the former, absence of photoreactivation and strong oxygen dependence in the latter – to some of the principal chemical and physical differences between the two radiations. These would include the preferential absorption of UV in certain sites, the random absorption of energy from ionizing radiation, and the charge separation involved in the act of ionization. The possible importance of the instantaneous creation of positive and negative charges in a polar medium, which will result in very strong electric fields during the interval before the polar molecules become orientated, has been pointed out very cogently by PLATZMAN and FRANCK (1958). These authors conclude that the formation of a single ion pair might very well dislocate as many as 20 hydrogen bonds in the immediate neighbourhood. It would seem that this degree of dislocation of secondary structure might well be instrumental in providing the circumstances for a re-arrangement of base pairs in the DNA molecule. Most, if not all, reversible mutations which can be induced by radiation also occur naturally. The function of radiation may be simply to increase the number of occasions on which such re-arrangements can occur.

Rather than occupy you with speculations of this kind, I prefer in conclusion to draw attention to the remarkable effectiveness of ionizing radiation in destroying the ability of genes to initiate enzyme synthesis, because I think that a better understanding of this process is probably a prerequisite to the formulation of a model for radiation mutagenesis.

It might have been supposed that in order to destroy the capacity of a cell to synthesize a particular enzyme, it would only be necessary to damage that small portion of a genome which genetical analysis shows to be concerned with the synthesis of the enzyme in question, and that the required dose would be large in proportion to the smallness of the target. The contrary turns out to be the case. The ability of the cell to initiate the synthesis of those enzymes so far tested seems to be destroyed by a dose equal to, or not much larger than, that which destroys reproductive integrity. A number of recent studies throws light on this observation.

We now have evidence that in phage, micro-organisms, higher plant cells and mammalian cells genetic material sometimes exerts an immediate control over enzyme synthesis. We know that RNA is concerned with this control, and recent experiments with phage (BRENNER, JACOB, and MESELSON, 1961) and bacteria (GROS, HIATT, GILBERT, KURLAND, RISEBROUGH, and WATSON, 1961) have provided very convincing evidence that the "messenger" RNA which, on biochemical grounds, appears to be the carrier of the information encoded in the base sequence of DNA, has a very short life, measured in seconds.

The results of a recent investigation into the in vitro synthesis of an inducible enzyme β-galactosidase agree well with this conclusion, since NOVELLI (1961) (EISENSTADT and NOVELLI, 1961) has shown that in addition to the cell components such as the ribosomes, which are known to be the actual seat of enzyme synthesis, it is essential to have present also DNA from a strain of organism inducible for the enzyme in question. Essential information contained in the ribosomes at the time of their separation from the cell had obviously been lost. The abrupt cessation of capacity for adaptive enzyme synthesis, when *E. coli* is infected with phage T2 which inactivates the bacterial genome (but not phage λ which depends to some extent on the bacterial genome for phage synthesis), also supports the same conclusion.

The evidence from experiments with ^{32}P, X-rays and UV is that an early step in this genetic control of enzyme synthesis is very sensitive of damage by radiation. This is most clearly seen in cases in which enzyme synthesis has to be initiated.

Phage: In phage, this situation arises immediately after infection, and the sensitivity of this step is inferred from (a) the fall in sensitivity to ^{32}P decay during the first few minutes after infection, (b) the much greater scope for cross reactivation and multiplicity reactivation of phage exposed intracellularly to ionizing radiation as compared with phage irradiated before infection, (c) by the existence of the "vulnerable centre" to UV inactivation, and (d) probably by the decline in photo-reactivability during the first few minutes after infection.

Bacteria: In bacteria, it is most evident from studies of inducible enzyme synthesis. Thus, as a result of ^{32}P decay in *E. coli*, loss of capacity for inducible (and constitutive) enzyme synthesis occurs at the same rate as loss of colony-forming ability (MCFALL, PARDEE, and STENT, 1958; PARDEE, 1958). After exposure to X-radiation, the capacity of several microorganisms to synthesize adaptive enzymes has been measured and found to be impaired at doses equal to, or not much above, those which cause loss of colony-forming ability; e. g. Lysine decarboxylase in *B. cadaveris* (PAULY, 1959), catalase in *Rhodopseudomonas* (CLAYTON and ADLER, 1961), β-galactosidase in *E. coli* (NOVELLI, 1961), and L-histidase in *Pseudomonas* (D. L. DEWEY; personal communication). DEWEY's data are particularly striking. Loss of capacity to synthesize induced L-histidase was exponentially related to dose with $D_0 \sim 3000$ rad for aerobic irradiation. From this a minimum target size of $\sim 6.10^{-16}$g (m. wt. $\sim$ 360 million) may be inferred – a not inappreciable fraction of the genome. This is about 100 times larger than the target for the non-specific incorporation of labelled amino acids into protein which POLLARD and his collaborators (POLLARD, 1961) estimated at 6.10^{-18}g (for histidine), and suggested might be the 70 S ribosomes.

Both PAULY (1959) and DEWEY (personal communication) have observed that the inhibition of capacity for inductive enzyme synthesis is exponentially related to dose, and oxygen dependent to about the same extent as colony forming ability ($O_2/N_2 \sim 4$ for lysine decarboxylase and ~ 3.5 for lysine histidase).

The ^{32}P experiments (McFall, Pardee, and Stent 1958; Pardee, 1958) indicate that in micro-organisms the important damage is either in the DNA itself or in some material which is within reach of the disturbance caused by the ^{32}P disintegration within the DNA.

Mammalian cells: I have been unable to find any evidence that a deliberate attempt has been made to investigate the effect of irradiation on adaptive enzyme synthesis in mammalian cells, but I believe that the delayed appearance of thymidylic kinase when partial hepatectomy is performed on irradiated liver is in fact an example of the information we seek. In the resting liver this enzyme is present in very small amount, and its rapid rise, preceding DNA synthesis, suggests that it is serving the role of an adaptive enzyme for the cells undergoing transition from the non-proliferating to the proliferating state. Here again the dose required to affect enzyme synthesis is comparable with that which destroys reproductive integrity of the proliferating cells.

I remarked in my opening paragraph that the ease with which radiation destroys the reproductive integrity of cells was not only an obstacle to the study of mutagenesis, but also introduced complications into the task of interpreting the observations.

In the last few paragraphs I have dwelt at greater length on what seem to me some very grave complications, namely the ability of low doses of radiation to impair the capacity for the synthesis of enzymes under immediate genetic control, including enzymes concerned with the duplication of DNA itself. Unless the loss of capacity to synthesize certain classes of enzyme is an all or none phenomenon for any one cell, which invariably leads to loss of reproductive integrity, it seems likely that the effect of radiation on enzyme synthesis will affect the yield of mutations per surviving cell, since any mutation from a deficient state to the wild type requires the initiation of enzyme synthesis before it can be expressed.

Summary

The forms of damage to the genome which may be expressed as mutation range from gross visible re-arrangements down to an unknown lower limit is probably a change in sequence affecting a few nucleotides. It seems that we have to revise our ideas concerning structural re-arrangements induced by X- and γ-radiation to take account of the fact that the induction of a simple break in chromosomes of plants and animals is a much rarer event than was formerly supposed and that, as proposed by Revell, most of the breaks seen at a metaphase have arisen in association with the formation of exchanges. We lack data upon which to decide whether breaks seen after exposure to densely ionizing radiation have arisen in association with exchanges or independently. α-radiation is extremely efficient compared with X- or γ-radiation. Almost every transit of an α-particle through a plant nucleus yields an aberration visible at metaphase.

Some at least of the mutations induced in micro-organisms by X-rays as well as UV appear to be changes in the genetic material which are strictly reversible. They are invariably initiated by one ionizing particle but may not be due to energy deposition in a single site. Mutations induced by UV frequently increase in proportion to the square of the dose when the organisms are plated on to a rich medium. Recent work, however, shows that on other plating media, both those which depress the yield and those which augment it, the mutation frequency rises almost linearly with dose. It is suggested therefore that the two sites of energy deposition indicated by the square law dependence may be concerned with quite different functions.

Recent work has shown that mutations in spermatogonia and oocytes by X-ray are dose-rate dependent, and also that the induction of somatic mutations in proliferating plant cells is greatly modified by exposure to a dose of 100 rad or less two hours before the test dose. These facts are examined in the light of a number of investigations which show that the ability of a gene to initiate enzyme synthesis is impaired by a dose which is not much greater than that which causes complete loss of reproductive integrity; a dose, moreover, which corresponds to the deposition of energy by a single ionizing particle anywhere in a region which represents an appreciable fraction of the genome.

References

Alper, T., and N. E. Gillies, 1958: Dependence of the observed oxygen effect on the post-irradiation treatment of microorganisms. Nature (London) **181**, 961—963.

Alper, T., and P. Howard-Flanders, 1956: Role of oxygen in modifying the radiosentitivity of *E. coli* B. Nature (London) **178**, 978—979.

Anderson, E. H., 1951: The effect of oxygen on mutation induction by X-rays. Proc. nat. Acad. Sci. U. S. **37**, 340—349.

Baikie, A. G., W. M. Court-Brown, K. E. Buckton, D. G. Harnden, P. A. Jacobs, and I. M. Tough, 1960: A possible specific chromosome abnormality in human chronic myeloid leukaemia. Nature (London) **188**, 1165—1166.

Beatty, A. V., and J. W. Beatty, 1960a: Post-irradiative effects on chromosomal aberrations in *Tradescantia* microspores. Genetics **45**, 331—344.

Beatty, A. V., and J. W. Beatty, 1960b: Potassium gluconate and ATP effects on chromosome aberration yield. Proc. nat. Acad. Sci. U. S. **46**, 1488—1492.

Beatty, A. V., J. W. Beatty, and C. Collins, 1956: Effects of various intensities of X-radiation on chromosomal aberration. Amer. J. Botany **43**, 328—332.

Belgovsky, M. L., 1958: The shape of frequency-dosage curve for recessive lethals in *Drosophila* in relation to differential radiosentitivity of different stages in germ cell development (Abstr.). Proc. 10th Internat. Congress of Genetics **2**, 19—20.

Bender, M. A., and P. C. Gooch, 1961: Somatic chromosome aberrations in normal and irradiated humans (Abstr.). Rad. Res. **14**, 451.

Bender, M. A., and S. Wolff, 1961: X-ray induced chromosome aberrations and reproductive death in mammalian cells. Amer. Naturalist **95**, 39—52.

Brenner, S., F. Jacobs, and M. Meselson, 1961: An unstable intermediate carrying information from genes to ribosomes for protein synthesis. Nature (London) **190**, 576—581.

Catcheside, D. G., and D. E. Lea, 1943: Effect of ionization distribution on chromosome breakage by X-rays. J. Genet. **45**, 186.

Clayton, R. K., and H. I. Adler, 1961: Protein synthesis and viability in X-irradiated *Rhodopseudomonas spheroides*. In preparation.

Conger, A. D., M. L. Randolph, C. W. Sheppard, and H. J. Luippold, 1958: Quantitative relation of RBE in *Tradescantia* and average LET of gamma-rays, X-rays and 1.3-, 2.5-, and 14.1-MEV fast neutrons. Rad. Res. **9**, 525—547.

Davidson, D., 1960: Protection and recovery from ionizing radiation: Mechanisms in seeds and roots. — Radiation protection and recovery. Editor: A. Hollaender. Pergamon Press, London, 175—211.

Davies, D. R., and E.T.Wall, 1961: Induced mutations at the V^{by} locus of *Trifolium repens*. II. Reduction below the additive base line by fractionated doses of gamma radiation. Genetics **46**, 787-798.

Demerec, M., and R. Latarjet, 1946: Mutations in bacteria induced by radiations. Cold Spring Harbor Symposia on Quant. Biol. **11**, 38—50.

DEMERC, M., and J. SAMS, 1960: Induction of mutations in individual genes of *Escherichia coli* by low X-radiation. Immed. and low level effects of ion. radiations. Proc. Venice Symp. 1959. Editor: A. A. Buzzati-Traverso. Taylor and Francis, London, 283–291.

DESCHNER, E. E., and L. H. GRAY, 1959: Influence of oxygen tension on X-ray induced chromosome damage in Ehrlich ascites tumour cells irradiated in vitro and in vivo. Rad. Res. **11**, 115–146.

DE SERRES, F. J. jun., 1958: Studies with purple adenine mutants in *Neurospora crassa*. III. Reversion of X-ray-induced mutants. Genetics **43**, 187–206.

DEWEY, D. L., and J. W. BOAG, 1960: Inaktivierung von Bakterien mittels einzelner Elektronenimpulse. 2. Naturforsch. **6**, 372–374.

EDDINGTON, C. W., and M. L. RANDOLPH, 1958: A comparison of the relative effectiveness of different average linear energy transfer on the induction of dominant and recessive lethals in *Drosophila*. Genetics **43**, 715–727.

EISENSTADT, J., and G. D. NOVELLI, 1961: The requirement for specific DNA in the cell-free synthesis of β-Galactosidase. In preparation.

GILES, N. H. jun., 1951: Studies on the mechanism of reversion in biochemical mutants of *Neurospora crassa*. Cold Spring Harbor Symposia on Quant. Biol. **16**, 283–313.

GILES, N. H., jun., 1954: Radiation-induced chromosome aberrations in *Tradescantia*. – Radiation Biology. Vol. 1: High energy radiation. Editor: A. Hollaender. McGraw-Hill Book Co., New York, 713–761.

GLASS, B., and L. E. METTLER, 1958: The oxygen effect in respect to point mutations in *Drosophila melanogaster*. Proc. 10th Internat. Congress of Genetics (Abstr.) **2**, 97–98.

GOODAL, S. H., C. S. RUPERT, and R. M. HERRIOTT, 1957: Photoreactivation of *Hemophilus influenzae* transforming factor for streptomycin resistance by an extract of *Escherichia coli* B. – The chemical basis of heredity. Editors: W. D. McElroy and B. Glass. The John Hopkins Press, Baltimore; 341–350.

GRAY, L. H., 1959: The influence of oxygen on the response of cells and tissues to ionizing radiation. Lectures on the Scientific Basis of Medicine 1957/58 **7**, 314–317.

GRAY, L. H., 1961: The modification by chemical agents of biological response to irradiation. Proc. Conf. on Research on the Radiotherapy of Cancer, Madison 1960. American Cancer Society, New York; 70–96.

GROS, F., H. HIATT, W. GILBERT, C. G. KURLAND, R. W. RISEBROUGH, and J. D. WATSON, 1961: Unstable ribonucleic acid revealed by pulse labelling of *Escherichia coli*. Nature (London) **190**, 581–585.

JAGGER, H., 1958: Photoreactivation. Bact. Rev. **22**, 99–142.

KAPLAN, R. W., U. WINKLER, and H. WOLF-ELLMAUER, 1960: Induction and reversion of c-mutations by irradiation of the extracellular kappa phage of *Serratia*. Nature (London) **186**, 330–331.

KIHLMAN, B. A., 1959: Studies on the production of chromosomal aberrations by visible light: The effects of cupferron, nitric oxide and wavelength. Exp. Cell. Res. **17**, 490–593.

KIMBALL, R. F., N. GAITHER, and S. M. WILSON, 1959: Reduction of mutation by post-irradiation treatment after ultraviolet and various kinds of ionizing radiations. Rad. Res. **10**, 490–497.

KIRBY-SMITH, J. S., and D. L. CRAIG, 1957: The induction of chromosome aberrations in *Tradescantia* by ultraviolet radiation. Genetics **42**, 176–187.

KOTVAL, J. P., and L. H. GRAY, 1947: Structural changes produced in microspores of *Tradescantia* by α-radiation. J. Genet. **48**, 135–154.

LEA, D. E., 1946: Actions of Radiations on living cells. Cambridge University Press, Cambridge.

LEE, H. H., and T. T. PUCK, 1960: The action of ultraviolet radiation on mammalian cells as studied by single cell techniques. Rad. Res. **12**, 340–348.

LERMAN, L. S., 1961: Structural considerations in the interaction of DNA and acridines. J. Molec. Biol. **3**, 18–30.

LOVELACE, R., 1954: Chromosome shattering by ultraviolet radiation (2650 Å). Nature (London) **181**, 961–963.

MARCOVICH, H., and R. LATARJET, 1958: Radiobiological aspects of the induction of lysogenic bacteria to produce phage with X-ray, gamma-ray, and ultraviolet radiations. Advances in biological and medical physics. Editors: C. A. Tobias and J. H. Lawrence. Academic Press, New York; **6**, 75–94.

MCFALL, E., A. B. PARDEE, and G. S. STENT, 1958: Effects of radiophosphorus decay on some synthetic capacities of bacteria. Biochim. biophysica Acta **27**, 282–297.

MUTZUTANI, M., Y. OHNUKI, Y. H. NAKANISHI, and C. M. POMERAT, 1961: Changes in modal values of chromosomes after irradiation of human amnion cells. Proc. Conf. on Research on the Radiotherapy of Cancer, Madison 1960. American Cancer Society, New York; 51–69.

Muller, H. J., 1954: The nature of the genetic effects produced by radiation: The manner of production of mutations by radiation. — Radiation biology. Vol. 1: Editor: A. Hollaender. McGraw-Hill, New York; 351—626.

Novelli, G. D., 1961: Mechanisms involved in the control of enzyme synthesis. 15th Annual Symposium of Fundamental Cancer Research 1961. See also Novelli, G. D., T. Kameyana, and J. M. Eisenstadt, 1961, J. Cell. Comp. Physiol **58** (Suppl.), 225–244.

Oster, I. I., 1958: Frequency-dosage relations for mutations following irradiation of sensitive and resistant germ cells (Abstr.). Proc. 10th Internat. Congress of Genetics **2**, 210—211.

Oster, I. I., 1959: The spectrum of sensitivity of *Drosophila* germ cell stages to X-irradiation. — Radiation biology. Proc. 2nd Australasia Conf. on Radiation Biology 1958. Editor: J. H. Martin. Butterworths Scientific Publications, London; 253—271.

Pardee, A. B., 1958: Experiments on the transfer of information from DNA to enzymes. Exper. Cell Research Suppl. **6**, 142—151.

Pauly, H., 1959: X-ray sensitivity and target volume of enzyme induction. Nature (London) **184**, 1570.

Platzman, R., and J. Franck, 1958: A physical mechanism for the inactivation of proteins by ionizing radiation. Symp. Information Theory in Biology 1956. Pergamon Press, London; 262—275.

Pollard, E. C., 1961: Effects of ionizing radiation on nucleoproteins. Proc. Conf. on Research on the Radiotherapy of Cancer, Madison, 1960. American Cancer Society, New York; 38—44.

Read, J., 1958: The breakage of *Tradescantia* chromatids by ionizing radiations. Physics in Medicine and Biology **2**, 258—270.

Read, J., 1959: Radiation biology of *Vicia faba* in relation to the general problem. Blackwell Scientific Publications, Oxford.

Read, J., 1960: Chromosome size, structure, and radiation damage. Proc. Symposium on the Effects of Ionizing Radiation on Seeds and Their Significance for Crop Improvement, Karlsruhe 1960.

Revell, S. H., 1959: The accurate estimation of chromatid breakage, and its relevance to a new interpretation of chromatid aberrations induced by ionizing radiations. Proc. Roy. Soc. (London) B **150**, 563—589.

Rotblat, J., and P. Lindop, 1961: British Empire Cancer Campaign Annual Report for 1960. **38**, 276–278.

Rupert, C. S., 1961: The mechanism of photoreactivation. — Comparative effects of radiation. Editors: M. Burton, J. S. Kirby-Smith, and L. J. Magee. John Wiley & Sons Inc., New York; 49—71.

Russell, W. L., L. B. Russell, and E. M. Kelly, 1960: Dependence of mutation rate on radiation intensity. — Immed. and low level effects of ion. radiations. Editor: A. A. Buzzati-Traverso. Taylor & Francis, London; 311—320.

Stapleton, G. E., A. J. Sbarra, and A. Hollaender, 1955: Some nutritional aspects of bacterial recovery from ionizing radiations. J. Bact. **70**, 7—14.

Swanson, C. P., 1942: Effects of ultra-violet light and X-rays on pollen tube chromosomes of *Tradescantia*. Genetics **27**, 491.

Thoday, J. M., 1942: The effects of ionizing radiations on the chromosomes of *Tradescantia bracteata*. A comparison between neutrons and X-rays. J. Genet. **43**, 189.

Thoday, J. M., 1951: The effect of ionizing radiations on the broad bean root. IX. Chromosome breakage and the lethality of ionizing radiations to the root meristem. Brit. J. Radiol. N.S. **24**, 572—622.

Thoday, J. M., and J. Read, 1949: Effect of oxygen on the frequency of chromosome aberrations produced by alpha rays. Nature (London) **163**, 133.

Thomlinson, R. H., and L. H. Gray, 1955: The histological structure of some human lung cancers and the possible implications for radiotherapy. Brit. J. Cancer **9**, 539.

Timoféeff-Ressovsky, N. W., and M. Dellbrück, 1936: Radiation researches on visible mutations and the mutability of single genes in *Drosophila*. Z. indukt. Abstamm.-Vererb. lehre **71**, 322.

Trumbore, C. N., and E. J. Hart, 1959: α-ray oxidation of ferrous sulfate in 0.4 M sulfuric acid solutions; effect of 0—0.4 M oxygen. J. physic. Chem. **63**, 867—873.

Witkin, E. M., 1961: Modification of mutagenesis initiated by ultraviolet light through post-treatment of bacteria with basic dyes. Proc. Oak Ridge National Laboratory 14th Ann. Res. Conference, Gatlinburg. J. Cell. Comp. Physiol. **58** (Suppl.), 135–144.

Wolff, S., 1959: Interpretation of induced chromosome breakage and rejoining. Rad. Res. (Suppl. 1), 453—462.

Wolff, S., K. C. Atwood, M. L. Randolph, and H. E. Luippold, 1958: Factors limiting the number of radiation-induced chromosome exchanges. 1. Distance: Evidence from non-interaction of X-ray and neutron-induced breaks. J. Biophys. Biochim. Cytol. **4**, 365—372.

Wright, E. A., 1961: British Empire Cancer Campaign Annual Report for 1960. **38**, 306–308.

Radiation Induced Mutagenesis — A Biochemist's Viewpoint on some Aspects of Mutation

M. Errera

Laboratoire de Biophysique et Radiobiologie, Université Libre de Bruxelles

I. Introduction

Since the pioneer work of Muller about induced mutagenesis, our concept of mutation has considerably improved and become more complicated. Furthermore our notions concerning the structure of the genes and chromosomes have made a great deal of headway, but most important, the macromolecules carrying genetic information from one cell generation to the next one have been identified, their physical structure and their overall chemical constitution determined. It would be too long to give a historical background of these fundamental discoveries, but one should mention the names of Miescher who discovered the presence of nucleic acids in white blood cells and in their nuclei, of Levene who gave their fundamental chemical analysis, and of Brachet who demonstrated the general occurrence of both ribo- and desoxyribonucleic acid in all living cells, which allowed him to establish their fundamental importance. The procedures of isolation of DNA were considerably perfectioned and enabled Avery and his colleagues as well as Boivin to demonstrate its physiological activity as a transforming principle to bacteria, and many physical chemists to measure its macromolecular dimensions. Wilkins was able to produce most beautiful X-ray diagrams of oriented DNA fibres which allowed the exact determination of the magnitude of the crystalline units along the fibre axes, studies which culminated in the model proposed by Watson and Crick which not only postulated a double helical structure but had the immense merit of leading to a logical theory of replication.

At this point, I will have to apologize for giving this lecture: I have not been working directly in the field I mean to develop, and my only excuse for preparing this talk is my own need of attempting to get an integration of recent advances. The chemical understanding of mutagenesis will have to deal with a certain number of problems, some of which we will attempt to describe. The points we should like to bring into focus are:

a) the chemical understanding of the DNA molecules, the mechanism of their replication, and their genetic significance;
b) the way in which the molecules can be altered by physical or chemical means in vitro or in vivo;
c) the way this alteration can be duplicated and fixed in the new mutated species;
d) some of the aspects of the expression of the normal or altered information in the new species;
e) some extension of concepts acquired at the molecular level to the cellular or subcellular structures.

If one of the fundamental discoveries of classical genetics is that of the linearity of the genes along the chromosomes, its counterpart at the molecular level – as it is very probable from work on viruses and bacteria – is the notion that the genes are linearly disposed along the DNA molecules. The concept of the chromosome at the level of the virus or bacteria is certainly not as well defined morphologically as it is in higher organisms.

In bacteriophage for instance, the total amount of DNA per phage particle is not very big compared to the molecular weight and it is practical to visualize the phage chromosome as a long DNA molecule coiled upon itself inside the protein shell of the virus. Furthermore, one knows that DNA is very susceptible to shearing forces which are unavoidable during the purification of the material for molecular weight analysis, it is therefore very probable that the actual molecular weight of the DNA is of the order of magnitude of the DNA content of the phage particles which makes this visualization quite legitimate (KOZLOFF, 1960). In fact the heigher estimates obtained for the molecular weight of DNA of phages T_2 and T_4 reach a value nearing 10^8.

II. Present concepts of gene structure

But before talking further about genes and genetics, one should remember, as BENZER pointed out in 1956, that when one uses the word gene for the unit of recombination, of mutation, or of genetic function one is not necessarily talking of the same thing (BENZER, 1957). If the model of the DNA molecule as proposed by WATSON and CRICK (1953) is a double helix formed of two complementary strands of opposite polarity linked to each other by H bonds between purine and pyrimidine bases (A–T and G–C) one should then be able to assign to these three concepts of "gene" a number of nucleotides. BENZER, studying the hereditary material of the T 4 bacteriophage of *E. coli* and focussing his attention on recombination of the rII region and using an ingenious method of testing either spontaneous or induced mutant phages, was able to obtain a map of the various mutated genes. As the total lengths of the hereditary material of T 4 is of the order of 10^5 nucleotide pairs, then the size of a recombination unit is of no more than 2 nucleotide pairs, and a mutation can be caused by the alteration within a very small number of nucleotides (probably not more than 5). As mutations in different points of the chain or of the same region of the chain are capable of producing the same phenotypic alteration, the unit of function of gene activity will be somewhat more difficult to define and will be larger than the recombination and mutation units. These very important experiments of BENZER have led him to draw a very complex but linear map of the rII region of T 4 to which we will come back to later on. The cistron (or unit of genetic function) rII *B* could contain approximately 1500 nucleotides and would be approximately only 0.75 p. c. of all the genetic material of this phage. The linear arrangement of markers along the bacterial chromosomes also comes from the elegant conjugation experiment in *E. coli* of WOLLMAN, JACOB, and HAYES (1956) in which it is shown that the chromosome from the donor (Hfr) bacteria cell penetrates the receptor (H^+) in a linear fashion and that the genetic markers always penetrate in the order of their arrangement along the chromosome as determined by recombination tests. Only the DNA of the donor bacteria is needed for the recombination with the receptor genome, thus, the ordered and unidimensional organisation of the genetic material along the DNA is also highly probable in this more complex material. Attempts to determine the minimum genetic unit of transforming DNA have led to the conclusion that if the pneumococcal DNA molecules of $M = 10 \cdot 10^6$ as starting material are sheared, they can be cut into small fragments and retain their transforming activity. However, for some yet unknown mechanism, the penetration of DNA appears to be impaired by fragmentation into shorter double helical fragments (ROSENBERG et al., 1959) and estimations become difficult. Furthermore the action of DNA ase on pneumococcal DNA results in a "one hit" inactivation process from which it can be deduced that a single phosphodiester bond broken in a region of 500 000 M is sufficient to inactivate the DNA (LERMAN and TOLMACH, 1959).

III. Replication of the DNA molecule

The experiments of KORNBERG have almost demonstrated the validity of the concept of WATSON and CRICK on DNA replication. These have not only shown that a DNA model or primer is needed for the 4 deoxytriphosphonucleotides to condense into a macromolecular structure, but that the deoxynucleotides in the DNA newly formed have a unique and non random pattern which is strikingly similar to that of the DNA used as primer and the analysis of the dinucleotides produced during degradation procedures of this newly formed DNA proves the opposite polarity of the 2 DNA partner strands in the double helix. These studies and those of BOLLUM have also shown that the priming of DNA synthesis is the function of only one strand of DNA, the priming effect being considerably increased if the double helices of DNA added to the incubation mixture are dissociated by heating or other means. The WATSON-CRICK-scheme is further substantiated by the experiments of JOSSE, KAISER, and KORNBERG (1961) who showed with an in vitro system that the frequency of small segments of newly formed DNA is identical to what is found in the primer used to initiate synthesis and this is true for a variety of primers of viral, bacterial and mammalian origin.

The now classical experiments of MESELSON, STAHL, and VINOGRAD (1958) need hardly be described here. Let us just mention that if *E. coli* are labelled with ^{15}N their DNA can be distinguished from ^{14}N DNA in CsCl density gradients in the ultracentrifuge. If ^{15}N labelled *E. coli* are made to multiply in a medium containing a ^{14}N nitrogen source, the first generation gives DNA equilibrating in the gradient as a single band of $^{15}N^{14}N$ DNA as one could expect from the WATSON CRICK hypothesis. The next generation gives mixed $^{15}N^{14}N$ DNA and ^{14}N DNA as expected, and from there on, the mixed DNA becomes less and less frequent. The activity of a single strand of DNA as a primer, the mechanism of replication of the DNA molecules, and the existence of bacteriophage Øx174 containing a single standed DNA of molecular weight of $1.7.10^6$ (SINSHEIMER, 1959), all points to the concept that a single polynucleotide chain of DNA is sufficient to carry the genetic information. This is in line with the fact that RNA viruses have their genetic information coded in a RNA molecule which is a single stranded unit. The linearity of the DNA structure compels one to think of a unidimensional code made of various combination of very short sequences of deoxynucleotides carrying the necessary information to sort out the 20 or so amino acids and align them in a specific manner to make proteins functional.

Then any alteration of this code will lead to the replication of an equally altered code. Now there are two ways by which one can imagine the code could be altered: either by introducing into the template molecule some unnatural building block or by changing the molecular structure of one or a small number of nucleotides. Both these methods appear to be possible ways of altering the DNA and both must therefore be considered as probable ways of inducing mutations. The important fact being here that mutagenic agents only need to induce an alteration in a single strand of DNA and in only one or two nucleotides at most if the "muton" of BENZER is only 4 or 5 nucleotides long which can be "rescued" when mutated, by recombination mechanisms involving approximately 2 nucleotides.

IV. Mutagenic reactions of nucleic acids

a) Induced by chemical action on DNA or RNA

1. Nitrous acid. The mutagenic effect of nitrous acid aquired its precise meaning when SCHUSTER and SCHRAMM (1958) and GIERER and MUNDRY (1958) demonstrated that tobacco mosaic virus RNA, when treated with NO_2H at a pH slightly above pH4 became partly deaminated in its purine and pyrimidine rings and when rubbed on tobacco leaves gave lesions some of which were phenotypically different from those produced by non treated RNA.

SIEGEL (1960), in studying NO_2H mutation in TMV, came to the conclusion that one mutagenic deamination is obtained for 2 or 3 lethal ones, but that the loss of each amino group resulted in a detectable biological effect.

In *E. coli*, NO_2H (KAUDEWITZ, 1959) is also lethal or mutagenic but with a 2 hit dose effect curve. The mutants have different phenotypes to the ones produced by UV, and apparently both strands of DNA have to be altered for mutations to occur. Isolated DNA of *Pneumococcus* has also been submitted to NO_2H by LITMAN and EPHRUSSI-TAYLOR (1959) and mutations are obtained for resistance to aminopterin, streptomycin and optochin. Quantitative aspects are difficult to assess at present on account of the little which is known of the effect of the agent on uptake of the transforming agent by the cells.

Nitrous acid has also proved to be mutagenic for phage T4 (FREESE, 1959) and for the single stranded phage Øx174 (TESSMAN, 1959). TESSMAN (1959) has shown by comparing the progeny of the single stranded Øx174 and of the double stranded T4 that action of nitrous acid leads to pure clones of phage; in the case of T4, one obtains mixed clones containing non mutated phage originating from the duplication of the non mutated strand, and mutant phage originating from the duplication of the other strand of the same DNA molecule which has been deaminated somewhere by NO_2H treatment (TESSMAN, 1959).

VIELMETTER and WIEDER (1959) demonstrated an exponential effect of the chemical with increasing concentration for the inactivation of T2 bacteriophage and for mutant production in the surviving fraction: 3 percent of survivors (in experiments when only one survivor remained for every 10^4 particles) had mutated to produce altered plaque morphology – some of the mutants where also shown to revert. If NO_2H treated phage T2 particles are made to infect *E. coli* in conditions where not more than one particle infects a bacteria, then some of the phage progeny give equal number of wild and mutant types showing that, in some cases at least, only one of the DNA strands has been affected. The authors postulate a certain number of possible events: a) inactivation by a one hit event, b) mutation of one DNA strand by a one hit event producing a "heterozygotic" type of molecule, c) inactivation of one strand and survival or mutation of the other.

SCHUSTER (1960) showed that in DNA of various origins guanine was more rapidly deaminated than adenine and cytosine which are probably better protected by H bonding and not as well as in RNA where these bases react equally well with NO_2H (SCHUSTER, 1960). However, in vitro reactions show that guanine reacts more rapidly with nitrous acid than does adenine or cytosine, but if one changes the pH of the reaction mixture from pH 4.2 to pH 5.0 the velocity of the reaction for adenine and cytosine decreases by a factor of 90 as does also the velocity of mutation induction in phage T2 in contrast to the reaction with guanine which decreases by a factor of only 35. These arguments lead to the conclusion that in vivo, adenine and cytosine are the sensitive bases the deamination of which is mutagenic to the bacteriophage.

Very recently NO_2H has proved mutagenic for Newcastle virus. If the latter are treated at pH 5.8, a relatively slow lethal effect is observed without mutations taking place. At lower pH (4.26) the higher lethality is accompanied by a high mutation rate – the progeny having altered plaque morphology and having a different type of virulence. The fact that in mixed infections with different mutants no recombination takes place seems to indicate a very localised lesion along the genome (GRANOFF, 1961).

2. Formaldehyde. Reaction of formaldehyde, known to be mutagenic for many species, also appears to be related to the reactivity of this agent with purines and pyrimidines. Formaldehyde can react directly with aminated purines and pyrimidines and with DNA either double or single stranded, and this reaction with the double stranded DNA leads to an important drop of the melting temperature of the hydrogen bonds of the double helix. But the major macromolecular effect of formaldehyde appears to prevent the reconstitution of the double helix which has been melted by heating (GROSSMAN, 1961). Similar reactions have been demonstrated for polyadenylic and polycytidylic acid (ALDERSON, 1960) which indicates that a priori there is no reason to postulate that the agent is reacting more with DNA than with RNA in the living cell. The chemical reaction is probably an hydroxymethylation of the amino group ($R–NH–CH_2OH$ or perhaps $R–N = (CH_2OH)_2$) with higher concentration of formaldehyde. The first compound would be susceptible to react with another amino group ($R–NH–CH_2–NH–R$) which would become blocked and also unavailable for hydrogen bonding, and this would prevent reconstitution of a melted double helix (ALDERSON, 1960). This author has also shown that in the case of *Drosophila*, RNA or simply adenylic acid needs to be present in the culture medium to be mutagenic so that the effect of this agent might not be a direct one on DNA (ALDERSON, 1960). Adenylic acid is not as effective as RNA when it is supplemented alone to *Drosophila*, but it becomes as active in the presence of other nucleosides or nucleotides if they are added with it. The formation of a compound such as $A–NH–CH_2–NH–A$ ($\overline{AA}$) is therefore very plausible and this "dimer" of adenine might also conceivable be taken up into a DNA chain, this compound having dimensions compatible with hydrogen bonding with complementary thymine and cytosine of a twin DNA strand; during the next replication the $\overline{AA}$ portion of this new DNA strand would replicate a TT pair, the sequence of the daughter DNA molecules having then been very locally changed.

3. Alkylating agents. Various alkylating agents have been shown to produce mitotic inhibition, chromosome breakage and gene mutation (see BACQ and ALEXANDER, 1961). These agents have been shown to affect DNA in vitro by changing its molecular asymmetry. The chemical mechanism of this effect is under investigation and several possibilities have been envisaged and summarised in BACQ and ALEXANDER's recent book (1961), these are:

1. alkylation of a phosphate group weakening the chain at this point of attack;
2. alkylation of a purine which becomes unstable and breaks off the main DNA chain;
3. crosslinking of DNA molecules in the case of bifunctional agents.

The third mechanism or something very closely related to what might take place in pure DNA solution, appears to operate on many types of living cells treated with chlorambucil ($COOH–(CH_2)_3–N<>(CH_2CH_2Cl)_2$) (see BACQ and ALEXANDER, 1961).

The alkylating agent, ethylmethane sulfonate (EMS), has been shown to produce mutations (or back mutations) in bacteriophage T4 treated in vitro: MCDONALD, GREEN, and

KRIEG (1961) have shown that in a 50 p. c. lethally inactivated phage preparation, 1 p. c. of the plaques formed are phenotypically altered, characteristic of mutation in the rII region – it is likely that the mutation process per se becomes definitive only after reproduction of the phage EMS complex inside the host bacteria.

BROOKS and LAWLEY (1961) have recently shown that the N_7 of guanine became alkylated and that 25 p. c. of the molecules could become cross-linked and BAUTZ-FREESE (1961) has obtained reversions of mutations of the rII region of the bacteriophages which appeared to be accompanied by the liberation of guanine residues. In TMV, FRAENKEL-CONRAT (1961) has obtained lethal effects resulting from guanine alkylation, but did not observe mutations except with dimethylsulfate.

4. Radiation. We will not undertake to describe radiation effects on nucleic acids, they will be dealt with in detail by Drs. SHUGAR and SCHOLES. What happens in the cells is not certain. DNA is probably crosslinked, as shown by ALEXANDER and STACEY (1959) in sperm heads after X-rays and by KAPLAN in *Serratia* after UV where it has been shown that similar doses are needed to produce cross-linking, killing and mutation by irradiation in vivo. ORD and STOCKEN (1960) have found that DNA extracted immediately after irradiation of thymocytes in vivo give an elution pattern on Ecteola columns which is different to that of the control, indicating changes in structure. One of the present problems is that of the mechanism of DNA alteration by radiation. We will not enter into great details but it seems at present certain that DNA can be altered by direct or indirect action of the DNA molecule itself or of a possible precursor which could become incorporated by normal metabolic processes into duplicating DNA and thus make it either unstable or abnormal at that point.

Nucleoproteins have been known for a long time to be susceptible to X-rays either by in situ or in vitro irradiation (ERRERA, 1946); since then many instances have confirmed this fact but the molecular understanding of the process is still very uncertain (see ERRERA, 1947 and 1958; ORD and STOCKEN, 1961; BACQ and ALEXANDER, 1961). Although it has been claimed that intracellular DNA is very radioresistant if its transforming ability is determined after in vivo irradiation of *Pneumococci* (DREW, 1955), it was calculated by GUILD that the assay method used was not sufficiently accurate to be able to detect any changes.

One puzzling fact had been for a long time the impossibility of inducing mutations in bacteriophage by irradiating extracellular preparations of bacteriophages. These have at present been obtained with UV and X-rays (KAPLAN et al., 1960) for the x phage of *Serratia marcescens*, this mutated phage being susceptible to reversion to the normal state by visible light or by heating at 40°C within the next hour following the UV.

In the lysogenic strain HY(x), intracellular phage mutates with UV irradiation at the same rate as extracellular irradiated x phage; it seems probable that the effect is not indirect and the reversibility of the mutagenic effect of UV light by mild heat treatment suggests a mechanism very similar to that resulting in a drop of the extinction coefficient observed in irradiated DNA, which is also reversed by heat. In the latter case, SINSHEIMER (1957) was able to demonstrate a specific hydration of the double bond of the cytosine ring.

There are not many indications whereby one can tell which specific changes are induced in DNA chains by radiation in vitro; it is probable that both for UV radiation and X-rays, pyrimidine bases are more susceptible than purines (SCHOLES and WEISS, 1961, present symposium; ERRERA, 1957), but there are also breaks which occur in the DNA double or single strands, which could also be mutagenic. Whether these alterations take place in vivo has not yet been analysed. Irradiation of nucleoproteins (DNP) in vivo leads to a breaking of DNP gels (ERRERA, 1946), to dissociation of nucleic acid and protein bonds (COLE and ELLIS). UV irradiation of bacteria has shown that thymine residues could become linked and form dimers in vitro, and there are some arguments to believe that dimerisation also takes place in vivo (Wacker, 1961). However, it must be emphasized that radiations never act as "cleanly" as chemicals (or at least as we think they do) and that many alterations can be expected to occur simultaneously.

b) Uptake of anomalous precursors into DNA and RNA

The fact that normal cell enzymes are capable of utilising analogs of purine or pyrimidine bases was first described by Jeener and Rosseels (1953) who demonstrated the replacement of uridylic acid by thiouridylic in the RNA of TMV when viruses were grown in the presence of thiouracil. Since then many studies on the uptake and even on the viability of various types of cells having integrated an anomalous nitrogen base into their DNA or RNA have been reported, and Dr. Szybalski will be talking during this conference on the increased radiosensitivity of tissue culture cells which have taken up bromodeoxyuridine into their DNA. The fact that this type of abnormal metabolism could be mutagenic was demonstrated when Litman and Pardee (1956) showed that 5-bromouracil (BrU) was taken up in the DNA of microorganisms and that when *E. coli* was infected with bacteriophage T2 in the presence of BrU, this compound was integrated in great proportions in the phage DNA and a large number of mutants appeared among the progeny (Litman and Pardee, 1956, 1960; Brenner and Smith, 1959).

In addition to being mutagenic, BrU also caused a great number of phages to become non infective and unable to kill the bacteria, and these abnormal phage particles were also more sensitive to the action of UV light.

Structurally bromouracil is an analog of thymine, the voluminous bromine atom having approximately the same volume as the methyl group of thymine and it is logical to think that the mutation is induced by the substitution of BrU for thymine. Bromouracildeoxyriboside which is taken up faster in DNA as is its normal counterpart, thymidine, is also a better mutagenic agent in bacteriophage (Freese, 1959) and BrU is mutagenic only in cases when no thymidine is formed in the host bacteria and cannot compete with the analog (for instance in the presence of sulfanilamide); only halogen derivatives of uracil which are analogs of thymine are mutagenic although many other derivatives are inhibitory of phage development. This is therefore a case by which thymine replacement in DNA appears to be the primary cause of mutation, and once the bromine derivative becomes incorporated in the nucleic acid, the production of mutants can no longer be halted, but mutagenesis actually begins only when phage DNA begins to replicate.

When 5-bromodeoxyuridine or 5-iododeoxyuridine were taken up into the DNA of streptomycin (SM) dependent mutant strain of *E. coli* (Sd4) in the presence of 5 fluorodeoxyuridine (FUDR) which inhibits specifically the synthesis of thymidilic acid, there was no evidence of any important back mutation effect of these analogs, although other loci of bacteria appeared to become mutated by the same agent (Zamenhof, 1959). But as well as becoming more sensitive to UV as a result of uptake of these analogs in their DNA, the surviving bacteria did show a definite increase of UV mutability (Lorkiewicz and Szybalski, 1960).

In addition to this type of effect, bromouracil labelled transforming DNA has a decreased activity for a variety of markers (Ephrati-Elizur and Zamenhof, 1959). This raises the question of the nature of the DNA molecules which are donated: Is the loss of transformation ability of certain markers due to the lack of penetration of whole labelled DNA molecules, carrying the markers in the competent bacteria, or are these molecules taken up but not integrated in the receptor's genome? The question remains open at the present time.

In multicellular organisms, 5-halogenated deoxyuridine derivatives have been shown to inhibit embryonic development of the sand dollar which is also very probably a genetic effect at the level of differentiating somatic cells, similar to the effect of radiation on these types of cells (KARNOFSKY and BASCH, 1960).

c) Thymine deficiency

It has been noted (COUGHLIN and ADELBERG, 1956; KANAZIR, 1958) that thymine starvation of thymine requiring bacteria induces the appearance of mutants before thymineless death appears. The cause of this effect is obscure but it is possible that the deficiency of thymine for DNA replication might cause "weak" points in the new DNA strand. The weakness could be caused by the total absence of a pyrimidine unit in the new strand or perhaps by the introduction in its place of an unsuitable precursor. Deficiency of nitrogen bases might also be the cause for mutations by heat shock as was demonstrated by GREER and ZAMENHOF (1959) who submitted bacteria to 60°C for 4 hours: this treatment, as suggested by the authors, might lead to the loss of purines from DNA.

d) ^{32}P decay

Lethality of bacteriophage and bacteria as the result of the decay of ^{32}P atoms introduced into the DNA is a well-known fact and has been reviewed in detail by STENT (1960). In this instance also, it has been shown that in bacteria besides killing the organisms, this decay was capable of producing a great variety of auxotrophic mutants like one finds in spontaneous or UV induced mutagenesis, and that the number of mutants increases linearly with the number of ^{32}P atoms which had decayed (KAUDEWITZ et al., 1958).

In experiments done with wild *E. coli*, the frequency of mutagenic ^{32}P disintegrations is 1500 times smaller than that of lethal disintegrations which have an efficiency of approximately 0.1 in a great variety of bacteriophage strains, and it could be calculated that approximately 1 out of 5×10^4 disintegrations is mutagenic. In this instance, it is highly probable that ^{32}P decay is efficient only when it concerns P atoms of the DNA molecule and STENT (1960) has suggested that lethal effects of ^{32}P might be caused by the decay of 2 atoms opposite each other in the polynucleotide double helix, the decay of one single atom causing non lethal damage and possibly mutations. This is borne out by the fact that single stranded bacteriophages are killed by ^{32}P decay with an efficiency of 1 (see STENT, 1960). Let us add that this picture of ^{32}P decay will still have to take into account that the lethal effect of ^{32}P on various T phages is decreased if this decay takes place in the presence of AET (2-aminoethyl isothiuronium bromide HBr), which might indicate that part of the process is indirect and not true suicide.

V. Replication of altered DNA

If bases of one DNA strand of a double helix have been altered by some chemical or radiation treatment, it can be expected that during the next DNA replications, as suggested by FREESE (1959, 1961) and BRENNER et al. (1961), an error may arise, the altered base having different hydrogen bonding possibilities. If some agent removes a base, an abnormal base might take its place and also lead to a stable, but abnormal DNA chain replication. For instance, bromouracil, if introduced in the place of thymine in a DNA molecule, will have the possibility of pairing with guanine, which will result after 3 division cycles by replacing a thymine-adenine pair by a cytidine-guanine pair. It has also been suggested that 2-aminopurine would either replace guanine in DNA or lead to the replacement of a guanine residue by another base, perhaps adenine; in this case a cytosine-guanine pair would be replaced by thymine-adenine and the effect of 2-amino-purine on a bromouracil induced mutant would lead to the reversion of this mutation. This has effectively been found by Freese (1959) in T4 phage, who also found that bromouridine could back mutate a 2-aminopurine induced mutant. Recently, Rudner (1961) has shown a correlation between DNA synthesis and the appearance of mutants in *Salmonella* when three DNA cycles were necessary before a bromodeoxyuridine induced mutation became expressed. With 2-aminopurine, the phenotypic expression of the mutant was already possible after the first DNA cycle.

In the case of bacteria (*E. coli*) grown in the presence of bromouracil, it is shown that this analog of thymine replaces the natural precursor to a certain extent (Shapiro and Chargaff, 1960). However, when the natural DNA and the bromouracil containing DNA are compared after partial hydrolysis and chromatographic analysis, it can be seen that the new molecules have, in the case of certain segments at least, been completely reorganised which leads the authors to believe that the substitution of BrU for thymine leads to the exclusion of some sequences. Of course in this experiment of Shapiro and Chargaff (1960) one thymine residue out of every three has been replaced by bromouracil and such changes in the sequence of some of the oligonucleotides could certainly prove to be much more difficult to demonstrate in the case of mutagenesis obtained more discretely.

VI. Biochemical processes

a) Role of nucleic acid and protein metabolism: mutation fixation

So far we have seen that the alteration of DNA by chemical or physical means is liable to cause the cells to mutate. However, using UV radiation, Demerec (1946) showed that the yield of bacteriophage resistant mutants increases during the cell divisions following irradiation and reaches a maximum after 10–13 generations. The mechanism of this delayed appearance of mutation – and many more examples have been discovered since Demerec's first experiment – has challenged investigations who have shown that DNA, RNA and protein metabolism all take some part in mutation fixation.

In 1956, Witkin demonstrated that for auxotrophic bacteria deficient in one amino acid, the yield of mutations was dependent on the amino acids present in the culture medium during the first hour after post irradiation incubation; as the addition of this amino acid at later times only increased the number of cell divisions, it was concluded that mutation fixation was dependent on amino acid metabolism and not on cell division (Witkin, 1956). As furthermore the presence of chloramphenicol, a known inhibitor of protein synthesis, also decreases the appearance of mutants when this agent is present immediately after irradiation, it was concluded that protein synthesis was the essential factor in the fixation of mutations. However, this is not an absolute requirement in all cases because (Witkin and Theil, 1960) mutations towards streptomycin resistance or independence are not inhibited by chloramphenicol, whereas other loci are highly affected. Witkin showed recently that the absence of thymine after irradiation of thymine requiring bacteria also prevented the fixation of mutations, so that mutation fixation is also related to DNA synthesis. These data lead to the concept that the processes necessary for induction of mutations need to occur prior to DNA synthesis, and that it is during this latter process that the induced mutation becomes stabilised and fixed in the genome.

In parallel studies, DOUDNEY and HAAS (1959) have come to a different but not necessarily incompatible conclusion: they suggest that mutations result from the uptake of abnormal nitrogen bases into the RNA of the bacteria; this concept results from the observation that a higher mutation yield is obtained by ultra violet light in bacteria containing a higher pool of nitrogen base precursors. These abnormal precursors should first have to be stabilised through some mechanism requiring amino acids or protein synthesis, and then fixed through a mechanism involving RNA synthesis – which ultimately leads to a state where chloramphenicol becomes ineffective. This modified RNA would then in some way lead to altered DNA formation; the final DNA synthesis being the mechanism by which the mutation is finally established.

The involvement of RNA in the mutation process as postulated by DOUDNEY and HAAS, rests on the fact that if high concentrations of cytidine or uridine are added to the culture medium prior to stabilisation and fixation, they can compete with the abnormal metabolites believed to be formed during irradiation and a smaller number of mutants are obtained. The argument would be strong if these nucleosides were specific RNA precursors, but it is known that they can be metabolised into deoxynucleosides and thus compete with DNA precursors. One might then imagine that the stabilisation of the abnormal DNA would be mediated by a process involving RNA synthesis.

It is at present very difficult to appreciate the role of RNA in the process of DNA synthesis or of its influence on the quality of DNA synthesized. STENT has postulated that, in bacteriophage multiplication, the genetic information of the DNA might be transferred for a short time to a ribonucleoprotein, a hypothesis which could explain that during this short time, the "phage" becomes resistant to ^{32}P decay (STENT, 1958). This hypothesis of STENT could explain the role of RNA metabolism in mutation processes. In any event it has to be remembered that alterations of DNA can be obtained in several ways and that each process may depend in a different manner on protein and nucleic acid metabolism.

The problem is certainly a complex one; but the dependency of mutation yield on subsequent protein metabolism has been found also to be true in the case of fungi. PHILIPPS (1961) has shown in irradiated *Actinomyces* spores, that mutation induction could result from the alteration in vivo of purines and pyrimidines, the induction process taking place before DNA synthesis but becoming stabilised as a result of DNA replication and of protein metabolism.

The picture is complicated by experiments of UAHARN (1961) who observed that in *Neurospora*, protein synthesis following UV treatment resulted in a smaller yield of mutants for low doses.

b) Phenotypic expression

The second important problem is how a mutated gene becomes expressed. If to simplify matters very much, one states that the phenotype of a cell is characterised by the chemical nature of its proteins (specific enzymes, "structure" proteins determining characteristic cell structures), one is led to ask the question: How do proteins get their information from the genes? Protein specificity might be determined solely by the sequence of its aminoacids which would in turn impose the secondary (degree of alpha helical structure for instance) and tertiary (characteristic folding of the polypeptide) structure which will determine protein specificity. It is believed at present that the primary structure of proteins is determined by some RNA template which has itself received genetic information from the DNA. One has effectively found that in the case of T bacteriophages, the introduction of the phage genome into the bacteria induces this host to make a new type of RNA which in its turn will preside to the synthesis of the proteins of the progeny phage (Volkin and Astrachan, 1956).

Hall and Spiegelman (1961) have been able to show that from phage infected cells it was possible to isolate hybrid helices containing one strand of the parent phage DNA hydrogen bonded to one of the newly formed RNA, thus demonstrating the complementarity of their structure. This strand of RNA could be the vehicle of genetic information from phage DNA to phage protein. How this RNA effects its mission is yet another problem to be solved – it could get incorporated into bacterial ribosomes which could then synthesize the proteins. We will not go into details of the mechanism of protein synthesis, but we would like to mention that this mechanism is now almost accessible to chemical analysis.

Nisman (1960) was able to synthesize in vitro the enzyme β galactosidase in the presence of ribosomes of a bacterial strain capable of making the enzyme, ribonucleoside triphosphates and the DNA of the same constitutive strain. The DNA of a non induced strain being ineffective, these experiments directly demonstrate the role of specific DNA in the synthesis of a specific protein (Nisman and Fukuhara, 1960). These experiments have recently been confirmed by Novelli (1961).

At present, very many instances are known where genetic deficiencies have been characterised by the change in a single amino acid of a specific protein: the example of hemoglobin is well known and several varieties characteristic of hereditary diseases are known. To take but one example, the replacement of only one glutamic acid residue by valine in one of the two polypeptide chains constituting the whole molecule is responsible for the sickle-cell syndrome. This is most probably due to a "mistake" in a small nucleotide sequence in a molecule of DNA. We would like at this point to give direct evidence that an induced change in the structure of an information carrying nucleic acid is liable to lead to abnormal proteins. When the ribonucleic acid of TMV is treated with nitrous acid and made to infect tobacco leaves, Tsugita and Fraenkel-Conrat (1960) have shown that one proline, one aspartic acid and one threonine residue of the virus protein become substituted for leucine, alanine and serine; it has further been shown that in the case of the proline residue, the change occurs in a residue near the carboxyl end of the polypeptide chain and that it is the additional leucine of the mutant which replaces it. This mutant is recognisable by the appearance of characteristic lesions on the tobacco plant which it infects.

VII. Some aspects of mutagenesis in multicellular organism

There is no a priori reason to doubt that the concepts described for mutagenesis in the case of microorganisms do not apply to the more complex types of organisms. However, the intricacy of cell structures in these organisms will certainly increase the difficulty of understanding mutagenic processes. The organisation of the genetic material into well-defined chromosomes has led to the distinction of point mutations and of chromosome mutations which are associated with the appearance of visible chromosome abnormalities. The term point mutation is in itself very vague and should be defined in terms of structure and function as suggested by BENZER (1956) for the microorganisms. The development of chemical concepts to complex organisms is unavoidably linked to the problems of chromosome structure, chromosome duplication and the induction of chromosome aberrations. Unfortunately, an extended discussion of these subjects would be the occasion for a complete new symposium, so that we cannot even attempt to summarize the present situation in a few pages. We will therefore only bring out some of the striking facts which should be considered in order to to analyse mutagenic processes in multicellular organisms.

a) Chemical organisation of the chromosomes

The presence of well-defined deoxyribo nucleoprotein (DNP) complexes in chromosomes is certainly of great importance. The X-ray diffraction studies of WILKINS on nucleproteins in vitro and in situ have led him to believe that histones (or protamines) are fitted into the "grooves" existing between the phosphate groups of the DNA double helix which enables the basic groups of the histones to salt link to the free acidic groups of the phosphates. This gives a greater stability to the double helix and histones have been suggested as playing an important role in the inhibition of gene action. The molecular weight of the whole complex for calf thymus is of the order of $18.5 \cdot 10^6$, half of which could be one molecule of DNA, the other half histones probably in the alpha helical configuration (ZUBAY and DOTY, 1959), but there are probably interruptions between the different histone molecules. The protein moiety of the complex might be crosslinked to other nucleohistone molecules in sperm heads.

To understand the structure of chromosomes, one will have to analyse the relationship of the nucleohistone molecules (90 p. c. of the chromosome) between themselves, and with the residual protein containing tryptophan (7 p. c.), the lipid components (2 p. c.), and the RNA (1 p. c.) (MIRSKY and RIS, 1951). How the DNA complexes are held together with the other constituents is the subject of speculation by various investigators. Electron micrographs so far obtained are still far from perfect, and cannot yet lead to a description of chromosome structure. Action spectra for chromosome breaking with UV is very similar to the spectrum of nucleic acids which would be the case if the DNA is continuous along the chromosome length and has to be broken for aberrations to appear (STADLER and UBER, 1942; KIRBY-SMITH and CRAIG, 1957). This seems also to be borne out by the observations of GALL (1958) showing the susceptibility of lampbrush chromosomes to DNA ase. To take care of these restrictive facts, FREESE (1950) has suggested that the interphase chromosomes are formed of double helices of DNA linked end to end by "connecting" units, one molecule of DNA could carry several different cistrons, which would only be carried by that molecule. A mutation would be produced by the alteration of a small segment or a break in one molecule or by breakage of connecting units which would lead to translocation, deletion, etc. During mitosis, these chains of DNA molecules which have duplicated during interphase would associate to form the complex coils observable in the microscope (FREESE, 1958).

Electron microscope data of metaphase chromosomes have been interpreted by RIS (1958) and have shown the multistranded structure of chromonemata, these being bundles of strands having a diameter of the order of 200 Å and which would contain the nucleohistone molecules, maybe coated by other proteins. These fibrils might be kept together by bridges of divalent cations (Ca^{2+}, Mg^{2+}) or by hydrogen bonds because agents known to break these linkages have been claimed to dissociate these fibrils into shorter units (BERNSTEIN and MAZIA, 1953; AMBROSE, 1956).

How all these chemical concepts of chromosome structures can be made to fit with the morphological coiling of chromosomes is a difficult thing to assess at present, but as stated by RIS (1958) "Chromosomes reveal a whole hierarchy of helical systems superimposed on one another", going right down to the molecular level of the double helix of DNA.

The answer to all these problems rests obviously on the future collaboration of biologists and physical chemists.

b) Chromosome duplication

However complex the chromosome structure, it is astonishing to find that chromosomes replicate as if they were behaving like a single double helix of WATSON and CRICK: the DNA synthesized during interphase (^{3}H thymidine uptake) becomes equally distributed between the two daughter chromosomes of the first division, non labelled chromosomes appearing in the next division when this is allowed to take place in non labelled medium. Actually, the pictures of TAYLOR and WOODS also demonstrate the occurrence of sister chromatid exchange in some cases (TAYLOR, 1959). However, some objections have been raised against the concept of TAYLOR by LA COUR and PELC (1959) on the basis of geometry artefacts occuring with the use of ^{3}H radioautographs. But if the concept of TAYLOR were correct, it could indicate that the chromatids could be formed of chains of DNA double helices and that chromatid exchange would be the result of exchanges between the 2 strands of one double helix. SCHWARTZ (1958) suggests that chromatids are in reality already double due to the fact that they separate at duplication, which is in agreement with cytological data referred to in his paper.

c) Production of chromosome aberrations

In a recent review (WOLFF, 1961), genetic effects of radiation have been classified as intragenic (true gene mutations) and intergenic (chromosome aberrations).

As stated earlier, there is no reason to assume that intragenic effects in higher organisms are not induced by similar processes than observed in microorganisms. We will only focus our attention on the second type of effects. Radiation, and chiefly ionizing radiation are capable of breaking chromosomes, the loose broken ends can be shown to reunite either in the fashion they were arranged originally (restitution) or with loose ends from other breaks, leading to various morphological aspects (deletion, rings, inversions, etc.). Besides radiation induced chromosome aberrations, these have been observed as a consequence of the uptake by the DNA of abnormal precursors like ^{3}H or ^{14}C thymidine where aberrations were scored in the anaphase and telophase chromosomes of root tips (MCQUADE and FRIEDKIN, 1960), and in root meristems *Vicia* (NATARAJAN, 1961). Bromodeoxyuridine produces chromosome aberrations (chromatid breaks and exchanges) in several mammalian cell strains in tissue culture, and the place of breaking along the chromosomes has a definite pattern, in many cases the aberration being the stretching of normally appearing constrictions along the chromosomes (HSU and SOMERS, 1961).

Alkylating agents have been shown to produce mutations and chromosome aberrations in many types of cells including mammalian (LIVINGSTONE and YERGANIAN, 1956). In general, it can be said that the distribution of breaks along the chromosome is much less random with alkylating agents than with X-rays, an observation which may depend on a series of factors besides the affinity of these agents for particular chromosome constituents (BACQ and ALEXANDER, 1961). Deficiency in Ca^{2+} and Mg^{2+} in the culture media of growing plants has also been shown to induce chromosome breakage, rearrangement and crossing over and to increase X-ray susceptibility of chromosomes (STEFFENSEN, 1955). Similar effects have been produced in young *Drosophila* (LEVINE, 1955) and in *Chlamydomonas* (EVERSOLE and TATUM, 1956) indicating the very great importance of salt linkages for chromosome stability. In this respect, it has been suggested by HAGEN (1960) that X-irradiation labilises the metal linkages (Ca^{2+}, Mg^{2+}) which appear to exist in nucleohistones and which, according to KIRBY, exist, besides electrovalent linkages, between the nucleic acid and the histone (KIRBY, 1958).

The role of metal ions in nucleohistone structure as suggested by KIRBY, BERNSTEIN and MAZIA and AMBROSE, will certainly have to be better known before their participation in chromosome damage can be adequately understood.

Studies of the effect of various biochemical agents on chromosome breakages have led WOLFF (1961) to recognise two kinds of chromosome breaks. This author showed that if oxydative phosphorylation inhibitors are given to *Tradescantia microspora* or to seeds of *Vicia faba* between two irradiations, the breaks produced during the first irradiation remained open and are capable of rejoining with breaks produced during the second irradiation and will increase the number of two hit aberrations (exchanges); in fact, one will find the same number of this type of aberrations as when the total dose of X-rays is given in a single irradiation. These experiments have led WOLFF to postulate the existence of an energy dependent rejoining system. If ATP is added to irradiated *Vicia*, restitution of breaks is increased and the number of exchanges is decreased. In contrast to this type of energy dependent and slow exchange process, WOLFF has observed another type of breaks which close much faster, and he has postulated that these might be breaks of ionic bonds of the type we have already described and whereby Ca^{2+} or Mg^{2+} ions could link DNA to histones or link together a certain number of deoxyribonucleoprotein (DNP) units. The situation is certainly very intricate and it has been suggested by JAYSON (1961) that the role of these ions might be more complex still: it is known that both X-rays and Ca^{2+} are capable of lowering ATP by breaking it down to ADP in spleen extracts; this means that the ionic effects obtained by STEFFENSEN could be interpreted on a metabolic basis: A low Ca^{2+} concentration in the cell would lead to increased ATP which would decrease the one hit aberrations by favouring restitutions and increase the two hit aberrations by favouring illegitimate reunions – mostly in low intensity or fractionated dose experiments. Although this recently expressed idea may not be found to apply in all cases, the exact role of metal ions both on oxydative phosphorylation and in chromosome structure will have to be known in order to understand the process of chromosome breakage and mutation.

VIII. Summarising Remarks

We have in many instances reasoned that many of the effects of radiation discussed here (mutation, chromosome damage) are due to alterations of the DNA molecule. That this could be the result of more complicated mechanisms than those envisaged comes from the following: Whereas the biological effects mentioned are strikingly reduced when the pressure of oxygen diminishes (oxygen effect), it has been observed in vitro on DNA or bacteriophage, that when irradiated in broth, there is no such dependency of the oxygen pressure and even the effect may be enhanced at lower O_2 pressure (EPHRUSSI-TAYLOR and LATARJET, 1955, BACHOFER and POTTINGER 1954; ALPER, 1954, 1961; MARCOVITCH, 1960).

DNA, even irradiated in broth, seems highly susceptible to indirect effects, but not to the presence of O_2 – and this has led Ephrussi-Taylor and Latarjet (1955) to state that X-rays do not alter DNA (genetic effects) by immediate action, but that some secondary processes, dependent on O_2 tension do take place before DNA is definitively damaged. What relation there is between these effects and "mutation fixation" is difficult to assess: the oxygen could very well increase damage to "recovery factors" like those discussed by Wolff (1961) for chromosome breakage; it might also reduce crosslinking in DNA as suggested by Alexander in the case of direct effects in DNA (Bacq and Alexander 1961).

More recently, Hutchinson (1961) has shown that in the presence of glutathione or "in vivo", DNA does show an oxygen effect when assayed for its transforming ability.

It is obvious that effects of specific chemicals on chromosomes will have to be studied with the refinement used in bacteriophage work in order to see if specific alterations of DNA, of proteins, or of other components of chromosomes lead to well-defined mutations and chromosome anomalies. These types of studies may prove to be very important to untangle the very complex structure of chromosome and help us to understand the possible mechanisms of radiation induced genetic effects.

References

Alderson, X. X., 1960: Mechanism of formaldehyde mutagenesis. The uniqueness of adenylic acid in the mediation of the mutagenic activity of formaldehyde. Nature **187**, 485–489.

Alexander, P., and K. A. Stacey, 1959: Cross-linking of DNA in sperm heads by ionizing radiations. Nature (London) **184**, 958–960.

Alper, T., 1954: Bacteriophage inactivation under changing conditions of irradiation. Intern. Conf. Radiobiol., Liège, pp. 39–45.

Ambrose, E. J., 1956: The structure of chromosomes. Progress in Biophysics **6**, 26–55.

Bachofer, C. S., and A. Pottinger, 1954: Role of high ionic concentrations in protection against X irradiation. J. Gen. Physiol. **37**, 663–667.

Bacq, Z., and P. Alexander, 1961: Fundamentals in radiobiology. Pergamon Press.

Bautz-Freese, E., 1961: Transitions and transversions induced by depurinating agents. Proc. nat. Acad. Sci. U. S. **47**, 540.

Benzer, S., 1958: The chemical basis of heredity. John Hopkins University Press, Baltimore, pp. 70–93.

Bernstein, M. H., and D. Mazia, 1953: The desoxyribonucleoproteins of sea-urchin sperm. Biochim. biophysica Acta **11**, 58–68.

Bernstein, M. H., and D. Mazia, 1954: Desoxyribonucleoprotein of cell nuclei and their sensitivity to X-rays. Nature (London) **174**, 463–464.

Bollum, F. J., 1959: The thermal conversion of non priming DNA to primer. J. biol. Chemistry **234**, 2733.

Brenner, S., and J. D. Smith, 1959: Induction of mutations in the DNA of phage T 2 synthesized in the presence of chloramphenicol. Virology **8**, 124–125.

Brenner, S., L. Barnett, F. H. C. Crick, and A. Orgel, 1961: The theory of mutagenesis. J. Mol. Biol. **3**, 121–124.

Brooks, P., and P. D. Lawley, 1961: A comparison of the reaction of mono- and difunctional alkylating agents with nucleic acids in vivo and in vitro. Biochem. J. **78**, 2 P.

Coughlin, C. A., and E. A. Adelberg, 1956: Bacterial mutations induced by thymine starvation. Nature (London) **178**, 531–532.

Dekker, Ch. A., 1960: Nucleic acids. Selected topics related to their embryology and chemistry. Annual Rev. Biochem. **29**, 453–474.

Demerec, M., 1946: Induced mutations and possible mechanisms of the transmission of heredity in *E. coli*. Proc. nat. Acad. Sci. U. S. **32**, 36–46.

Doudney, C. O., and F. L. Haas, 1959: Gene replication and mutation induction in bacteria. J. Mol. Biol. **1**, 81–83.

Drew, R. M., 1955: Irradiation of *Pneumococcus* D.N.A. Radiation Research **3**, 116–120.

Ephrati-Elizur, E., and S. Zamenhof, 1959: Biological effects of incorporation of bromodeoxyuridine into *B. subtilis* on its transformation. Nature (London) **184**, 472—473.

Errera, M., 1947: Action of radiations on nucleoprotein complexes. Cold Spring Harbor Symp. **12**, 60.

Errera, M., 1952: Étude photochimique de l'ADN. Biochim. biophysica Acta **8**, 30—37, et **8**, 115—124.

Errera, M., 1957: Mécanismes de l'action biologique des radiations. Aspects biochimiques. In: Protoplasmatologia, Handbuch der Protoplasmaforschung X, 3, p. 241, Springer Verlag, Vienne.

Errera, M., 1958: Effects of radiation on cells. In: The cell; edited by A. Brachet and A. Mirsky, Chap. 15, pp. 695—740.

Eversole, R. A., and E. L. Tatum, 1956: Chemical alterations of crossing-over frequency in *Chlamydomonas*. Proc. nat. Acad. Sci. U. S. **42**, 68—93.

Fraenkel-Conrat, H., 1961: Chemical modification of viral RNA. Biochim. biophysica Acta **49**, 169—180.

Freese, E., 1958: The arrangement of DNA in the chromosome. Cold Spring Harbor Symp. **23**, 13—18.

Freese, E., 1959: The differences between spontaneous mutations and base analog induced ı ıtations in phage T 4. Proc. nat. Acad. Sci. U. S. **45**, 622—633.

Gall, J. G., 1958: Chromosomal differentiation. Symposium on the chemical basis of development; edited by W. D. McElroy and B. Glass, John Hopkins University Press, Baltimore, pp. 103—135.

Granoff, A., 1961: Induction of Newcastle virus mutants with NO_2H. Virology **13**, 402—408.

Greer, S., and S. Zamenhof, 1959: Loss of purine from DNA heated in mutagenic conditions at physiological pH. Fed. Proc. **18**, 238.

Grossman, L., S. S. Levine, and W. S. Allison, 1961: The reaction of formaldehyde with nucleotides and T 2 bacteriophage DNA. J. Mol. Biol. **3**, 47—60.

Guild, W. R., 1957: Evidence related to the effects of moderate doses on DNA in cells. Rad. Res. Soc. Abstracts **55**, 28.

Hagen, U., 1960: Metal linkages in the thymus nucleohistone and their damage by X radiation. Biochem. J. **76**, 56 P—57 P.

Hall, B. D., and S. Spiegelman, 1961: Sequence complementarity of T 2 DNA and T 2 RNA. Proc. nat. Acad. Sci. U. S. **47**, 137—146.

Hutchinson, F., 1961: Molecular basis for action of ionizing radiations. Science (Washington) **134**, 533—538.

Hsu, T. C., and C. E. Somers, 1961: Effect of 5 bromodeoxyuridine on mammalian chromosomes. Proc. nat. Acad. Sci. U. S. **47**, 396—403.

Jayson, G., 1961: Bivalent metal ions as the coupling factor between cell metabolism and the rate of cell mutation. Nature (London) **190**, 144—146.

Jeener, R., and Rosseels J., 1953: Incorporation of 2-H thiouracil ^{35}S in the ribose nucleic acid of TMV. Biochim. biophysica Acta **11**, 435—438.

Josse, J., A. D. Kaiser, and Kornberg A., 1961: Enzymatic synthesis of DNA. VIII. Frequencies of nearest neighbor base sequences in DNA. J. biol. Chemistry **236**, 864—875.

Kanazir, D., 1958: The apparent mutagenicity of thymine deficiency. Biochim. biophysica Acta **30**, 20—23.

Kaplan, R. W., V. Winkler, H. Wolff-Ellmauer, 1960: Induction and reversion of C-mutations by irradiation of the extracellular x phage of *Serratia marcescens*. Nature (London) **186**, 330—331.

Karnofsky, D. A., and R. S. Basch, 1960: The effect of 5 fluorodeoxyuridine and related halogenated pyrimidines on the sand dollar embryo. J. Biophys. Biochem. Cytol. **7**, 61—71.

Kaudewitz, F., W. Vielmetter, and H. Friedrich-Freksa, 1958: Mutagene Wirkung des Zerfalles von radioaktivem Phosphor nach Einbau in Zellen von *E. coli*. Z. Naturforsch. **13b**, 793—802.

Kaudewitz, F., 1959: Inaktivierende und mutagene Wirkung salpetriger Säure auf Zellen von *E. coli*. Z. Naturforsch. **14b**, 528—537.

Kaudewitz, F., 1961: Bakteriengenetik, Methoden und Ergebnisse. Naturwissenschaften **48** (8), 276—282.

Kirby-Smith, J. S., and D. L. Craig, 1957: The induction of chromosome aberrations in *Tradescantia* by UV radiation. Genetics **42**, 176—187.

Kirby, K. S., 1958: Preparation of some DNA protein complexes from rat liver homogenates. Biochem. J. **70**, 260—265.

Kozloff, L. M., 1960: Biochemistry of animal viruses. Annual Rev. Biochem. **29**, 475—502.

La Cour, L. F., and S. R. Pelc, 1959: Effect of colchicine on the utilization of thymidine labelled with tritium during chromosomal reproduction. Nature **183**, 1455—1456.

Lerman, L. S., and L. S. Tolmach, 1959: Genetic transformation. II. The significance of damage to the DNA molecule. Biochim. biophysica Acta **33**, 371–387.

Levine, R. P., 1955: Chromosome structure and the mechanism of crossing over. Proc. nat. Acad. Sci. U. S. **41**, 727–730.

Litman, R. L., and H. Ephrussi-Taylor, 1959: Inactivation et mutation des facteurs génétiques de l'ADN du pneumocoque par l'ultra violet et l'acide nitreux. C. R. hebd. Séances Acad. Sci. **249**, 838–840.

Litman, R. L., and A. B. Pardee, 1956: Production of bacteriophage mutants by a disturbance of DNA metabolism. Nature (London) **178**, 529–531.

Litman, R. L., and A. B. Pardee, 1960: The induction of mutants of bacteriophage T2 by 5 bromouracil. III. Nutritional and structural evidence regarding mutagenic action. Biochim. biophysica Acta **42**, 117–130.

Lorkiewicz, Z., and W. Szybalski, 1960: Genetic effects of halogenated thymidine analogs incorporated during thymidylate synthetase inhibition. Biochem. Biophys. Res. Comm. **2**, 413–418.

Marcovitch, H., 1960: International symposium on primary and initial effects of ionizing radiations on living cells, Moscow. Abstracts, p. 54.

Marmur, J., W. F. Anderson, L. Matthews, K. Berns, E. Gajewska, D. Lane, and P. Doty, 1961: The effects of UV light on the biological and physical properties of desoxyribonucleic acids. J. Cell. Comp. Physiol. **58** (Suppl.), 33–55.

Matheson, A. T., and C. A. Thomas, 1960: The indirect effects accompanying ^{32}P suicide of bacteriophage. Virology **11**, 289–291.

McDonald Green, D., and D. R. Krieg, 1961: Delayed origin of mutants induced by exposure to ethylmethanesulfonate. Proc. nat. Acad. Sci. U. S **47**, 64–72.

McQuade, H. A., and M. Friedkin, 1960: Radiation effects of thymidine ^{3}H and thymidine ^{14}C. Exper. Cell Res. **21**, 118–125.

Meselson, M., F. W. Stahl, and J. Vinograd, 1958: The replication of DNA in *E. coli*. Proc. nat. Acad. Sci. U. S. **44**, 671–682.

Mirsky, A. E., and H. Ris, 1951: The composition and structure of isolated chromosomes. J. Gen. Physiol. **34**, 475–492.

Mundry, K. W., and A. Gierer, 1958: Die Erzeugung von Mutationen des TMV durch Behandlung seiner Nucleinsäure in vitro. Z. Vererbungslehre **89**, 614–630.

Mundry, K. W., 1959: The effect of nitrous acid on TMV: mutation, not selection. Virology **9**, 722–726.

Natarajan, A. T., 1961: Chromosome breakage and mitotic inhibition induced by tritiated thymidine in root meristems of *Vicia faba*. Exper. Cell Res. **22**, 275–281.

Nisman, B., and B. Fukuhara, 1960: Rôle de l'ADN dans la synthèse des protéines par les deux fractiom enzymatique liées à des particules de *E. coli*. C. R. hebd. Séances Acad. Sci. **250**, 410–412.

Novelli, G. D., T. Kameyama, and J. M. Eysenstadt, 1961: The effect of ultraviolet light and X rays on an enzyme forming system. J. Cell. Comp. Physiol. **58**, (Suppl.), 225–244.

Ord, M. G., and L. A. Stocken, 1960: Structural changes in DNA after X radiation in vitro. Biochim. biophysica Acta 37, 352–353.

Ord, M. G., and L. A. Stocken, 1961: In: Mechanisms in radiobiology. Editors: M. Errera and A. Forssberg, Academic Press, N. Y.

Phillips, J. N., 1961: Modification of mutation frequency in microorganisms. I. Preirradiation and postirradiation treatments. Genetics **46**, 317–322.

Ris, H., 1958: Chromosome structure. The chemical basis of heredity. John Hopkins University Press, Baltimore, pp. 23–62.

Rosenberg, B. H., F. M. Sirotnak, and L. F. Cavalieri, 1959: Size of the genetic determinants in *Pneumococcus* and the nature of the variables. Proc. nat. Acad. Sci. U. S. **45**, 144–156.

Rudner, R., 1960: Mutation as an error in base pairing. Biochem. Biophys. Res. Comm. **3**, 275–286.

Scholes, G., and J. Weiss, 1961: This symposium.

Schuster, H., and G. Schramm, 1958: Bestimmung der biologisch wirksamen Einheit in der RNS des TMV auf chemischem Weg. Z. Naturforsch. **13b**, 697–704.

Schuster, H., 1960: Die Reaktionsweise der DNS mit salpetriger Säure. Z. Naturforsch. **15b**, 298–304.

Schuster, H., 1960: The reaction of DNA with nitrous acid. Biochem. Biophys. Res. Comm. **2**, 320–323.

Schwartz, D., 1958: Desoxyribonucleic acid side chain model of chromosomes. Nature (London) **181**, 1149–1150,

Shapiro, A., and E. Chargaff, 1960: Severe distortion by 5-BrU of the sequence characteristics of bacterial DNA. Nature (London) **188**, 62–64.

Siegel, A., 1960: Studies on the induction of TMV mutants with nitrous acid. Virology 11, 156–167.

Sinsheimer, R., 1959: A single stranded DNA from bacteriophage ⌀x 174. J. Mol. Biol. 1, 43–53.

Stadler, L. J., and F. M. Uber, 1942: Genetic effects of UV radiation in maize. IV. Comparison of monochromatic radiations. Genetics 27, 84–118.

Steffensen, D., 1955: Chromosome rupture in *Tradescantia* by calcium deficiency. Proc. nat. Acad. Sci. U. S. 41, 155–160.

Stent, G. S., 1958: Mating in the reproduction of bacterial viruses. Advances in Virus Research 5, 95–160.

Stent, G. S., 1960: Genetic and physiological effects of the decay of incorporated radioactive phosphorus in bacterial viruses and bacteria. Advances Biol. Med. Phys. 7, 2–76, Acad. Press, New York/London.

Taylor, J. H., 1959: Symposium Molecular Biology. Editor: R. Zirkle, University of Chicago Press, Chicago, Ill., pp. 304–320.

Tessman, I., 1959: Mutagenesis in phage ⌀x 174 and T4. Virology 9, 375–389.

Tsugita, A., and H. Fraenkel-Conrat, 1960: The amino-acid composition and C terminal sequence of a chemically evoked mutant of T.M.V. Proc. nat. Acad. Sci. U. S. 46, 636–642.

Vaharu, T., 1961: Modifications of UV induced mutation frequency in *Neurospora crassa*. Genetics 46, 247–256.

Van Dyke, J. G., 1959: Oxygen effect of X irradiation in intracellular phage T2Hr. Biochem. Biophys. Res. Comm. 1, 54–56.

Vielmetter, W., and H. Schuster, 1960: The base specificity of mutation induced by nitrous acid in phage T2. Biochem. Biophys. Res. Comm. 2, 324–328.

Vielmetter, W., und C. M. Wieder, 1959: Mutagene und inaktivierende Wirkung salpetriger Säure auf freie Partikel der Phagen T2. Z. Naturforsch. 14b, 312–317.

Volkin, E., and L. Astrachan, 1956: Phosphorus incorporation in *E. coli* RNA after infection with bacteriophage T2. Virdogy 2, 433–437.

Weygand, F., A. Wacker, and H. Dellweg, 1952: Stoffwechseluntersuchungen bei Mikroorganismen mit Hilfe radioaktiver Stoffe. Z. Naturforsch. 7b, 19–25.

Witkin, E., 1956: Time, temperature and protein synthesis in study of UV induced mutation in bacteria. Cold Spring Harbor Symp. 21, 123–138.

Witkin, E., and E. C. Theil, 1960: The effect of post irradiation posttreatment with chloramphenicol on various UV induced mutations in *E. coli*. Proc. nat. Acad. Sci. U. S. 46, 226–231.

Wolff, Sh., 1961: Radiation genetics; in Mechanisms in Radiobiology. Editors: M. Errera and A. Forssberg, Academic Press Inc., N. Y.

Wollman, E. L., F. Jacob, and W. Hayes, 1956: Conjugation and genetic recombination in *E. coli* K12 Cold Spring Harbor Symp. 21, 141–162.

Zamenhof, S., and Griboff, G., 1954: Incorporation of halogenated pyrimidines into the DNA of *Bacterium coli*. Nature (London) 174, 306–307.

1954: *E. coli* containing 5-BrU in its DNA. Nature (London) 174, 307–308.

Zamenhof, S., 1959: Further studies on the unstable and unnatural deoxyribonucleic acids. Ann. N. Y. Acad. Sc. 81, 784–787.

Zubay, G., and P. Doty, 1959: The isolation and properties of desoxyribonucleoprotein particles containing single nucleic acid molecules J. Mol. Biol. 1, 1–20.

Straleninduzierte Mutagenese – Gesichtspunkte des Genetikers

R. Rieger und H. Böhme

Institut für Kulturpflanzenforschung Gatersleben der Deutschen Akademie der Wissenschaften zu Berlin

I. Einleitung

Seit den Pionierarbeiten von Muller (1927, 1928) und Stadler (1928) in den USA, Nadson und Phillipov (1925) in der UdSSR, Stubbe (1930) in Deutschland und einer ganzen Reihe anderer Autoren in verschiedenen Ländern steht es fest, daß die Einwirkung energiereicher Strahlen Veränderungen des genetischen Materials auszulösen vermag. Diese Veränderungen werden im allgemeinen in zwei nicht scharf voneinander abzugrenzende Kategorien, die Genmutationen und Chromosomenaberrationen, eingeteilt, wobei die letztgenannten mikroskopisch sichtbare Strukturveränderungen an den Chromosomen darstellen. Mutationen und Aberrationen, auf die sich die folgenden Ausführungen im wesentlichen beziehen, stellen nur eine Komponente aus der Vielzahl strahleninduzierter Veränderungen bei biologischen Objekten dar, die untereinander ohne Zweifel in vielfachen Wechselbeziehungen stehen. Eine Beschränkung auf die genetische Komponente stellt also unter allen Umständen eine starke Vereinfachung der tatsächlichen Geschehnisse dar, ist aber hier unvermeidlich. Darüber hinaus soll im folgenden auch nicht der Versuch unternommen werden, eine balancierte Übersicht über alle Aspekte der strahleninduzierten Mutagenese zu geben. Wir wollen statt dessen jene Phänomene und Probleme etwas ausführlicher darstellen, die uns im Zusammenhang mit einer möglichen Deutung des Mutationsprozesses von besonderer Bedeutung erscheinen. Dabei mußten verschiedene Faktoren und ihr Einfluß auf das Mutationsgeschehen (etwa die Bedeutung des Wassergehaltes und der Temperatur) unberücksichtigt bleiben. Im Mittelpunkt unserer Ausführungen werden die Abhängigkeiten des Mutationsprozesses von der Art und Applikationsweise der energiereichen Strahlung und seine Beeinflussungsmöglichkeiten durch genetische und nicht-genetische Faktoren stehen. Abschließend soll der Versuch gemacht werden, einige Schlußfolgerungen über mögliche Ablaufformen des Mutationsprozesses aus der Sicht des Genetikers zu ziehen.

II. Einflüsse auf den strahleninduzierten Mutationsprozeß: Strahlenquantität und -qualität

Die Beziehungen zwischen eingestrahlter Strahlendosis, Dosisrate (Intensität) sowie Strahlenqualität und den ausgelösten strahlengenetischen Effekten lassen sich etwa wie folgt zusammenfassen:

1. Unabhängig von der Strahlenqualität ist im Bereich nicht allzu hoher Dosen ionisierender Strahlenarten die Rate der induzierten Genmutationen und Chromosomenbrüche im

allgemeinen der Dosis direkt proportional. Diese lineare Dosis-Effekt-Beziehung dürfte auch im Bereich sehr niedriger Strahlendosen gelten, obwohl hier nur sehr wenige und zudem nicht widerspruchsfreie Ergebnisse vorliegen (SPENCER und STERN, 1948; BONNIER und LÜNING, 1949; DEMEREC, 1959). Bei Einwirkung hoher Strahlendosen können die Dosis-Effekt-Kurven einem Sättigungsniveau zustreben (ATWOOD und MUKAI, 1954; KAPLAN, 1956; GWATKIN und GOTTLIEB, 1956; RUSSEL und RUSSEL, 1958) oder nach einem Maximum auch wieder abfallen (NEWCOMBE und MCGREGOR, 1954; ALIKHANIAN et al., 1957). Nicht lineare Dosis-Effekt-Beziehungen sind besonders häufig nach Einwirkung von UV zur Beobachtung gekommen (s. KAPLAN, 1959).

Für jene Gruppe induzierter Chromosomenaberrationen, an denen nach der Bruch-Reunions-Hypothese mehrere Brüche beteiligt sind, gilt ein linearer Anstieg mit steigender Strahlendosis, wenn dicht ionisierende Strahlenarten wie Neutronen und α-Strahlen zur Anwendung kommen (GILES, 1940, 1943; THODAY, 1942). Erfolgt demgegenüber ihre Auslösung durch Strahlenarten mit niedriger Ionisationsdichte (Röntgen- und γ-Strahlen), ergibt sich eine exponentielle Dosis-Effekt-Beziehung, wobei die Häufigkeit der Aberrationen mit einem Exponenten von 1,5 bis 2 zunimmt, wenn die Dosis in kurzer Zeit appliziert wird (SAX, 1940; THODAY, 1951; GILES, 1954; KAUFMANN, 1954). Lineare Dosis-Effekt-Beziehungen für den gleichen Aberrationstyp einerseits und exponentielle andererseits machen es wahrscheinlich, daß zur Einleitung der Aberrationen bei Einwirkung von Strahlenarten mit niedriger Ionisationsdichte zwei Primärereignisse (Brüche) unabhängig voneinander erforderlich sind, während im Falle dicht ionisierender Strahlen der Durchgang eines ionisierenden Partikels zur gleichzeitigen Einleitung beider Primärereignisse ausreicht (SAX, 1940; RICK, 1940; MARINELLI et al., 1942).

2. Die Veränderung der Dosisrate durch Protrahierung oder Fraktionierung bleibt im allgemeinen ohne Einfluß, wenn es sich um induzierte Genmutationen handelt (MULLER, 1954, 1957). Das gleiche gilt für Mehrbruch-Chromosomenaberrationen dann, wenn sie durch dicht ionisierende Strahlenarten induziert werden. Kommen statt dessen Strahlenarten mit niedriger Ionisationsdichte zur Einwirkung, wird ein Dosisrateneinfluß erkennbar, der sich darin ausdrückt, daß protrahierte oder fraktionierte Bestrahlung im allgemeinen (s. FETTNER, 1956; ABRAHAMSON, 1959) zu einer geringeren Ausbeute an „Mehrfach-Aberrationen“ führt (SAX, 1939; FABERGÉ, 1940; GILES, 1943, 1954; WOLFF et al., 1954, 1955, 1956; BEATTY et al., 1956). Der Verdünnungs- und Fraktionierungseffekt läßt sich nach der Bruch-Reunions-Hypothese so deuten, daß in diesem Falle auf Grund von Restitutionsvorgängen zur Vorbruchstruktur weniger Bruchflächen zur Reunion zur Verfügung stehen, da ein Teil der zuerst induzierten Brüche bereits restituiert ist, wenn die zur Kooperation notwendigen weiteren Brüche in der zur Reunion erforderlichen Entfernung eintreten. Von WOLFF und ATWOOD (1954) und DESCHNER und SPARROW (1955) konnte in diesem Zusammenhang zusätzlich noch ein dosisabhängiger Strahleneinfluß auf die Reunionsrate der Brüche nachgewiesen werden, so daß ein erholungsfähiges Reunionssystem postuliert wird, das unter Strahleneinfluß geschädigt werden kann (WOLFF und LUIPPOLD, 1956).

3. Die relative biologische Wertigkeit (RBW) von Strahlenarten ist mit der unterschiedlichen Ionisationsdichte bzw. unterschiedlichen linearen Energieübertragung pro Einheit Weglänge der ionisierenden Partikel korreliert. Es liegen heute wohl genügend brauchbare Daten vor, die die Schlußfolgerung gestatten, daß die RBW der Strahlenarten im allgemeinen mit steigender Ionisationsdichte im Hinblick auf alle genetischen Effekte ansteigt (BAKER und v. HALLE, 1951; KING und WOOD, 1955; EDINGTON, 1956; EDINGTON und RANDOLPH, 1958; ALEXANDER, 1958). Im Falle dominanter Letalfaktoren und Chromosomenaberrationen ist die RBW der dichter ionisierenden Strahlen im allgemeinen höher als bei der Auslösung rezessiver Letalfaktoren und Genmutationen. Das Wirkungsverhältnis variiert in

Abhängigkeit vom Einzelfall nach Gray (1959) in weiten Grenzen etwa um einen Faktor 1000, wenn die schnellsten und langsamsten im strahlenbiologischen Experiment eingesetzten Partikel zum Vergleich kommen. Nur in wenigen Fällen wurde festgestellt, daß die RBW bei steigender linearer Energieübertragung absank oder unbeeinflußt blieb (Übersichten bei Zirkle, 1954; Bacq und Alexander, 1958). Qualitative Unterschiede in den ausgelösten strahlengenetischen Effekten treten bei Verwendung von Strahlenarten verschiedener Ionisationsdichte im allgemeinen nicht auf.

Innerhalb des für Mutationsexperimente verwendeten UV-Wellenbereiches von etwa 2000 bis 3000 Å sind klare quantitative, mit der Wellenlänge korrelierte Unterschiede in bezug auf die Auslösung von Genmutationen und Chromosomenaberrationen nachgewiesen worden. Im allgemeinen wurde gefunden, daß das Wirkungsspektrum mit dem UV-Absorptionsspektrum der Nukleinsäuren (Maximalwirkung bei 2600 Å), in einigen Fällen auch mit dem der Proteine (McAuley et al., 1947) oder DNS-Basen (Wirkungsmaximum bei 2800 Å) zusammenfällt. Im Bereich der Wellenlänge von 2000 bis 3000 Å bewirkt UV bei einer Quantenenergie von nur einigen eV lediglich Anregungen und keine Ionisationen, während die ionisierenden Strahlenarten ihre Gesamtenergie beim Eintritt in die Materie etwa zur Hälfte in Anregungen, zur Hälfte in Ionisationen umsetzen. Bei Verwendung von UV zur Auslösung von Genmutationen und Chromosomenaberrationen hängt die Energieabsorption vom Molekulargewicht der durchstrahlten Materie ab und variiert mit der chemischen Konstitution des biologischen Systems. In energetischer Hinsicht ist UV im Vergleich mit ionisierenden Strahlenarten nur wenig wirksam (Setlow, 1957).

Dieser notwendigerweise sehr kurze Überblick über den Einfluß von Strahlenquantität und -qualität auf die ausgelösten strahlengenetischen Effekte läßt eine Reihe von Schlußfolgerungen zu, die im wesentlichen auch der Treffer- und Treffbereichstheorie zugrunde liegen. Diese „Theorien" stellen auch heute noch, trotz mancher Diskussionen, das einzige wirklich geschlossene Hypothesensystem zur Deutung quantitativer Strahleneffekte dar und dürften unter Berücksichtigung einer Reihe neuer Erkenntnisse weiterhin die Grundlage strahlenbiologischer Interpretationen bleiben. Im Hinblick auf die Natur des Mutationsprozesses ergibt sich auf Grund der bisher referierten Fakten nur eine begrenzte Aussagemöglichkeit, die im wesentlichen die Einleitung des Geschehens und seine Kinetik betrifft und sich etwa so formulieren ließe: Lineare und exponentielle Dosis-Effekt-Beziehungen lassen unter Berücksichtigung des Einflusses der Dosisrate und der Ionendichte der Strahlung den Schluß zu, daß der Strahlenwirkung primär quantenhafte Elementarprozesse zugrunde liegen. Der Durchgang eines ionisierenden Partikels durch sensible Bereiche (Treffbereiche), bzw. die Partikelpassage in deren unmittelbarer Umgebung, gibt zu physikalischen Elementarakten (Anregungen, Ionisationen) Anlaß, denen chemische Änderungen der Treffbereiche folgen, die über mehr oder weniger lange Reaktionsketten zum erkennbaren genetischen Strahlenschaden führen. Ob ein Elementarereignis zur Einleitung des Strahlenschadens genügt oder ob mehrere kooperieren müssen, hängt vom Einzelfall ab und ist mit dem Typ des Effektes und der linearen Energieübertragung der benutzten Strahlenart deutlich korreliert.

Offen bleibt nach den bisherigen Erörterungen die Frage, welche Vorgänge durch die lokalisierte Energiezufuhr im Treffbereich (direkte Strahlenwirkung) oder in dessen Nähe (indirekte Wirkung) ausgelöst werden und welche Zeit zu ihrer Komplettierung zur Verfügung steht. Klare Zeitabhängigkeiten ergeben sich bei Einwirkung von Strahlenarten mit niedriger Ionisationsdichte bei der Auslösung komplexer Chromosomenaberrationen, die im allgemeinen nach der Bruch-Reunions-Hypothese gedeutet werden. Es muß aber darauf hingewiesen werden, daß der primäre Eintritt von „Brüchen" nicht als eindeutig bewiesen gelten darf. Er wird durch die von Revell (1955, 1959) aufgestellte Austauschhypothese verneint, und die Fülle der von Revell zusammengetragenen Argumente gegen die Bruch-Reunions-

Hypothese macht es sehr deutlich, mit welchen Schwierigkeiten die Deutung strahlengenetischer Befunde verknüpft ist.

Unabhängig davon, ob der Interpretation des Zeitfaktors bei der Entstehung von sog. Mehr-Bruch-Aberrationen die Bruch-Reunions- oder die Austauschhypothese zugrunde gelegt wird, leitet sich daraus ein Zweiphasensystem ab: 1. Die Chromosomen erfahren beim Durchgang ionisierender Partikel direkt oder indirekt einen Primärschaden, der mehr oder weniger instabil ist. 2. Paarweise auftretende Schädigungen dieser Art können unter der Voraussetzung, daß sie zeitlich und räumlich dicht genug beieinander eintreten, kooperieren und einen sekundären, stabilen Zustand, die in der Metaphase faßbaren Chromosomenaberrationen, erreichen. Zeitfaktor-Effekte sind immer dort zu erwarten, wo der Strahlenschaden schon mit Beginn seiner Einleitung Restaurierungen erfährt oder wo der realisierte sichtbare Effekt von der Wechselwirkung zwischen den Ergebnissen von zwei oder mehr unabhängigen Primärereignissen abhängt, die schon während der Einstrahlung der Restaurierung unterliegen. Die Bedeutung des Zeit-Intensitätsfaktors für die Interpretation der Mutationsprozesse liegt auf der Hand.

Im Zusammenhang mit der Tatsache, daß sich die strahlengenetischen Effekte mit steigender Ionisationsdichte der verwendeten Strahlenart pro Einheit der Weglänge des Partikels verstärken, dürfte sich andeuten, daß eine gewisse Mindestzahl gleichzeitig eintretender Elementarprozesse (Anregungen, Ionisationen) pro Raumeinheit die Auslösung genetischer Strahlenschäden positiv beeinflußt. Zur Zeit der starken Betonung radiochemischer Befunde zur Wasserzersetzung und ihrer Bedeutung für die Strahleneffekte in vivo schien sich eine relativ zwanglose Erklärung des Einflusses der Ionisationsdichte auf die biologischen Reaktionen in der unterschiedlichen Anordnung der strahleninduzierten OH- und H-Radikale des Wassers auf Grund verschiedener Ionisationsverteilung anzubieten. In der Zwischenzeit sind aber u. a. von ALPER (1956), ZIMMER (1956, 1960) und ALEXANDER (1960) ernsthafte Einwände gegen diese Interpretation erhoben worden. So wichtig eine genaue Kenntnis der RBW von Strahlen mit hoher und niedriger Ionisationsdichte vom Gesichtspunkt möglicher Deutungen des strahlengenetischen Schadens auch sein mag, so schwierig erscheint ihre Deutung gegenwärtig ohne genauere Vorstellungen von der Natur, der relativen Bedeutung und dem Anteil direkter und indirekter Strahlenwirkungen in vivo.

Das letztgenannte Problem macht u. a. auch den Versuch, aus den quantitativen Befunden zur Strahlenbiologie auf die Größe, Zahl und schließlich Form der Treffbereiche schließen zu wollen, fragwürdig. Dieser Versuch setzt voraus, daß die Treffer in der unmittelbaren submikroskopischen Umgebung des genetischen Materials bzw. in ihm selbst erfolgen, daß jede Ionisation im betreffenden Treffbereich zur Mutation oder Aberration führt, daß Anregungen derartige Effekte nicht auslösen und alle Mutationen identifizierbar sind. Da keine dieser Prämissen vollgültig sein dürfte, müssen derartige Kalkulationen fragwürdig bleiben, zumal außerdem Gen und sensibles Volumen nicht identisch zu sein brauchen.

In jüngster Zeit ist die Kenntnis über den Mutationsprozeß insofern über Aussagen zum physikalischen Primärereignis hinausgekommen, als intermediäre Reaktionen zwischen „Treffer" und Mutationskomplettierung nachgewiesen werden konnten, die neue Wege zur experimentellen Erforschung der dem Primärereignis folgenden Phasen des Mutationsprozesses freilegten, unabhängig davon, ob der Treffer direkt oder indirekt erfolgte. Es ist heute sicher, daß verschiedene Faktoren die Strahlenreaktionen verändern können. Diese Tatsache findet ihren Ausdruck darin, daß die Wirkung der gleichen Menge absorbierter Energie der eingestrahlten Photonen oder Partikel je nach den Bestrahlungsbedingungen verschieden sein kann und vom biologischen System beeinflußt wird. Erste Hinweise in dieser Richtung gehen schon auf die Anfänge strahlenbiologischer Forschung zurück mit der Feststellung, daß verschiedene Organismen, Organe, Zellen, Zellstadien und Loci unterschiedlich strahlensensibel sind. Dieses Phänomen ist das Thema der folgenden Erörterungen, wobei zunächst die genetischen und dann die nicht-genetischen, den Mutationsprozeß beeinflussenden Faktoren dargestellt werden sollen, ohne daß angestrebt wird, eine vollständige Übersicht zu geben.

III. Einflüsse auf den strahleninduzierten Mutationsprozeß: Genetische Faktoren

Relativ starke Unterschiede in der mutagenen Wirksamkeit der verschiedenen Strahlenarten ergeben sich beim Vergleich der Raten strahleninduzierter Mutationen zwischen Angehörigen verschiedener systematischer Kategorien. Aus den Untersuchungen von RUSSEL (1951, 1956) an 7 Loci der Maus ergibt sich z. B. für röntgenbestrahlte Spermatogonien eine durchschnittliche Mutationsrate von 25×10^{-8} pro r pro Locus. Dieser Wert ist wesentlich höher, als entsprechende Untersuchungen für *Drosophila melanogaster* ergaben. ALEXANDER (1954) fand in Versuchen mit 8 Loci an diesem Objekt bei der Bestrahlung reifer Spermien 6×10^{-8} Mutationen pro r pro Locus. Unter Berücksichtigung der Tatsache, daß die Strahlensensibilität von Spermatogonien etwa halb so groß ist wie die der Spermien, ergibt sich für die Maus eine etwa 6- bis 8mal höhere mutagene Wirksamkeit der Röntgenstrahlen. So wichtig diese Angaben auch für die Übertragung der Ergebnisse experimenteller strahlengenetischer Untersuchungen etwa auf den Menschen sind, so lassen sie in dem hier diskutierten Zusammenhang doch keine Deutung der möglichen Ursachen anderer Sensibilitätsunterschiede und keine Abschätzung des relativen Anteiles genetischer und nicht-genetischer Faktoren zu.

In ähnlichem Maße trifft dies auch für die bei *Drosophila* beobachteten Unterschiede in der induzierten Mutationsrate zwischen den beiden Geschlechtern zu. Generell findet man eine höhere strahleninduzierte Mutabilität im männlichen Geschlecht. Die Stärke des Unterschiedes ist, wie z. B. aus den Untersuchungen von GLASS (1956) hervorgeht, vom Stadium der Gametogenese abhängig, in dem die Bestrahlung vorgenommen wurde. OSTER (1958) fand bei Verwendung der empfindlichsten Stadien der Spermiogenese, den Spermatiden, und der resistentesten Stadien der Oogenese, den Oogonien, eine um etwa das 12fache höhere Rate induzierter rezessiver Letalfaktoren im männlichen Geschlecht. Eine Reihe von Faktoren kann zur Erklärung dieser Sensibilitätsunterschiede der Geschlechter herangezogen werden. Für Chromosomenaberrationen ist zunächst der unterschiedliche meiotische Entwicklungszustand, in dem sich z. B. Spermatozoen und reife Oozyten befinden, von Bedeutung. Nach GLASS (1956) läßt sich jedoch nur ein kleiner Anteil der Differenz in der induzierten Mutabilität beider Gametentypen mit diesem Unterschied erklären. Hinzu kommen – jedoch nur im Hinblick auf Chromosomenaberrationen – die unterschiedlichen räumlichen Verhältnisse in Spermatozoen und Oozyten sowie die unterschiedlich starken Chromosomenbewegungen, die beide zu einer höheren Wahrscheinlichkeit von Translokationen im männlichen Geschlecht führen. Inwieweit neben diesen und ähnlichen sekundären Erscheinungen eine primär verschieden starke Mutabilität des genetischen Materials beider Geschlechter vorliegt, läßt sich aus den vorliegenden experimentellen Befunden nicht eindeutig ableiten.

Der Einfluß des genetischen Materials selbst auf die mutagene Wirksamkeit der verschiedenen Strahlenarten läßt sich in den Ermittlungen der induzierten Mutationsrate spezifischer Loci gut erkennen. Aus den Ergebnissen der Versuche an höheren Organismen (Übersicht über die älteren Arbeiten bei MULLER, 1952) sowie vor allem an Mikroorganismen lassen sich zusammenfassend drei Erscheinungskomplexe ableiten:

1. **Die unterschiedliche Sensibilität verschiedener Loci eines Genotyps gegenüber der mutagenen Wirkung der gleichen Strahlenart und -dosis.** Valencia und Muller (1949) wiesen relativ starke Unterschiede in der Mutationsbereitschaft zwischen verschiedenen Genen bei *Drosophila melanogaster* nach. In ihrer Untersuchung von 15 Loci des X-Chromosoms ergaben sich nach Bestrahlung mit 5000 r Röntgenstrahlen als Extremwerte (umgerechnet in Mutationsrate/r): für das Gen *garnet* ($+ \rightarrow g$) $5 \cdot 10^{-8}$ und für das Gen *raspberry* ($+ \rightarrow r$) $1{,}6 \cdot 10^{-9}$. Unterschiede in ähnlicher Größenordnung lassen sich auch aus den Angaben über die bereits zitierten Arbeiten an den 7 Loci der Maus von Russel entnehmen.

Methodisch einfacher und in größerem Maßstab lassen sich die Versuche an spezifischen Loci bei Mikroorganismen durchführen. Die von Kølmark angestellten Rückmutationsversuche an einer Adenin- und Inositol-Doppelmutante von *Neurospora crassa* ergaben eine sehr schwache mutative Reaktion des *inos*$^-$-Allels auf Röntgenstrahlen (0,2 *inos*$^+$/10^6 bestrahlter Makrokonidien) gegenüber einer um mehr als das 15fache höheren Mutantenhäufigkeit beim *ad*$^-$-Allel. Auf UV-Bestrahlung reagieren die beiden Allele umgekehrt: Am *inos*$^-$-Allel wird nach Bestrahlung mit gleicher Dosis die doppelte Mutantenhäufigkeit (7,1/10^6) beobachtet wie beim *ad*$^-$-Allel (3,5/10^6) (s. Westergaard, 1960).

Die größten Erfahrungen liegen in dieser Beziehung bei *Escherichia coli* vor (Demerec, 1955; Glover, 1956). Etwa 10 bis 15% der bisher untersuchten Allele dieses Objekts sind absolut mutagenstabil, d. h., weder mutagene Strahlen noch Chemikalien induzieren Mutationen, obwohl eine spontane Mutabilität beobachtet wird. Bei *Salmonella typhimurium* ist der Anteil mutagenstabiler Allele noch höher (Hartman, 1956). Die quantitativen Unterschiede verschiedener Gene ein und desselben Stammes gegenüber der mutagenen Wirkung mutagener Strahlenarten sind bei *E. coli* besonders stark ausgeprägt. So schwankt die röntgenstrahleninduzierte Mutantenhäufigkeit bei 5 untersuchten Genen eines Stammes z. B. zwischen 2680/10^8 Überlebenden für das Allel *leu-6* und 54/10^8 für das Allel *ar-2*. Bei UV-Bestrahlung beobachtete Glover Unterschiede in analoger Größenordnung.

Zu dem gleichen Phänomen, nämlich der unterschiedlichen Sensibilität oder Reaktionsbereitschaft verschiedener Gene eines Genotyps gegenüber dem gleichen Mutagen, zählen wir ebenfalls die in den letzten Jahren in zunehmendem Umfang beobachtete Erscheinung des Auftretens verschiedener Mutantenspektren nach Anwendung verschiedener Mutagene. Die meisten Beobachtungen hierzu liegen zwar auf dem Gebiet der chemischen Mutagenese vor, jedoch konnten in einzelnen Fällen auch Unterschiede zwischen verschiedenen Strahlenarten ermittelt werden. Das gilt z. B. für die von Newcombe (1952) bei *E. coli* festgestellten Unterschiede im Verhältnis dreier – sicher alleler – Mutantentypen nach UV- und γ-Bestrahlung und für analoge andere Fälle, bei denen z. T. auch Unterschiede im Spektrum in Abhängigkeit von der Dosis der angewandten Strahlung mitgeteilt werden (Plough, Young und Grimm, 1950; Alikhanian und Klepikova, 1957).

Was die Verteilung der strahleninduzierten Chromosomenaberrationen über die Einzelchromosomen und den Chromosomensatz anbelangt, so wird in vielen Arbeiten (z. B. Ford, 1949; Revell, 1953) eine zufallsgemäße, längenproportionale Verteilung angegeben. Im Gegensatz hierzu steht eine Reihe neuerer Arbeiten, in denen u. a. Röntgen- und γ-Strahlen zur Einwirkung kamen (Rieger und Michaelis, 1959; Evans und Bigger, 1961; ältere Literatur bei Sparrow, 1951). Wie weit eine in der Metaphase nach der Bestrahlung festgestellte, nicht-zufallsgemäße Aberrationsverteilung über die Chromosomen Ausdruck einer nicht-zufallsgemäßen Verteilung der strahleninduzierten Primärereignisse ist, muß als offenes Problem bezeichnet werden. Auch die Verteilung der strahleninduzierten Aberrationen zwischen den Zellen wird im allgemeinen als zufallsgemäß angegeben und soll der Poisson-Verteilung folgen (Übersicht bei Lea, 1955). Klare Abweichungen von dieser Regel sind in jüngster Zeit bekannt geworden (JACKSON und BARBER, 1958; ATWOOD und WOLFF, 1959; BENDER und WOLFF, 1961; EVANS und BIGGER, 1961).

2. Die unterschiedliche Sensibilität des gleichen Allels gegenüber der mutagenen Wirkung verschiedener Strahlenarten. Die bereits zitierten Untersuchungen an *Neurospora*, *E. coli*, *Salmonella typhimurium* und anderen Objekten erbrachten auch den Nachweis dafür, daß das gleiche Gen sehr unterschiedlich auf die verschiedenen mutagenen Reize reagieren kann. Während z. B. das Allel *inos*$^-$ der von KØLMARK verwendeten Doppelmutante auf Röntgenstrahlen nur sehr schwach reagiert (0,2/10^6), werden durch UV-Bestrahlung 7,1 *inos*$^+$ Mutationen pro 10^6 bestrahlter Konidien induziert. Das Gen *ar-1* von *E. coli* (GLOVER, 1956) ist gegenüber Röntgenstrahlen mutagen-stabil, reagiert auf UV jedoch mit einer Mutantenhäufigkeit von 718/10^8. Ohne weitere Beispiele anzuführen, sei auf einen wichtigen Befund hingewiesen, der sowohl bei *Neurospora*, *E. coli* als auch *Salmonella* gemacht wurde: In ihrer Reaktion auf verschiedene Mutagene verhalten sich nicht identische Allele (mutational sites oder mutons) ein und desselben Genlocus verschieden. Die sog. Mutagenspezifität ist daher nicht eine Eigenschaft des Genlocus, sondern des mutierenden Allels.

3. Die unterschiedliche Sensibilität des gleichen Allels gegenüber der mutagenen Wirkung derselben Strahlenart und -dosis in Abhängigkeit vom genotypischen Milieu. Die Beeinflussung der strahleninduzierten Inaktivierung durch das genotypische Milieu ist aus Untersuchungen an verschiedenen Organismen bekannt. So wird darüber z. B. in den Arbeiten von LASKOWSKI und STEIN (1960) an Hefe und LAMPRECHT an *Phaseolus* (1958) berichtet. In unserem Zusammenhang sind besonders die Ergebnisse von GLOVER (1956) von Bedeutung, die zeigen, daß die mutative Reaktion eines Allels bereits durch das Fehlen oder die Anwesenheit eines Allels eines anderen Genlocus grundlegend verändert werden kann. Das bereits mehrfach erwähnte Allel *ar-1*$^-$ von *E. coli* ist in Kombination mit dem Normalallel *try-5*$^+$ gegenüber Röntgenstrahlen mutagen-stabil. In Verbindung mit dem mutierten Allel *try-5*$^-$ werden durch Röntgenstrahlen jedoch Mutationen mit einer Häufigkeit von 39/10^8 Überlebenden induziert. Das Allel *ar-3*$^-$ reagiert in der Kombination *ar-3*$^-$; *leu-3*$^+$ auf UV-Bestrahlung mit 486 Mutationen/10^8; bei gleicher UV-Dosis beträgt die Anzahl von Mutationen bei der Kombination *ar-3*$^+$; *leu-3*$^-$ jedoch 2900/10^8. Analoge Einwirkungen anderer Gene auf die mutative Reaktion nach UV bzw. Röntgenbestrahlung wurden u. a. am Allel *str-d-4* festgestellt.

Die wenigen hier angeführten Beispiele zeigen deutlich, daß die einzelnen Loci eines Genotyps quantitativ unterschiedlich auf die mutagene Strahlenwirkung reagieren. Über die Ursachen dieser Reaktionsunterschiede können zum gegenwärtigen Zeitpunkt nur Vermutungen geäußert werden. Zunächst ist bei den Fällen, in denen sich die Mutationsrate zweier Loci nach Behandlung mit der gleichen Strahlenart und -dosis eindeutig unterscheidet, an Unterschiede in der Struktur dieser Loci zu denken. Beweiskräftig können in diesem Sinne jedoch nur Experimente sein, in denen entweder Mutationsschritte vom Normalallel zum mutierten Allel geprüft werden oder – im Falle von Reversionen – wenn Doppelmutanten Verwendung finden. Die oben angedeuteten Strukturunterschiede wären damit auch gleichzeitig die Ursache für das Auftreten unterschiedlicher „Mutantenspektren“ nach Behandlung mit verschiedenen mutagenen Strahlen oder chemischen Mutagenen. Neben möglichen Strukturunterschieden im genetischen Material spielen bei dem hier besprochenen Problem der sog. elektiven Mutabilität zweifellos auch Differenzen in der Wirkungsweise der mutagenen Faktoren eine deutlich zu erkennende Rolle. Diese manifestieren sich in Sensibilitätsunterschieden der mutativen Reaktion des gleichen Locus auf verschiedene mutagene Reize. Wie z. B. die Untersuchungen von DE SERRES (1958, 1959) am *ad-3A*- und *ad-3B*-Cistron von *Neurospora* zeigen, muß in derartigen Fällen jedoch zunächst an die Existenz mehrerer Mutationsorte (mutational sites) innerhalb eines „Locus“ gedacht werden, die, durch Strukturdifferenzen bedingt, auf die verschiedenen Mutagene unterschiedlich reagieren.

Die Verwendung von chemischen Verbindungen gestattet es infolge der Vielfalt der strukturell nicht verwandten, mutagenen Chemikalien (Basenanaloge, alkylierende Verbindungen, Peroxyde usw.), den Mutationsprozeß auf stärker verschiedenartigen Wegen (Wirkungsmechanismen) einzuleiten, als dies bei der Verwendung verschiedener Strahlenarten möglich ist. Wenn der Wirkungsmechanismus eines Mutagens für die unterschiedliche, mutationsauslösende Potenz verschiedener mutagener Faktoren gegenüber ein und demselben Mutationsort verantwortlich zu machen ist, dann ist a priori das Auffinden von Fällen elektiver Mutabilität in Versuchen mit chemischen Mutagenen häufiger zu erwarten als beim Vergleich verschiedener Strahlenarten. Die bisherigen Erfahrungen bestätigen diese Erwartung völlig.

Schließlich sind die erwähnten Beispiele des Einflusses des genotypischen Milieus – in einigen Fällen nur eines mutierten Allels – auf die Mutabilität eines Locus ein Zeichen für die starke Sensibilität der Prozesse, die nach dem Eintritt des Primärereignisses bis zur Fixierung des neuen allelen Zustandes ablaufen. Wenn zwar auch noch in keinem Falle experimentell nachgewiesen, so ist doch mit hoher Wahrscheinlichkeit damit zu rechnen, daß der von dem modifizierenden Gen herbeigeführte veränderte physiologische Zustand der Zelle oder einzelner ihrer Bereiche für die Beeinflussung des an einem anderen Locus ablaufenden Mutationsprozesses verantwortlich ist.

Zusammenfassend zeigt die Betrachtung des Einflusses genetischer Faktoren auf den strahleninduzierten Mutationsvorgang, daß dieser das Ergebnis des wechselseitigen Zusammenwirkens zwischen mutagenem Agens und seiner für die Wirkung charakteristischen Eigenschaften, der spezifischen Struktur des mutierenden Locus und des durch Genotyp und Außenbedingungen definierten Zustandes der Zelle ist.

IV. Einflüsse auf den strahleninduzierten Mutationsprozess: Nicht-genetische Faktoren

a) Der Einfluß von Entwicklungsstadium und Wachstumsphase

Die nicht-genetischen Faktoren, die über eine Veränderung des physiologischen Zustandes der Zelle einen Einfluß auf den Ablauf des strahleninduzierten Mutationsprozesses ausüben, sind außerordentlich zahlreich, und es kann hier nur eine relativ unvollständige Übersicht gegeben werden. Allen die Strahlenwirkung quantitativ modifizierenden nicht-genetischen Faktoren ist gemeinsam, daß ihr Modifizierungsvermögen mit steigender linearer Energieübertragung der Strahlung abnimmt.

Die unterschiedlich starke mutagene Wirkung von Strahlen auf Zellen in verschiedenen Entwicklungsstadien ist bei *D. melanogaster* intensiv untersucht worden (Auerbach, 1954; Kishin, 1955; Lüning, 1952; Oster, 1958). Ohne auf Einzelheiten einzugehen, herrscht Übereinstimmung zwischen den Autoren, daß Spermatiden die stärkste Sensibilität gegenüber der letalen und mutagenen Wirkung ionisierender Strahlen aufweisen. Die Sensibilität steigt, angefangen von den Spermatogonien, als dem resistentesten Stadium der Spermatogenese, über die primären Spermatozyten bis zu den Spermatiden an; reife Spermatozoen sind jedoch wieder deutlich resistenter gegenüber der Strahlenwirkung. Diese Abstufung in der Reaktionsempfindlichkeit der verschiedenen Entwicklungsstadien ist für Chromosomenaberrationen deutlicher ausgeprägt als für Genmutationen. Als mögliche Ursache für die hohe Sensibilität der Spermatiden ist wohl vor allem an den von Oster (1956, 1958) und

Fritz-Niggli (1958) in Betracht gezogenen hohen intra- oder interzellulären Sauerstoffdruck im Vergleich mit anderen Stadien der Spermatogenese zu denken. Nach Sobels (1961) liegt möglicherweise in diesem Stadium auch eine geringere Restitutionschance für die eingetretenen Chromosomenbrüche vor, da das Reunionssystem in den Spermatiden leichter zu schädigen ist. Daneben spielen sicher auch die germinale Selektion (für Chromosomenaberrationen) und die unterschiedlichen Stoffwechselverhältnisse der verschiedenen Stadien eine gewisse Rolle. Die letztgenannte Ursache ist vor allem aus den Befunden von Bonnier und Lüning (1953) sowie Abrahamson und Telfer (1954) abzuleiten, nach denen die Strahlensensibilität von reifen Spermatozoen bei Bestrahlung im Rezeptakulum des Weibchens größer ist als bei Bestrahlung im männlichen Organismus. In diesem Zusammenhang sei auch die bereits seit längerer Zeit bekannte Tatsache erwähnt, daß die Strahlensensibilität der verschiedenen Stadien der Mitose, gemessen an der Häufigkeit induzierter Chromosomenaberrationen, unterschiedlich ist.

Auch bei Mikroorganismen läßt sich ein deutlicher Einfluß des Entwicklungsstadiums, in dem die Bestrahlung vorgenommen wird, auf die Inaktivierung und Mutationsauslösung feststellen. Zunächst ist ein deutlicher Sensibilitätsanstieg gegenüber der letalen und mutagenen Wirkung von ionisierenden Strahlen und UV-Licht während der Sporenkeimung festzustellen (Meffert und Wyss, 1951; Newcombe, 1956). Weiterhin verändert sich der Sensibilitätsgrad gegenüber der inaktivierenden Wirkung ionisierender Strahlen innerhalb des Wachstumszyklus einer Bakterienkultur (Stapleton, 1955); Zellen aus der logarithmischen Wachstumsphase, d. h. solche Zellen, die sich intensiv teilen und eine hohe Stoffwechselaktivität aufweisen, sind wesentlich sensibler als Zellen aus der lag-Phase und der stationären Phase. Sowohl aus den Untersuchungen von Kimball, Gaither und Wilson (1957) an *Paramaecium* mit Röntgenbestrahlung als auch aus den Angaben von Haas und Doudney (1957) über die Induktion von Mutationen durch UV bei *E. coli* geht eine Beziehung zwischen der Sensibilität der mutativen Reaktion auf Bestrahlung und der Kernteilung bzw. der Nukleinsäuresynthese hervor. Bei *Paramaecium* liegt das Maximum der Sensibilität in der ersten Hälfte der Interphase, und bei *E. coli* wird die höchste Mutantenrate bei Bestrahlung vor der Nukleinsäure-Verdopplung („Kernteilung") erzielt.

Neben den inneren Bedingungen wie Alter, Entwicklungsstadium und Wachstumsphase haben äußere Faktoren einen bedeutenden Einfluß auf die Strahlensensibilität des gleichen Zelltyps. Das zeigt sich besonders klar im Falle des sog. Sauerstoffeffektes.

b) Der Einfluß von Sauerstoff während der Bestrahlung

Erhöht sich nach Bestrahlung unter normaler Belüftung oder in reinem Sauerstoff das Ausmaß der induzierten strahlengenetischen Effekte, wenn mit dem Bestrahlungsergebnis in Stickstoff verglichen wird, wird von einem „Sauerstoffeffekt" gesprochen. Seit seiner Wiederentdeckung durch Thoday und Read (1947) und der Feststellung des Ausbleibens einer sensibilisierenden Wirkung des Sauerstoffes bei Einwirkung von α-Strahlen (Thoday und Read, 1949) sind zahlreiche Untersuchungen zum Problem Sauerstoffeffekt durchgeführt worden, deren wesentlichste Ergebnisse sich etwa so zusammenfassen lassen:

1. Die Sensibilität biologischer Systeme gegenüber Strahlenarten mit niedriger Ionisationsdichte wird in Sauerstoffgegenwart unabhängig von der Temperatur im Bereich von 0 bis 37°C etwa um den Faktor 2,5 bis 3 erhöht; bei steigender Ionisationsdichte schwächt sich der Effekt ab (Neutronen) und fehlt bei Verwendung von α-Strahlen vollständig (Howard-Flanders, 1958). Ebenso wie für α-Strahlen ist für UV kein Sauerstoffeffekt nachweisbar (Kirby-Smith und Craig, 1957). Bei Verwendung von Strahlenarten mit niedriger Ioni-

sationsdichte blieb ein Sauerstoffeffekt im Sinne einer Strahlensensibilisierung nur in wenigen Fällen aus (s. GRAY, 1959). Sauerstoffeffekte sind sowohl für Genmutationen (z. B. GLASS und METTLER, 1958) als auch für Chromosomenaberrationen nachgewiesen worden, wobei die Sensibilisierung durch Sauerstoffgegenwart im ersten Falle wohl im allgemeinen kleiner ist.

2. Die Tatsache, daß es keine Rolle spielt, ob der Sauerstoff z. B. durch Stickstoff, Stickoxydul, CO_2, Azetylen oder Wasserstoff ersetzt wird, zeigt an, daß das Fehlen von Sauerstoff und nicht die Verwendung von Stickstoff für die Reduktion strahlengenetischer Effekte verantwortlich ist. Aus den Untersuchungen von KIHLMAN (1955, 1959) und anderen Autoren geht hervor, daß eine Unterdrückung des Sauerstoffeffektes für strahleninduzierte Chromosomenaberrationen durch Respirationsinhibitoren nicht möglich ist. Daraus ergibt sich, daß der Sauerstoffeffekt im Gegensatz zu Befunden mit bestimmten Radiomimetika (KIHLMAN, 1955, 1956; RIEGER und MICHAELIS, 1960) nicht mit oxydativen Stoffwechselprozessen verbunden ist, die über die Respirationskette verlaufen. Durch die Benutzung von Respirationsinhibitoren zur Ausschaltung von Sauerstoffgradienten war es außerdem möglich, Sauerstoffeffekte in Geweben schon bei sehr niedrigen, unter 21% liegenden Sauerstoffkonzentrationen nachzuweisen. Liegt keine totale Anaerobiose vor, kann auch die Temperatur über indirekte Einflüsse auf den Respirationskoeffizienten die Strahlensensibilität beeinflussen, wie DESCHNER und GRAY 1959 zeigen konnten.

3. Nach ALPER und HOWARD-FLANDERS (1956) läßt sich für Mikroorganismen die Sensibilität gegenüber Röntgenstrahlen bei verschiedenen Sauerstoffkonzentrationen zur Zeit der Bestrahlung durch die folgende Gleichung ausdrücken: $\frac{S}{S_N} = \frac{m\,(O_2) + K}{(O_2) + K}$. Dabei bedeuten S die Sensibilität bei der vorliegenden Sauerstoffkonzentration (O_2), S_N die Sensibilität bei Sauerstoffabschluß. Der Faktor m symbolisiert das Verhältnis der zur Auslösung eines bestimmten Effektes benötigten Dosen unter Anaerobiose und Sauerstoff. K ist diejenige Sauerstoffkonzentration in μM, die eine Sensibilität von $(m + 1)/2 \times$ Stickstoffsensibilität ergibt. Nach Ausschaltung von Sauerstoffgradienten sind die für Mikroorganismen und Gewebe höherer Formen gefundenen K-Werte größenordnungsmäßig gleich (HOWARD-FLANDERS, 1958; HOWARD-FLANDERS und JOCKEY, 1960; DESCHNER und GRAY, 1959; DEWEY, 1960; KIHLMAN, 1958, 1959).

4. Daß der Sauerstoffeffekt bei der Induktion von Chromosomenaberrationen auf eine erhöhte Bruchzahl und nicht auf unterschiedliche Reunionschancen bei Aerobiose und Anaerobiose (z. B. SCHWARTZ, 1952; BAKER und v. HALLE, 1955) zurückzuführen ist, wurde von GILES und RILEY (1950) festgestellt und sowohl in Fraktionierungsexperimenten von WOLFF et al. (1954, 1955, 1956) als auch bei einmaliger Dosisapplizierung von NEARY und EVANS (1958) bestätigt. Nach WOLFF et al. (1954, 1955) gilt der Sauerstoffeffekt für die Induktion von Brüchen und für die Schädigung des Reunionssystems, die sich im längeren Offenbleiben der Brüche ausdrückt. Die Wahrscheinlichkeit der Teilnahme an auswertbaren Aberrationen wird jedoch durch das geschädigte, aber erholungsfähige Reunionssystem nicht erhöht (WOLFF, 1957).

Die Korrelation zwischen Sauerstoffeffekt und Ionisationsdichte der verwendeten Strahlung löste eine breite Diskussion zum Problem der indirekten und direkten Strahlenwirkung aus, in deren Mittelpunkt die radiochemischen Befunde zur Wasserzersetzung durch ionisierende Strahlenarten standen.

In diesem Zusammenhang wiesen bereits THODAY und READ (1949) auf eine mögliche Bedeutung von strahleninduziertem H_2O_2 für die Auslösung genetischer Strahlenschäden hin. Ausgangspunkt dafür war der Befund, daß H_2O_2 bei Einwirkung ionisierender Strahlen mit niedriger linearer Energieübertragung nur in Sauerstoffgegenwart, bei Einwirkung von Strahlen mit hoher linearer Energieübertragung auch unter anaeroben Verhältnissen entsteht.

Außerdem stand die Bedeutung des Hydroperoxyl-Radikals (HO_2) zur Debatte, das nur in Sauerstoffgegenwart gebildet wird (Gray, 1953). Auf die in den letzten Jahren gegen die strahlenbiologische Bedeutung von H_2O_2 und HO_2 vorgebrachten Argumente wurde bereits hingewiesen. Die Frage einer Beteiligung strahlenchemisch erzeugter Wasserzersetzungsprodukte muß als völlig offen bezeichnet werden, da außerdem von Howard-Flanders (1957), Gray et al. (1958), Kihlman (1958, 1959), Dewey (1960), Capps (1961) und Dale et. al. (1961) gezeigt werden konnte, daß einerseits Stickoxyd (NO) den Sauerstoff zu vertreten vermag und der Effekt andererseits durch inerte Gase unter Druck unterdrückt wird. In diesem Zusammenhang weisen aber Dale et al. (1961) darauf hin, daß bei *Shigella flexneri* keine völlige Äquivalenz in der Wirkung beider Gase besteht, und nach Powers et al. (1959, 1960) sowie Sparrman et al. (1959) vermittelt NO bei Bestrahlung im trockenen Zustand einen Strahlenschutzeffekt.

Howard-Flanders (1958, 1959) sieht eine Deutung der Abhängigkeit des Sauerstoffeffektes von der Ionisationsdichte der Strahlung in der Annahme, daß in den Molekülen des genetischen Materials unter Strahleneinfluß entstehende kurzlebige Carbonradikale die primäre Veränderung darstellen. Diese Veränderung könnte bei Sauerstoffgegenwart durch eine Ionisation entstehen, während bei Anaerobiose mehrere Ionisationen notwendig wären. Die Inaktivierung des in Frage stehenden Moleküls soll dann durch Reaktion mit einem anderen Radikal erfolgen, bei dem es sich um Sauerstoff, NO oder ein strahleninduziertes OH-Radikal handeln könnte. Bei Strahlenarten mit niedriger Ionisationsdichte wäre die Entfernung zwischen dem Carbonradikal und den entstehenden OH-Radikalen im allgemeinen zu groß, um die notwendige Reaktion einzuleiten, so daß der Effekt von der Gegenwart von Sauerstoff oder Stickoxyd abhängig wäre. Die Inaktivierung des Moleküls könnte mit der Entstehung von C-O- oder C-N-Bindungen im Zusammenhang stehen (Howard-Flanders und Jockey, 1960).

So interessant diese nur angedeuteten und etwas simplifizierten Spekulationen sein mögen, so muß doch festgestellt werden, daß es gegenwärtig unbekannt ist, welcher oder welche Mechanismen dem O_2-Effekt zugrunde liegen und welche Faktoren über sein Ausbleiben entscheiden. Experimentell begründete Vorstellungen in dieser Richtung versprechen wichtige Aufschlüsse über den strahleninduzierten Mutationsprozeß zu bringen.

Nur am Rande sei erwähnt, daß der Sauerstoffeffekt sicherlich auch eine Rolle bei der Herabsetzung der Strahleneffekte durch sog. chemische Strahlenschutzstoffe, auf die nicht näher eingegangen werden kann, spielt (Burnett et al., 1951; Stapleton et al., 1952), obwohl hier in Abhängigkeit vom Einzelfall auch andere Interpretationen herangezogen werden müssen (Übersichten bei Eldjarn und Pihl, 1960; Hollaender, 1960). Die gleiche Deutung, nämlich eine Beeinflussung der Sauerstoffspannung, dürfte häufig auch dort gelten, wo die strahleninduzierte Mutations- oder Aberrationsrate mit steigender Temperatur absinkt, da das Gewebe bei niedriger Temperatur mehr Sauerstoff enthält. Hinweise in dieser Richtung liefert für geschlechtsgekoppelte rezessive Letalfaktoren bei *Drosophila* z. B. eine Arbeit von Baker und Sgourakis (1950). Auf alle Fälle ist aber bei Vorliegen eines Einflusses der Temperatur auf die Mutationsraten zu bedenken, daß Temperaturveränderungen vielgestaltige physiologische Veränderungen in den Zellen nach sich ziehen können.

Die bisher notwendigerweise sehr summarisch geschilderten Befunde zum Sauerstoffeffekt hatten zur Voraussetzung, daß der Sauerstoff zur Zeit der Bestrahlung zugegen war. Ohne Zweifel bezieht sich diese Sauerstoffwirkung auf die einleitende Phase der Mutationsprozesse, für die damit feststeht, daß sie umweltbeeinflußbar ist. Daß die beeinflußbare Reaktion oder Substanz sehr kurzlebig ist, geht klar aus Untersuchungen von Dewey und Boag (1959) hervor, die eine hohe Elektronendosis in 2 μ sec applizierten. Die aerob gehaltenen Bakterien (1% Sauerstoff in der Gasphase) reagierten auf diese Behandlung mit der Sensibilität annähernder Anaerobiose, woraus sich ergibt, daß der vorhandene Sauerstoff in den Zellen bereits während der Einstrahlung des ersten Teiles der Dosis aufgebraucht wurde und nicht schnell genug durch Diffusion ersetzt werden konnte, um aerobe Sensibilität aufrechtzuerhalten.

c) Photoreaktivierung und Photorevertierung

Der Befund, daß ein Teil der inaktivierenden und auch der mutagenen Wirkung der UV-Strahlen durch eine Nachbehandlung mit sichtbarem Licht rückgängig gemacht werden kann (KELNER, 1949), zeigte mit Eindeutigkeit, daß die UV-induzierte Mutationsauslösung nicht im Augenblick der Energieabsorption sofort zur stabilen Veränderung des genetischen Materials führt. Wir können auf die zahlreichen Einzelfragen, die mit der Photoreaktivierung und -revertierung zusammenhängen, hier nicht eingehen (Übersichten u. a. bei JAGGER, 1958, 1960; DULBECCO, 1955) und wollen nur auf einige Aspekte dieser Erscheinung hinweisen, die nach unserer Ansicht für das Verständnis des UV-induzierten Mutationsprozesses von Bedeutung sind.

Es ist zweckmäßig, bei der Erörterung des Mechanismus der Photoreaktivierung zwischen dem Chromophor, also der Struktur, die das reaktivierende Licht absorbiert, und dem „reaktivierbaren Zentrum" als dem Träger der UV-induzierten Veränderung, die durch die Photorevertierung rückgängig gemacht wird, zu unterscheiden (JAGGER, 1958). Photoreaktivierende Wirksamkeit zeigt der Wellenlängenbereich von 3130 bis 5490 Å, während die mutagene Wirkung des UV-Lichtes auf den Bereich von 2540 Å, das Absorptionsmaximum der Nukleoproteine, konzentriert ist. Dieses unterschiedliche Wirkungsspektrum ist nicht unbedingt als Beweis für die Annahme zu werten, daß das Chromophor nicht mit dem durch UV-Strahlen veränderten genetischen Material identisch ist, da zumindest der inaktivierende Prozeß im Sinne des Quantenbedarfs wesentlich effektiver ist als die Photoreaktivierung; auch nur schwach absorbierende Strukturen könnten daher in Betracht gezogen werden. Auf der anderen Seite zeigen Experimente wie die von SKREB und ERRERA (1957), die einen etwa gleichen photoreaktivierenden Effekt in kernhaltigen und kernlosen Bezirken von Amöben fanden, daß die Identität von Chromophor und Nukleoproteinen nicht notwendigerweise vorauszusetzen ist. Auch über die Lokalisation des reaktivierbaren bzw. revertierbaren Zentrums kann heute noch keine eindeutige Aussage gemacht werden. Ohne die Möglichkeit der Photorevertierbarkeit von Veränderungen anderer Zellbestandteile damit auszuschließen, steht zumindest fest, daß UV-induzierte Schädigungen bzw. Veränderungen von Nukleoproteinen und der DNS mit Licht der angegebenen Wellenlängen revertierbar sind. Das geht besonders aus den Untersuchungen von GOODGAL, RUPERT und HERRIOTT (1957) und RUPERT, GOODGAL und HERRIOTT (1958) hervor, die erstmalig die Möglichkeit der in-vitro-Photoreaktivierung von Transformations-DNS aus *Hemophilus influenza* nachwiesen. Die Tatsache, daß die in diesen Versuchen beobachtete Photoreaktivierung nur bei Anwesenheit eines proteinhaltigen Extraktes aus *E. coli* eintritt, läßt auf einen direkten, d. h. primär an der DNS ablaufenden Prozeß schließen, der jedoch enzymatisch kontrolliert wird. In diesem Zusammenhang ist ebenfalls wichtig, daß die in UV-bestrahlten Zellen gestoppte DNS- und RNS-Synthese photoreaktivierbar ist (IVERSON und GIESE, 1957; KELNER, 1953; u. a.). DUCHESNE und GARSON (1957) schlugen zur Deutung des Photoreaktivierungsmechanismus eine direkte Einwirkung des reaktivierenden Photons auf das DNS-Molekül vor, wobei der durch UV-Einwirkung induzierte metastabile Zustand zum Ausgangszustand zurückgeführt werden soll. Unter Berücksichtigung der Tatsache, daß eine in-vitro-Reaktivierung nur bei Anwesenheit eines Proteins möglich ist und die Photoreaktivierung in gefrorenen Zellen nicht eintritt, nimmt JAGGER (1960) einen „komplexen direkten Mechanismus“ an, wobei eine prosthetische Gruppe des Proteins die Lichtenergie absorbiert. Durch die Anregung des Moleküls wird die enzymatische Regulierung einer direkten chemischen Reaktion mit der UV-geschädigten DNS möglich, die zur Reaktivierung führt. Es sei darauf hingewiesen, daß es sich bei den hier besprochenen Deutungsversuchen um das Reaktivierungsphänomen und nicht um die Mutationsreversion handelt. Verschiedene Untersuchungen (s. KAPLAN, 1959), auf die wir hier nicht eingehen können, zeigen, daß beide Prozesse zwar sehr viele gemeinsame Charakteristika besitzen, jedoch nicht notwendigerweise auf den gleichen Mechanismus zurückzuführen sind.

d) Der Einfluß kontrollierter Stoffwechselbedingungen

Die bisher besprochenen Modifikationen des strahleninduzierten Mutationsprozesses werden durch Faktoren hervorgerufen, deren primäre Wirkungsart auf die Zelle weitgehend unbekannt ist; es ist daher schwer, aus der Feststellung des modifizierenden Einflusses dieser Faktoren Rückschlüsse auf den konkreten Ablauf der Mutagenese zu ziehen. In diesem Sinne sind die im folgenden zu besprechenden Versuche zur Beeinflussung des Mutationsprozesses mit solchen Faktoren, deren Primärwirkung auf bestimmte Stoffwechselvorgänge wir mehr oder weniger genau kennen, zweifellos von besonderem Interesse. Selbstverständlich muß bei der Deutung derartiger Versuchsergebnisse immer in Betracht gezogen werden, daß ein solcher Faktor neben der beabsichtigten Hauptwirkung in der Zelle (vielleicht auch nur in der bestrahlten Zelle) unbekannte Nebenwirkungen entfalten kann, die für das Zustandekommen der Modifikation des Mutationsprozesses von Bedeutung sind.

Der Einfluß der Veränderung spezifischer Stoffwechselprozesse auf den Mutationsvorgang ist am besten im Zusammenhang mit der UV-Bestrahlung bekannt. Wir werden uns daher hier im wesentlichen auf die zusammenfassende Darstellung dieser Befunde konzentrieren und die Ergebnisse, die aus Versuchen mit ionisierenden Strahlen vorliegen, nur soweit möglich an den entsprechenden Stellen einflechten.

Von starkem Einfluß auf die UV-induzierte Mutabilität ist der Nukleinsäurestoffwechsel der Zelle. Es wurde bereits darauf hingewiesen, daß die induzierte Mutationsrate am höchsten ist, wenn die Bestrahlung kurz vor der Zellteilung und damit kurz vor der Beendigung der Nukleinsäuresynthese erfolgt. HAAS und DOUDNEY (1957) fanden eine beträchtliche Steigerung des mutagenen Effektes der UV-Bestrahlung durch eine Erhöhung des Gehaltes an Purin- und Pyrimidinbasen der RNS bzw. ihrer Riboside in der Zelle. Die entsprechenden Purine und Pyrimidine der DNS sind in diesem Zusammenhang unwirksam. Da Cytidin und Uridin die entscheidenden Faktoren dieser Vorbehandlung sind (DOUDNEY und HAAS, 1958), gehen die Autoren bei der Deutung ihrer Versuche von der Annahme aus, daß die primäre, UV-induzierte Veränderung an Vorstufen der RNS erfolgt. Der Einbau dieser veränderten Vorstufen führt zur Synthese einer veränderten RNS, und schließlich erfolgt während der DNS-Replikation die Übertragung der Veränderung von RNS auf DNS. Die Autoren sind sich bewußt (DOUDNEY und HAAS, 1959 b), daß sie mit dieser Annahme eine Vermittlerrolle der RNS bei der Übertragung der genetischen Information während der DNS-Synthese postulieren; sie folgen damit einem von STENT (1958) geäußerten Vorschlag des Ablaufes der DNS-Synthese, der den Vorstellungen von WATSON und CRICK entgegengestellt wurde. Wir gehen auf diesen Aspekt der Deutungsversuche des induzierten Mutationsprozesses hier nicht ein. Diese Problematik ist der Inhalt einer der wichtigsten Fragen des Genetikers an den Biochemiker!

WITKIN (1956, 1958) sowie DOUDNEY und HAAS (1958) konnten zeigen, daß die Hemmung der Proteinsynthese durch Chloramphenicol oder durch Bebrütung der bestrahlten auxotrophen Kultur im Medium ohne Aminosäuren die Höhe der induzierten Mutationsrate wesentlich herabsetzt. Ein gleicher Effekt wurde kürzlich (KADA, DOUDNEY und HAAS, 1960) auch nach Röntgenbestrahlung beobachtet. Die empfindliche Phase, in der die Ausbeute an Mutationen durch Blockierung der RNS- oder Proteinsynthese beeinflußt werden kann, wird zeitlich durch die der Bestrahlung folgende DNS-Replikation begrenzt. Die Länge dieses Zeitraumes ist eine Funktion der UV-Dosis, und KELNER (1953), KANAZIR und ERRERA (1956), HAROLD und ZIPORIN (1958), DOUDNEY (1959) sowie DRACULIC und ERRERA (1959) zeigten, daß sofort nach der Bestrahlung zunächst nur die DNS-Synthese eingestellt wird, und weiterhin, daß zur Wiederaufnahme der DNS-Synthese eine vorherige Protein- bzw. RNS-Synthese erforderlich ist. Das Absinken der Mutationsrate bei Hemmung der Proteinsynthese könnte im Hinblick auf diese Befunde durch die Verlängerung der Zeitspanne bis zur nächsten DNS-Replikation verursacht werden; die Chance der Reversion von prämutativen Zuständen wäre dadurch erhöht (KIMBALL, GAITHER und WILSON, 1959). Eine alternative Deutung (DOUDNEY und HAAS, 1959) postuliert erstens die Notwendigkeit einer enzymatisch kontrollierten Stabilisierung der veränderten RNS-Vorstufen und zweitens die Möglichkeit ihres aktiven Abbaues durch einen ebenfalls enzymatisch geregelten Prozeß. DOUDNEY und HAAS (1960) konnten eine Reihe von Versuchsergebnissen vorlegen, die die zweite der angeführten Hypothesen stark stützen.

Über die Beeinflussung der durch ionisierende Strahlen induzierten Mutagenese durch spezifische Stoffwechselprozesse liegen weniger Untersuchungen vor. Aus den Arbeiten von KIMBALL und Mitarbeitern (KIMBALL, 1958; KIMBALL, GAITHER und WILSON, 1957, 1959) geht hervor, daß alle Einflüsse (Streptomycin, Aminosäuremangel, niedrige Temperatur u. a.), die das Wachstum der bestrahlten Individuen hemmen oder verzögern, die Zeitspanne bis zur nächsten Teilung bzw. Chromosomenverdopplung erweitern und hierdurch die Möglichkeit für spontane Reversionen der vormutativen Zustände erhöhen; mit Vollzug der der Bestrahlung folgenden Teilung bzw. DNS-Synthese ist auch hier, nach Röntgenbestrahlung, die Mutation fixiert. Für das Wachstum suboptimale Bedingungen nach der Betrahlung beeinflussen, wie ALPER und GILLIES (1958) zeigten, die letale Wirkung niedriger Dosen in der gleichen Richtung.

Eine Hemmung der Proteinsynthese in Spermatiden und vor allem Spermatogonien durch Chloramphenicol reduziert die Rate röntgeninduzierter Letalfaktoren bei *Drosophila* beträchtlich (SOBELS, 1961). SOBELS führt diesen Einfluß in Analogie zu den Verhältnissen bei Mikroorganismen auf einen von der Proteinsynthese abhängigen Prozeß der Mutationsfixierung zurück.

Auch für die durch Röntgenstrahlen ausgelösten Chromosomenaberrationen sind in einigen Fällen Einflüsse spezifischer Stoffwechselprozesse auf den Aberrationsvorgang bekannt geworden. So konnte durch WOLFF und Mitarbeiter (WOLFF und LUIPPOLD, 1955, 1958; WOLFF, 1960) zuerst für *Vicia faba*, dann auch für andere Objekte gezeigt werden, daß sich die Fraktionierungseffekte durch Zugabe von Respirationsinhibitoren oder durch Entkopplung von oxydativer Phosphorylierung und Atmung nach der ersten Halbdosis auslöschen lassen. Bei der Induktion von Zweibruchaberrationen drückt sich das darin aus, daß Brüche der ersten Strahlendosis mit jenen der zweiten in Wechselwirkung treten können. WOLFFS auf der Bruch-Reunions-Hypothese basierende Interpretation geht dahin, daß für Restitution und Reunion energiereiche Bindungen im Verlauf der oxydativen Phosphorylierung bereitgestellt werden müssen. Diese Vorstellung wird dadurch gestützt, daß Zugabe von ATP die Zeit, während der offene Brüche zur Verfügung stehen, verkürzt. Der Befund, daß zur Schließung dieser Brüche energiereiches Phosphat zur Verfügung stehen muß, führt zu der Vermutung, daß Kovalenzbindungen beteiligt sind.

Auch im Aberrationsprozeß spielt die Proteinsynthese eine wesentliche Rolle. So bleiben nach WOLFF (1959, 1960) bei Einwirkung von Chloramphenicol und Aureomycin zwischen den fraktioniert gegebenen Röntgendosen die Brüche länger offen. Daraus wird geschlossen, daß an der Restitution eines Teiles der Brüche Peptidbindungen beteiligt sind. Ein anderer Bruchtyp schließt sich ohne Energiebedarf und offensichtlich auch ohne Beteiligung der Proteinsynthese sehr schnell, und es wird angenommen, daß in diesem Falle Ionenbindungen beteiligt sind (WOLFF, 1956; COHN, 1958).

Im Anschluß an die Arbeiten von WOLFF und Mitarbeitern sind eine ganze Reihe von Autoren in der Deutung ihrer Befunde WOLFF gefolgt (COHN, 1958; BEATTY, u. BEATTY, 1957, 1959; ABRAHAMSON, 1959; FALK, 1959; SOBELS, 1960, 1961). Es sei nur vermerkt, daß die WOLFFsche Deutung den tatsächlichen Eintritt von Brüchen zur Voraussetzung hat, und nur unter dieser Vorbedingung ist ein Schluß auf die Natur der beteiligten Bindungen möglich.

Natürlich gibt es neben den angeführten Arbeiten, in denen modifizierende Faktoren mit Wirkung auf spezifische Stoffwechselprozesse eingesetzt wurden, eine Vielzahl von Befunden zur Modifikation der Strahleneffekte durch Veränderung des allgemeinen physiologischen Zustandes der Zelle. Eine kausalanalytische Deutung ist jedoch infolge der komplexen Natur der in der Zelle hierdurch herbeigeführten Veränderungen außerordentlich schwierig. In diesem Zusammenhang wären besonders die Arbeiten von ADAMS et al. (1955, 1958), NILAN (1955), EHRENBERG (1955, 1959), CALDECOTT et al. (1955, 1957), CURTIS et al. (1957) und JACKSON (1959) zu erwähnen, in denen u. a. der Einfluß von Temperatur und Sauerstoff nach der Bestrahlung auf das Ausmaß der induzierten genetischen Effekte geprüft wurde.

Aus dieser notwendigerweise fragmentarischen Übersicht über die Modifikationsmöglichkeiten des Mutationsprozesses geht zweierlei mit Deutlichkeit hervor:

1. Es ist bisher in keinem Falle möglich, eine Ablaufform des strahleninduzierten Mutationsprozesses lückenlos von der Energieabsorption bis zur Mutationsfixierung im Sinne definierter chemischer, biochemischer und physikochemischer Reaktionsschritte zu beschreiben.

2. Mit Sicherheit zeigen die Modifikationsmöglichkeiten jedoch, daß ein Teil der strahleninduzierten Mutationen auf einen aus mehreren Phasen bestehenden Mutationsprozeß zurückzuführen ist. Für den modifizierbaren Anteil von induzierten Genmutationen kann weiterhin festgestellt werden, daß die Modifikationsmöglichkeit durch die der Bestrahlung folgende DNS-Synthese begrenzt wird. Weiterhin ist die RNS- bzw. Proteinsynthese notwendige Voraussetzung zur Fixierung der Mutation.

Sowohl nach der UV-Bestrahlung als auch nach Einwirkung ionisierender Strahlenarten verbleibt ein im Einzelfall unterschiedlich großer Anteil nicht modifizierbarer Primärereignisse (z. B. ALPER und GILLIES, 1958; WITKIN und THIEL, 1960). Es muß dabei z. Z. noch offenbleiben, ob sich der modifizierbare und nichtmodifizierbare Anteil im Mechanismus der Primärreaktionen grundsätzlich unterscheiden.

V. Einige Überlegungen zum Ablauf des Mutationsprozesses

Die stufenweise Entstehung komplexer Chromosomenaberrationen ist ein schon seit langer Zeit bekanntes Phänomen. Eine der wichtigsten strahlenbiologischen Entdeckungen der letzten Zeit ist die Feststellung, daß die Induktion von Genmutationen zumindest durch Strahlenarten mit niedriger linearer Energieübertragung (LEÜ) und durch UV-Licht ebenfalls ein milieubeeinflußter, in Phasen ablaufender Prozeß ist. Nach den vorliegenden Befunden lassen sich zumindest drei Etappen erkennen: die Einleitungsphase, die Praemutationsphase und die Fixierungsphase. Die Dauer der Praemutationsphase ist dabei von der Art des induzierenden Agens sowie von den inneren und äußeren Milieufaktoren abhängig. Dabei betrifft die Modifizierbarkeit, die nur an einigen Beispielen erläutert werden konnte, nicht nur relativ spät liegende Abschnitte des Geschehens, sondern auch seine Einleitung, wie der O_2-Effekt und Vorbehandlungen zeigen. Hierdurch haben die Vorstellungen über den strahleninduzierten Mutationsprozeß einige einschneidende Wandlungen erfahren, ohne daß sich der Gesamtprozeß heute bereits in allen seinen Einzelheiten überschauen ließe.

Unter den folgenden Prämissen wollen wir versuchen, ein Bild vom möglichen Ablauf des Mutationsprozesses zu geben, das sich im wesentlichen auf biologische Gesichtspunkte stützt und nicht mehr als eine Diskussionsgrundlage aus der Sicht des Genetikers darstellen soll:

1. Die Ionisationsdichte der Strahlung ist von wesentlicher Bedeutung für die Aktivierung der sensiblen Bereiche oder den Verlauf der der Aktivierung folgenden Reaktionen.

2. Die für die Mutations- und Aberrationsentstehung entscheidenden sensiblen Bereiche sind mit dem genetischen Material identisch; die primäre Veränderung des genetischen Materials kann direkt oder indirekt über diffusible, strahleninduzierte Reaktionspartner induziert werden.

Um den Ablauf des Mutationsprozesses bei Einwirkung von Strahlenarten mit niedriger Ionisationsdichte und unter der Voraussetzung einer direkten Strahlenwirkung zu interpretieren, kann man von der Vorstellung ausgehen, daß in der einleitenden Phase eine nicht ausreichende „Aktivierung“ des Treffbereiches im genetischen Material erfolgt. Diese Annahme kann aus den Beziehungen zwischen RBW und Ionisationsdichte der Strahlung abgeleitet werden. Der Verlauf einer der „Aktivierung“ schnell folgenden Reaktion kann durch Sauerstoff und Strahlenschutzstoffe beeinflußt werden. Fehlender Sauerstoff oder zugegebene Strahlenschutzstoffe könnten die Weiterentwicklung der eingeleiteten Veränderungen verhindern oder ihre Restaurierung begünstigen. Bei Vorhandensein von Sauerstoff (oder Stickoxyd) nimmt demgegenüber die Folgereaktion eine Richtung, die zur Fixierung der Veränderung (Mutation) führen kann. Bis zur endgültigen Fixierung bleibt ein labiler „Vormutationszustand“ erhalten. Diese endgültige Fixierung ist vom Vorherrschen bestimmter Stoffwechselverhältnisse abhängig, bei deren Fehlen der Vormutationszustand zum Ausgangsstatus zurückkehrt. Die Dauer der Fixierungsperiode wird im allgemeinen durch die nächste der Bestrahlung folgende Zellteilung begrenzt.

Das für die Mutationsauslösung durch ionisierende Strahlenarten mit niedriger LEÜ unter Annahme direkter Strahlenwirkung skizzierte Bild läßt sich mit gewissen Einschränkungen auf den UV-induzierten Mutationsprozeß übertragen, sofern auch hier von einer direkten Wirkungsweise ausgegangen wird. Die Einschränkungen beziehen sich auf die Tatsache, daß im Falle UV-induzierter Mutationen und Aberrationen keine Sauerstoffabhängigkeit der einleitenden Phase des Mutationsprozesses nachweisbar ist. Darüber hinaus ist ein Teil der Primärveränderungen photoreaktivierbar.

Wenn sich auch bestimmte Phasen des durch UV-Licht oder durch ionisierende Strahlen mit niedriger LEÜ induzierten Mutationsprozesses in gleicher Weise und mit dem gleichen Ergebnis modifizieren lassen, kann man daraus weder auf die völlige Identität noch auf die gleiche Lokalisation des von den beiden verschiedenen Strahlenarten verursachten Primärereignisses schließen. So konnte z. B. durch Elektronen-Spin-Resonanzuntersuchungen von KIRBY-SMITH und RANDOLPH (zit. nach WOLFF, 1960) festgestellt werden, daß nach Röntgenbestrahlung die Proteine, nach UV-Bestrahlung die Nukleinsäuren der Chromosomen besonders starke Signale geben.

Die Situation ändert sich, wenn an Stelle einer direkten eine indirekte Wirkungsweise vorliegt. Im Falle ionisierender Strahlenarten mit niedriger LEÜ könnte die sensibilisierende

Wirkung des Sauerstoffes in einer Beeinflussung der Bereitstellung mutagener, strahlenchemischer Reaktionsprodukte gesehen werden. Dabei könnten etwa Radikale vom Typ HO_2 oder H_2O_2 bzw. organische Peroxyde eine Rolle spielen. Die Argumente von ALPER (1956), ZIMMER (1956, 1960), DAINTON (1959), EHRENBERG (1959), ALEXANDER (1960) u. a. legen es jedoch nahe, zumindest dem HO_2-Radikal und möglicherweise auch dem Wasserstoffperoxyd keine zu große Bedeutung bei der strahleninduzierten Mutagenese zuzumessen. In diesem Zusammenhang sollte z. B. auch an die von HOWARD-FLANDERS (1958, 1959) angestellten, bereits erwähnten Betrachtungen zur Deutung des Sauerstoffeffektes gedacht werden.

Durch die Untersuchungen von WITKIN (1956, 1958), HAAS und DOUDNEY (1957) sowie DOUDNEY und HAAS (1959, 1960) u. a. Autoren können einzelne Phasen des UV-induzierten Mutationsprozesses bei Mikroorganismen bereits besser verstanden werden. HAAS und DOUDNEY nehmen eine indirekte Wirkungsweise an, bei der zunächst eine strahleninduzierte Veränderung von RNS-Bausteinen eintritt; hierauf würde sich eine Übertragung dieser Veränderungen auf die DNS während deren Replikation anschließen. Auch bei diesem Mechanismus ist ein Einfluß des Stoffwechsels, insbesondere der Proteinsynthese, auf die Mutationsfixierung nachweisbar.

Bei Einwirkung von Strahlenarten mit hoher LEÜ konnte nach den bisher vorliegenden Befunden kein oder kein deutlich ausgeprägter Sauerstoffeffekt gefunden werden. Auch für mögliche spätere Phasen des Mutationsprozesses sind Beeinflussungsmöglichkeiten unter diesen Umständen nicht bekannt geworden. Im Falle direkter Strahlenwirkung wäre auf Grund dieses Tatbestandes bei sehr hoher Ionisationsdichte eine größere Wahrscheinlichkeit zur ausreichenden Aktivierung des Treffbereiches im genetischen Material anzunehmen. Ebenso könnte unter diesen Umständen die Revertierungswahrscheinlichkeit der der Aktivierung unmittelbar folgenden Primärreaktion sehr klein und praktisch unbeeinflußbar sein. Weiterhin wäre eine sofortige Fixierung der Veränderung im genetischen Material aus dem Ausbleiben der Modifizierungsmöglichkeit des Mutationsvorganges durch Milieufaktoren abzuleiten. Die Annahme einer indirekten Wirkung bei Verwendung dicht ionisierender Strahlen würde voraussetzen, daß auch in diesem Falle sowohl die primäre strahleninduzierte Reaktion als auch die Folgereaktionen nicht zu beeinflussen sind.

Zur Interpretierung jenes Anteiles genetischer Strahleneffekte, die nach Einwirkung von Strahlenarten mit niedriger linearer Energieübertragung auch bei Anaerobiose auftreten, wäre davon auszugehen, daß auch bei Einwirkung von Strahlen mit niedriger Ionisationsdichte neben den nicht ausreichenden Aktivierungen der Treffbereiche ein Anteil der Treffer zu ausreichenden Aktivierungen oder zu Reaktionen führt, die nicht auf Sauerstoff angewiesen sind. Für diese Fälle gilt das für dicht-ionisierende Strahlen Gesagte.

Die bisherigen Erörterungen lassen sich sowohl auf die Entstehung von Genmutationen als auch von Chromosomenaberrationen anwenden, wobei allerdings spezifische Modifikationen in Abhängigkeit davon erforderlich werden, ob im letzten Falle der Deutung die Bruch-Reunions-Hypothese oder die Austauschhypothese zugrunde gelegt wird. Eine eindeutige Entscheidung zugunsten der einen oder anderen Hypothese ist auf Grund der vorliegenden Befunde gegenwärtig nicht möglich. Außerdem ist im Falle von Chromosomenaberrationen, an denen zwei oder mehr Chromosomen beteiligt sind, bei ihrer Induktion durch Strahlenarten mit niedriger Ionisationsdichte die Bedeutung des Raumfaktors zu berücksichtigen, der beispielsweise durch Zentrifugierung vor, während und nach der Bestrahlung beeinflußt werden kann (SAX, 1943; WOLFF und v. BORSTEL, 1954; ANDERSON, 1955). Allgemein läßt sich feststellen, daß durch Strahlen mit niedriger Ionisationsdichte induzierte Chromosomenaberrationen durch jede Einwirkung nach Zahl und Art modifiziert werden können, die in der Lage ist, entweder die Zahl der primär induzierten Brüche und ihr

Verhalten (Bruch-Reunions-Hypothese) oder einen unbekannten primären Effekt bezüglich seiner Reversion oder Weiterentwicklung zu beeinflussen (Austauschhypothese).

Unsere bisherigen Erörterungen über den Mutationsprozeß sind ohne Zweifel sehr allgemein gehalten und zudem auf einige Annahmen gestützt. Eine stärkere Präzisierung fällt dem Genetiker deshalb schwer, weil einmal noch zahlreiche echte Lücken im Wissen um die strahlenbiologischen Prozesse bestehen und weil es andererseits außerordentlich schwierig ist, einen beträchtlichen Teil der gegenwärtig bekannten Einzelheiten richtig einzuordnen. Darüber hinaus ist es zwar heute sehr wahrscheinlich, daß die DNS (in Sonderfällen auch die RNS) Träger der Information ist, aber es ist auch bekannt, daß die DNS nur einen Bestandteil des Chromosoms ausmacht. Zur Interpretation der Mutations- und Aberrationsentstehung darf also die DNS nicht isoliert betrachtet werden, sondern als eine Komponente des Chromosoms, und dessen Aufbau ist trotz mancher Fortschritte und einer Reihe von Modellvorstellungen (Schwartz, 1955; Taylor, 1957; Freese, 1958; Nebel, 1959) noch weitgehend unbekannt. Die Postulierung zweier verschiedener Bruchtypen nach Einwirkung von Röntgenstrahlen durch Wolff und Mitarbeiter ist ohne Zweifel von großem Interesse. Ihre Ansicht, daß in dem einen Falle Ionenbindungen, im anderen Kovalenzbindungen beteiligt sein sollen, setzt aber voraus, daß primär tatsächlich Brüche eintreten, was nicht unzweifelhaft feststeht. Die von der Arbeitsgruppe Berends (Beukers, Ijlstra und Berends, 1958, 1959, 1960), von Wacker und Mitarbeitern (1960, 1961), Scholes und Weiss (1960) u. a. erhobenen Befunde zur chemischen Veränderung der DNS oder ihrer Basen in vitro erwecken größte Aufmerksamkeit. Offensichtlich geht aus der in den eben zitierten Arbeiten von Berends und Mitarbeitern sowie Wacker et al. mitgeteilten Umwandlung des Thymins in eine dimere Verbindung hervor, daß die unter dem Einfluß des UV-Lichtes eintretende photochemische Reaktion am DNS-Molekül selbst ablaufen kann. Hierfür könnte ebenfalls der Befund sprechen, daß der Ersatz des Thymins durch 5-Bromuracil zu einer verstärkten Strahlenempfindlichkeit der Zelle führt, wie es zuerst von Greer und Zamenhof (1957) gefunden und in der Folge von Lorkiewicz und Szybalski (1960) und anderen Autoren bestätigt wurde. Für den Genetiker ist es aber heute noch keineswegs einfach, diese Daten in sein Bild vom Mutationsprozeß einzuordnen und mit den biologischen Fakten in Einklang zu bringen. Hier ist der Genetiker ohne Zweifel auf die Hilfestellung des Physikers, Chemikers und Biochemikers angewiesen.

Die physikalisch definierte energiereiche Strahlung tritt bei der Einstrahlung in das biologische Objekt mit dem organischen Material in physikalische Wechselwirkung, die über chemische und biochemische Reaktionen zum biologisch faßbaren Effekt, etwa zur Genmutation oder Chromosomenaberration, führt. Die Art dieser Reaktionen ist leider noch weitgehend ungeklärt und kann nur mehr oder weniger spekulativ gedeutet werden, zumal die Übertragung von in vitro am Modellsystem erhaltenen Befunden auf biologische Objekte nur mit großer Vorsicht erfolgen darf. Die Fragen des Genetikers an den Physiker und Chemiker beziehen sich im wesentlichen auf die der Anregung oder Ionisation folgenden physikalisch-chemischen Prozesse, und hierbei besonders auf die Natur der chemischen Reaktionen, die der Strahlenabsorption folgen, und die in Abhängigkeit von der Ionisationsdichte der Strahlung unterschiedlich leicht durch die herrschenden Umweltbedingungen modifikabel sind. Zur Zeit der starken Betonung ausschließlich direkter Strahlenwirkungen ebenso wie in der Periode, in der mit der Entdeckung der Beeinflußbarkeit des Mutationsprozesses und unter dem Eindruck strahlen-chemischer Befunde zur Wasserzersetzung das Schwergewicht auf die indirekte Strahlenwirkung verlagert wurde, konnte der Eindruck entstehen, daß entweder nur der eine oder nur der andere Mechanismus wirksam sein könne. Mit zunehmender Kenntnis der Details stehen wir heute vor einem wesentlich komplexeren, aber sicher realeren Bild des Ursprungs strahleninduzierter Veränderungen des genetischen Materials; dabei deuten viele Ergebnisse darauf hin, daß sowohl die direkte als auch die indirekte Wirkungsweise nebeneinander mit wechselnden und im Einzelfalle schwer abschätzbaren Anteilen an der Entstehung dieser Veränderungen beteiligt sein können.

Literatur

Abrahamson, S., 1959: The influence of oxygen on the X-ray induction of structural changes in *Drosophila* oocytes. Genetics **44**, 173–185.

Abrahamson, S., and J. D. Telfer, 1954: Sex chromosome loss and translocation frequencies in *Drosophila melanogaster* after X-raying sperm in males or in females. Genetics **39**, 955–956.

Adams, I. D., R. A. Nilan, and H. M. Gunthardt, 1955: After-effects of ionizing radiation in barley. I. Modification by storage of X-rayed seeds in oxygen and nitrogen. Northwest Sci. **29**, 101–108.

Adams, I. D., and R. A. Nilan, 1958: After-effects of ionizing radiation in barley. II. Modification by storage of X-irradiated seeds in different concentrations of oxygen. Rad. Res. **8**, 111–122.

Alexander, M. L., 1954: Mutation rates at specific autosomal loci in the mature and immature germ cells of *Drosophila melanogaster*. Genetics **39**, 409–428.

Alexander, M. L., 1958: Radiation damage in the developing germ cells of *Drosophila virilis* from fast neutron treatment. Genetics **43**, 458–469.

Alexander, P., 1960: Protection of macromolecules in vitro against damage by ionizing radiations. In: A. Hollaender (Ed.), Radiation protection and recovery. Pergamon Press, 3–44.

Alikhanian, S. I., and F. S. Klepikova, 1957: A study of the effect produced by X-rays upon variation in *Actinomyces subtropicus*. Ber. Akad. Wiss. UdSSR **115**, 806–808.

Alper, T., 1956: The modification of damage caused by primary ionization of biological targets. Rad. Res. **5**, 573–586.

Alper, T., and P. Howard-Flanders, 1956: Role of oxygen in modifying the radiosensitivity of *E. coli* B. Nature (London) **178**, 978.

Alper, T., and N. E. Gillies, 1958: "Restoration" of *Escherichia coli* strain B after irradiation; its dependence on suboptimal growth conditions. J. Gen. Microbiol. **18**, 461–472.

Anderson, L., 1955: Effects of centrifugation on X-ray induced chromatid aberrations in *Tradescantia* irradiated in air and N_2. Genetics **40**, 563.

Atwood, K. C., and F. Mukai, 1954: Survival and mutation in *Neurospora* exposed at nuclear detonations. Amer. Nat. **88**, 295–314.

Atwood, K. C., and S. Wolff, 1959, in S. Wolff, 1959: Interpretation of induced chromosome breakage and rejoining. Rad. Res., Suppl. **1**, 453–462.

Auerbach, C., 1954: Sensitivity of the *Drosophila* testis to the mutagenic action of X-rays. Z. Vererbungslehre **86**, 113–125.

Bacq, Z. M., und P. Alexander, 1958: Grundlagen der Strahlenbiologie. Georg Thieme Verlag, Stuttgart.

Baker, W. K., and E. Sgourakis, 1950: The effect of oxygen concentration on the rate of induced mutations in *Drosophila*. Proc. nat. Acad. Sci. U. S. **36**, 176–184.

Baker, W. K., and E. S. v. Halle, 1953: The basis of oxygen effect on X-irradiated *Drosophila* sperm. Proc. nat. Acad. Sci. U. S. **39**, 152–161.

Baker, W. K., and E. S. v. Halle, 1954: The production of dominant lethals in *Drosophila* by fast neutrons from cyclotron irradiation and nuclear detonations. Science (Washington) **119**, 46–49.

Baker, W. K., and E. S. v. Halle, 1955: Evidence on the mechanism of the oxygen effect by use of a ring chromosome. J. Cellul. Comp. Physiol. **45** (Suppl. 2), 299–307.

Beatty, A. V., and J. W. Beatty, 1957: Chromosome breakage and rejoining in *Tradescantia* microspores. Amer. J. Bot. **44**, 778–783.

Beatty, A. V., and J. W. Beatty, 1959: Metabolic inhibitors and chromosome rejoining. Amer. J. Bot. **46**, 317–323.

Beatty, A. V., J. W. Beatty, and C. Collins, 1956: Effects of various intensities of X-radiation on chromosomal aberrations. Amer. J. Bot. **43**, 328–332.

Bender, M. A., and S. Wolff, 1961: X-ray induced chromosome aberrations and reproductive death in mammalian cells. Amer. Nat. **95**, 39–52.

BEUKERS, R., and W. BEHRENDS, 1960: Isolation and identification of the irradiation product of thymine. Biochim. biophysica Acta **41**, 550—551.

BEUKERS, R., J. IJLSTRA, and W. BEHRENDS, 1958: The effect of ultraviolet light on some components of the nucleic acids. II. In rapidly frozen solutions. Recueil Trav. chim. Pays-Bas **77**, 729—732.

BEUKERS, R., J. IJLSTRA, and W. BEHRENDS, 1959: The effect of ultraviolet light on some components of the nucleic acids. III. Apurinic acid. Recueil Trav. chim. Pays-Bas **78**, 247—251.

BEUKERS, R., J. IJLSTRA, and W. BEHRENDS, 1960: The effect of ultraviolet light on some components of the nucleic acids. VI. The origin of the U. V. sensitivity of deoxyribonucleic acid. Recueil Trav. chim. Pays-Bas **79**, 101—104.

BONNIER, G., and K. G. LÜNING, 1949: Studies on X-ray mutations in the white and forked loci of *Drosophila melanogaster*. I. A statistical analysis of mutation frequencies. Hereditas **35**, 163—189.

BONNIER, G., and K. G. LÜNING, 1953: Sex linked lethals in *Drosophila melanogaster* from different modes of X-ray irradiation. Hereditas **39**, 193—200.

BORNET, W. T., G. E. STAPLETON, M. L. MORSE, and A. HOLLAENDER, 1951: Reduction of X-ray sensitivity of *Escherichia coli* B/r by sulfhydryl compounds, alcohols, glycols and sodium hydrosulfite. Proc. Soc. Exper. Biol. Med. **77**, 636—638.

CALDECOTT, R. S., 1955: Effects of hydration on X-ray sensitivity in *Hordeum*. Rad. Res. **3**, 316—330.

CALDECOTT, R. S., E. B. JOHNSON, D. I. NORTH, and C. F. KONZAK, 1957: Modification of radiation-induced injury by posttreatment with oxygen. Proc. nat. Acad. Sci. U. S. **43**, 975—983.

CAPPS, A. S., 1961: The effects of nitric oxide on radiation damage in *Drosophila virilis* and *Drosophila melanogaster*. Genetics **46**, 123—128.

COHN, N., 1958: An analysis of the rejoining of X-ray-induced broken ends of chromosomes in the root tips of *Allium cepa*. Genetics **43**, 362—373.

CONGER, A. D., 1960: Genetical protection. In: A. HOLLAENDER (Ed.), Radiation protection and recovery. Pergamon Press, Oxford, 212—241.

CURTIS, H. J., R. S. CALDECOTT, N. DELIHAS, and C. F. KONZAC, 1958: Modification of radiation damage in dormant seeds by storage. Rad. Res. **8**, 526—534.

DAINTON, F. S., 1959: Chemical effects of radiation. Rad. Res., (Suppl. **1**), 1—25.

DALE, W. M., J. V. DAVIES, and C. RUSSEL, 1961: „Nitric oxide and oxygen effect" in X-irradiated *Shigella flexneri*. Nature (London) **189**, 851.

DEMEREC, M., 1955: What is a gene? — Twenty years later. Amer. Nat. **89**, 5—20.

DEMEREC, M., and J. SAMS, 1959: Induction of mutations in individual genes of *Escherichia coli* by low X-radiation. Intern. J. Rad. Biol. (Suppl.), 283—291.

DESCHNER, E., and A. H. SPARROW, 1955: Chromosome rejoining capacity with respect to breakage sensivitity to X-rays and thermal neutrons. Genetics **40**, 460—475.

DEWEY, D. L., 1960: Effect of oxygen and nitric oxide on the radiosensitivity of human cells in tissue culture. Nature (London) **186**, 780—782.

DEWEY, D. L., and J. W. BOAG, 1959: Dose-rate effects in bacteria. Rad. Res. **9**, 106.

DEWEY, D. L., and J. W. BOAG, 1959: Modification of the oxygen effect when bacteria are given large pulses of radiation. Nature (London) **183**, 1450—1451.

DOUDNEY, C. O., 1959: Macromolecular synthesis in bacterial recovery from ultraviolet light. Nature (London) **184**, 189—190.

DOUDNEY, C. O., and F. L. HAAS, 1958: Postirradiation modification of ultraviolet induced mutation frequency and survival in bacteria. Proc. nat. Acad. Sci. U. S. **44**, 390—398.

DOUDNEY, C. O., and F. L. HAAS, 1959a: Mutation induction and macromolecular synthesis in bacteria. Proc. nat. Acad. Sci. U. S. **45**, 709—722.

DOUDNEY, C. O., and F. L. HAAS, 1959b: Gene replication and mutation induction in bacteria. J. Mol. Biol. **1**, 81—83.

DOUDNEY, C. O., and F. L. HAAS, 1960: Some biochemical aspects of the post-irradiation modification of ultraviolet-induced mutation frequency in bacteria. Genetics **45**, 1481—1502.

DRACULIC, M., and M. ERRERA, 1959: Chloramphenicol sensitive DNA synthesis in normal and irradiated bacteria. Biochim. biophysica Acta **31**, 459—463.

DUCHESNE, J., et J. GARSON, 1957: Photo-restauration in biologie. J. Chim. physique Physico-Chim. biol. **54**, 789—791.

DULBECCO, R., 1955: Photoreactivation. In: Radiation biology (Editor: A. HOLLAENDER), Vol. II, 455—486.

EDINGTON, C. W., 1956: The induction of recessive lethals in *Drosophila melanogaster* by radiations of different ion density. Genetics **41**, 814—821.

EDINGTON, C. W., and M. L. RANDOLPH, 1958: A comparison of the relative effectiveness of radiations of different average linear energy transfer on the induction of dominant and recessive lethals in *Drosophila*. Genetics **43**, 715—727.

EHRENBERG, L., 1955: The influence of post-radiation factors on effects produced in barley. Radiobiol. Symp. Proc. Liège, 1954, 285—289.

EHRENBERG, L., 1959: Radiobiological mechanisms of genetic effects. A review of some current lines of research. Rad. Res. (Suppl. 1), 102—123.

ELDJARN, L., and A. PIHL, 1960: Mechanisms of protective and sensitizing action. In: ERRERA, M., and A. FORSSBERG, Mechanisms in radiobiology. Academic Press, New York, Vol. II, 231—296.

EVANS, H. J., and T. R. L. BIGGER, 1961: Chromatid aberrations induced by gamma irradiation. II. Non randomness in the distribution of chromatid aberrations in relation to chromosome length in *Vicia faba* root-tip cells. Genetics **46**, 277—289.

FABERGÉ, A. C., 1940: An experiment on chromosome fragmentation in *Tradescantia* by X-rays. J. Genet. **39**, 229—248.

FALK, R., 1959: Delay in joining of X-ray-induced breaks by anoxia in *Drosophila melanogaster*. Genetics **44**, 509 (Abstr.).

FRITZ-NIGGLI, H., 1958: Mögliche Ursachen der verschiedenen Strahlenempfindlichkeit des Erbmaterials in Keimzellen unterschiedlichen Alters. Naturwissenschaften **45**, 557—564.

FRITZ-NIGGLI, H., 1959: Strahlenbiologie, Grundlagen und Ergebnisse. Georg Thieme Verlag, Stuttgart.

GILES, N. H., 1940: The effect of fast neutrons on the chromosomes of *Tradescantia*. Proc. nat. Acad. Sci. U. S. **26**, 567—575.

GILES, N. H., 1943: Comparative studies of the cytogenetical effects of neutrons and X-rays. Genetics **28**, 398—418.

GILES, N. H., 1954: Radiation-induced chromosome aberrations in *Tradescantia*. In: A. HOLLAENDER (Ed.), Radiation biology. McGraw-Hill, New York, Vol. 1, 713—761.

GILES, N. H., and H. P. RILEY, 1950: Studies on the mechanism of the oxygen effect on the radiosensitivity of *Tradescantia* microspores. Proc. nat. Acad. Sci. U. S. **36**, 337—344.

GLASS, B., 1956: Differences in mutability during different stages of gametogenesis in *Drosophila*. Brookhaven Symp. in Biol. 8 (Mutation), 148—170.

GLASS, B., and L. METTLER, 1958: The oxygen effect in respect to point mutations in *Drosophila*. Proc. 10th Internat. Congress of Genetics, Montreal, 97.

GLOVER, S. W., 1956: A comparative study of induced reversions in *Escherichia coli*. In: Carnegie Inst. Wash. Publ. **612**, 121—136.

GOODGAL, S. H., C. S. RUPERT, and R. M. HERRIOTT, 1957: Photoreactivation of *Hemophilus influenza* transforming factor for streptomycin resistance by an extract of *E. coli* B. In: W. D. MCELROY and B. GLASS (Ed.), The chemical basis of heredity, 341—350.

GRAY, L. H., 1953: Characteristics of chromosome breakage by different agents. Heredity **6** (Suppl.), 311—315.

GRAY, L. H., 1959: Radiation dose — macroscopic, microscopic and sub-microscopic aspects. In: I. H. MARTIN (Ed.), Radiation biology. Butterworth, London, 1—13.

GRAY, L. H., F. O. GREEN, and C. A. HAWES, 1958: Effect of nitric oxide on the radiosensitivity of tumour cells. Nature (London) **182**, 952—953.

GREER, S., and S. ZAMENHOF, 1957: Effect of 5-bromouracil in deoxyribonucleic acid of *E. coli* on sensitivity to ultraviolet irradiation. Amer. chem. Soc. Abstr. (131st meeting), 3 C.

GWÁTKIN, R. B., and D. GOTTLIEB, 1956: Mutants recovered after exposure of *Streptomyces venezuelae* to X-rays. J. Bact. **71**, 328—332.

HAAS, F. L., and C. O. DOUDNEY, 1957: A relation of nucleic acid and protein synthesis on ultraviolet induced mutation in bacteria. Proc. nat. Acad. Sci. U. S. **43**, 871—883.

HAROLD, F. M., and Z. Z. ZIPORIN, 1958: Synthesis of protein and of DNA in *Escherichia coli* irradiated with ultraviolet light. Biochim. biophysica Acta **29**, 439—440.

HARTMAN, Z., 1956: Induced mutability in *Salmonella typhimurium*. In: Carnegie Inst. Wash. Publ. **612**, 107—120.

HOLLAENDER, A. (Ed.), 1960: Radiation protection and recovery. Pergamon Press, Oxford.

HOWARD-FLANDERS, P., 1957: Effect of nitric oxide on the radiosensitivity of bacteria. Nature (London) **180**, 1191—1192.

HOWARD-FLANDERS, P., 1958: Physical and chemical mechanisms in the injury of cells by ionizing radiations. In: Biological and medical physics. Vol. 6, 544—603.

HOWARD-FLANDERS, P., 1959: Primary physical and chemical processes in radiobiology. In: „Radiation Biology and Cancer". Univ. of Texas, 29—40.

HOWARD-FLANDERS, P., and T. ALPER, 1957: The sensitivity of microorganisms to irradiation under controlled gas conditions. Rad. Res. 7, 518.

HOWARD-FLANDERS, P., and D. MOORE, 1958: The time interval following pulsed irradiation within which injury to bacteria can be modified by dissolved oxygen. Rad. Res. 9, 422.

HOWARD-FLANDERS, P., and P. JOCKEY, 1960: Similarities in the effects of oxygen and nitric oxide on the rate of inactivation of vegetative bacteria by X-rays. Rad. Res. 13, 466—478.

IVERSON, R. M., and A. C. GIESE, 1957: Synthesis of nucleic acids in ultra-violet-treated *Escherichia coli*. Biochim. biophysica Acta 25, 62—68.

IVES, P. T., R. P. LEVILE, and H. T. YOST, 1955: The production of mutation in *Drosophila melanogaster* by the fast neutron radiation of an atomic explosion. Proc. nat. Acad. Sci. U. S. 40, 165—171.

JACKSON, W. D., 1959: The life-span of mutagens produced in cells by irradiation. In: I. M. MARTIN (Ed.), Radiation biology. Butterworth, London, 190—208.

JACKSON, W. D., and H. N. BARBER, 1958: Patterns of chromosome breakage after irradiation and ageing. Heredity 12, 1—25.

JAGGER, J., 1958: Photoreactivation. Bact. Rev. 22, 99—138.

JAGGER, J., 1960: Photoreactivation. In: A. HOLLAENDER, Radiation protection and recovery. Pergamon Press, 352—377.

KADA, T., C. O. DOUDNEY, and F. L. HAAS, 1960: Some biochemical factors in X-ray induced mutation in bacteria. Genetics 45, 995.

KANAZIR, D., and M. ERRERA, 1956: Alterations of intracellular deoxyribonucleic acid and their biological consequence. Cold Spring Harbor Symposia on Quant. Biol. XXI, 19—29.

KAPLAN, R. W., 1956: Dose-effect curves of s-mutation and killing in *Serratia marcescens*. Arch. Mikrobiol. 24, 60—79.

KAPLAN, R. W., 1959: Strahlengenetik der Mikroorganismen. In: H. SCHINZ (Ed.), Strahlenbiologie, Nuklearmedizin und Krebsforschung. Georg Thieme Verlag, Stuttgart, 109—156.

KAUFMANN, B. P., 1954: Chromosome aberrations induced in animal cells by ionizing radiations. In: A. HOLLAENDER (Ed.), Radiation biology. McGraw-Hill, New York, Vol. 1, 627—712.

KELNER, A., 1949: Effect of visible light on the recovery of *Streptomyces griseus* conidia from ultraviolet irradiation injury. Proc. nat. Acad. Sci. U. S. 35, 73—79.

KELNER, A., 1953: Growth, respiration and nucleic acid synthesis in ultra-violet-irradiated and photoreactivated *Escherichia coli*. J. Bact. 65, 252—262.

KIHLMAN, B. A., 1955: Oxygen and the production of chromosome aberrations by chemicals and X-rays. Hereditas 41, 384—404.

KIHLMAN, B. A., 1955: Studies on the effect of oxygen on chromosome breakage induced by 8-ethoxycaffein. Exper. Cell Res. 8, 404—407.

KIHLMAN, B. A., 1956: Factors affecting the production of chromosome aberrations by chemicals. J. Biophys. Biochem. Cytol. 2, 543—555.

KIHLMAN, B. A., 1958: The effect of oxygen, nitric oxide, and respiration inhibitors on the production of chromosome aberrations by X-rays. Exper. Cell Res. 14, 639—642.

KIHLMAN, B. A., 1958: Respiration and radiosensitivity of broad bean roots. Nature (London) 182, 730—731.

KIHLMAN, B. A., 1959: Effect of nitric oxide on the production of chromosomal aberrations by X-rays. Exper. Cell Res. 17, 588—590.

KIHLMAN, B. A., 1959: The effect of respiratory inhibitors and chelating agents on the frequencies of chromosomal aberrations produced by X-rays in *Vicia*. J. Biophys. Biochem. Cytol. 5, 481—490.

KIHLMAN, B. A., 1961: Biochemical aspects of chromosome breakage. Advances in Genetics 10, 1—59.

KIMBALL, R. F., 1958: Post-irradiation modification of mutation after various radiations. Rad. Res. 9, 138.

KIMBALL, R. F., N. GAITHER, and S. M. WILSON, 1957: Postirradiation modification of mutagenesis in *Paramecium* by streptomycin. Genetics 42, 661—669.

KIMBALL, R. F., N. GAITHER, and S. M. WILSON, 1959: Reduction of mutation by post-irradiation treatment after ultraviolet and various kinds of ionizing radiations. Rad. Res. 10, 490—497.

KING, R. C., and E. M. WOOD, 1955: Sex-linked lethal mutations induced by thermal neutrons in male and female *Drosophila melanogaster*. Genetics 40, 490—499.

KIRBY-SMITH, I. S., and D. L. CRAIG, 1957: The induction of chromosome aberrations in *Tradescantia* by ultraviolet radiation. Genetics 42, 176—187.

Kishin, A. F. E., 1955: The response of the immature testis of *Drosophila* to the mutagenic action of X-rays. Z. Vererbungslehre **87**, 97—112.

Lamprecht, H., 1958: Röntgen-Empfindlichkeit und genotypische Konstitution von *Phaseolus*. Agri. Hort. Genetica XVI, 196—208.

Laskowski, W., 1960: Inaktivierungsversuche mit homozygoten Hefestämmen verschiedenen Ploidiegrades. Z. Naturforsch. **15b**, 495—506.

Lea, D. E., 1955: Actions of radiations on living cells. Cambridge University Press, 2nd Ed.

Lorkiewicz, Z., and W. Szybalski, 1960: Genetic effects of halogenated thymidine analogs incorparated during thymidylate synthetase inhibition. In: Biochemal and Biophysical Research Communications, Vol. 2, No. 6, 413—418.

Lüning, K. G., 1952: X-ray induced dominant lethals in different stages of spermatogenesis in *Drosophila*. Hereditas **38**, 91—107.

Marinelli, L. D., B. R. Nebel, N. H. Giles, and D. R. Charles, 1942: Chromosomal effects of low X-ray doses on five-day *Tradescantia* microspores. Amer. J. Bot. **29**, 866—874.

McAuley, A. L., I. M. Ford, and D. L. Dobie, 1949: Production of lethal mutations in the fungus *Chaetomium globosum* by monochromatic ultraviolet irradiation. Heredity **3**, 109—120.

Meffert, R. B., and O. Wyss, 1951: The mutability of *Bacillus anthracis* spores during germination. J. Bact. **61**, 357—364.

Mickey, G. H., 1954: Visible and lethal mutations in *Drosophila*. Amer. Nat. **88**, 241—255.

Muller, H. J., 1927: Artificial transmutation of the gene. Science (New York) **66**, 84—87.

Muller, H. J., 1928: The production of mutations by X-rays. Proc. nat. Acad. Sci. U. S. **14**, 714—726.

Muller, H. J., 1954: The nature of the genetic effects produced by radiation. In: A. Hollaender (Ed.), Radiation biology. McGraw-Hill, New York, Vol. 1, 351—473.

Muller, H. J., 1954: The manner of production of mutations by radiation. In: A. Hollaender (Ed.), Radiation biology. McGraw-Hill, New York, Vol. 1, 475—626.

Muller, H. J., 1954: The relation of neutron dose to chromosome changes and point mutations in *Drosophila*. I. Translocations. Amer. Nat. **88**, 437—459.

Muller, H. J., 1957: Damage from point mutations in relation to radiation dose and biological conditions. In: Effect of radiation on human heredity. World Health Org., Geneva, 25—47.

Nadson, G. A., and G. S. Phillipov, 1925: Influence des rayons X sur la sexualité et la formation des mutantes chez les champignons inférieurs (*Mucoracéae*). Compt. rend. Soc. biol. **93**, 473—475.

Neary, G. J., and H. J. Evans, 1958: Chromatid breakage by irradiation and the oxygen effect. Nature (London) **182**, 890—891.

Nebel, B. R., 1959: Observations of mammalian chromosome fine structure and replication with special reference to mouse testis after ionizing radiation. Rad. Res. (Suppl. 1), 431—452.

Newcombe, H. B., 1952: A comparison of spontaneous and induced mutations of *Escherichia coli* to streptomycin resistance and dependence. J. Cellul. Comp. Physiol. **39** (Suppl. 1), 13—26.

Newcombe, H. B., 1956: The timing of induced mutations in *Streptomyces*. Brookhaven Symp. Biol. **8**, 88—102.

Newcombe, H. B., and J. F. McGregor, 1954: Dose-response relationships in radiation induced mutation. Saturation effects in *Streptomyces*. Genetics **39**, 619—627.

Nilan, R. A., 1955: Post-irradiation storage effect on chromosomes in barley seeds X-rayed at normally ineffective dosages. Genetics **40**, 588 (Abstr.).

Oster, J. J., 1956: Modification of X-ray mutagenesis in *Drosophila*. II. Relative sensitivity of spermatids and mature spermatozoa. Advances in Radiobiology, Proc. V. Internat. Conf. on Radiobiology, Stockholm, 475—480.

Oster, J. J., 1958: Radiosensitivity. Genen en Phaenen **3**, 53—66.

Plough, H. H., H. N. Young, and M. R. Grimm, 1950: Penicilline-screened auxotrophic mutations in *Salmonella typhimurium* and their relation to X-ray dosage. J. Bact. **60**, 145—157.

Powers, E. L., B. F. Kaleta, and R. B. Webb, 1959: Nitric oxide protection against radiation damage. Rad. Res. **11**, 461.

Powers, E. L., R. B. Webb, and B. F. Kaleta, 1960: Oxygen and nitric oxide as modifiers of radiation injury in spores of *Bacillus megaterium*. Proc. nat. Acad. Sci. U. S. **46**, 984—993.

Revell, S., 1955: A new hypothesis for chromatid changes. Proceedings of the Radiobiol. Symp., Liège, 1954. Butterworth, London, 243—253.

Revell, S., 1959: The acurate estimation of chromatid breakage and its relevance to a new interpretation of chromatid aberrations induced by ionizing radiations. Proc. Roy. Soc., Biol. Sc., **150**, 563—589.

RICK, C. M., 1940: On the nature of X-ray induced deletions in *Tradescantia* chromosomes. Genetics **25**, 466–482.

RIEGER, R., und A. MICHAELIS, 1959: Vergleichende Untersuchungen zur Verteilung durch verschiedene Mutagene induzierter Brüche über den Chromosomensatz von *Vicia faba* L. Chromosoma **10**, 163–178.

RIEGER, R., und A. MICHAELIS, 1960: Der Einfluß von Stoffwechselinhibitoren auf die radiomimetische Wirkung von Äthylalkohol. Mber. d. DAdW **2**, 290–297.

RIEGER, R., und A. MICHAELIS, 1960: Über die radiomimetische Wirkung von Äthylalkohol bei *Vicia faba*. In: H. STUBBE, (Hrsg.), Chemische Mutagenese. Abh. dtsch. Akad. Wiss. Berlin, Kl. med. Wiss., Jahrg. 1960, Nr. 1, 54–65.

RUPERT, C. S., 1960: Photoreactivation of transforming DNA by an enzyme from baker's yeast. J. Gen. Physiol. **43**, 573–595.

RUPERT, C. S., S. H. GOODGAL, and R. M. HERRIOTT, 1958: Photoreactivation in vitro of ultraviolet inactivated *Hemophilus influenza* transforming factor. J. Gen. Physiol. **41**, 451–471.

RUSSEL, W. L., 1951: X-ray-induced mutations in mice. Cold Spring Harbor Symposia on Quant. Biol. **16**, 327–336.

RUSSEL, W. L., 1956: Genetic effects of radiation in mice and their bearing on th eestimation of human hazards. Proc. Internat. Conf. on the Peaceful Uses of Atomic Energy, Geneva, Vol. 11, 382.

RUSSELL, W. L., and L. B. RUSSEL, 1958: Radiation-induced genetic damage in mice. Second United Nat. Internat. Conf. on the Peaceful Uses of Atomic Energy, A/Conf. 15/P/897 U. S. A.

SAX, K., 1939: The time factor in X-ray production of chromosome aberrations. Proc. nat. Acad. Sci. U. S. **25**, 225–233.

SAX, K., 1940: An analysis of X-ray induced chromosomal aberrations in *Tradescantia*. Genetics **25**, 41–68.

SAX, K., 1943: The effect of centrifuging upon the production of X-ray induced chromosomal aberrations. Proc. nat. Acad. Sci. U. S. **29**, 18–21.

SCHOLES, G., and J. WEISS, 1960: Organic hydroxy-hydroperoxides: a class of hydroperoxides formed under the influence of ionizing radiation. Nature (London) **185**, 305–306.

SCHWARTZ, D., 1952: The effect of oxygen concentration on X-ray induced chromosome breakage in maize. Proc. nat. Acad. Sci. U. S. **38**, 490–494.

SCHWARTZ, D., 1955: Studies on crossing over in maize and *Drosophila*. J. Cellul. Comp. Physiol. **45** (Suppl. 2), 171–183.

DE SERRES, F. J., 1958: Studies with purple adenine mutants in *Neurospora crassa*. III. Reversion of X-ray-induced mutants. Genetics **43**, 187–206.

DE SERRES, F. J., 1960: Studies with purple adenine mutants in *Neurospora crassa*. IV. Lack of complementation between different *ad-3 A* mutants in heterokaryons and pseudowild types. Genetics **45**, 555–566.

SETLOW, R. B., 1957: Action spectroscopy. Advances Biol. Med. Phys. **5**, 37–71.

SETLOW, R. B., G. WATTS, and C. DOUGLAS, 1959: Inactivation of proteins by vacuum ultraviolet radiation. Proc. First Nat. Biophys. Conf. Yale Univ. Press, New Haven, Conn., 174–183.

SKREB, Y., and M. ERRERA, 1957: Action des rayons U. V. sur des fragments nucliés et anucliés d'amibes. Exper. Cell Res. **12**, 649–656.

SOBELS, F. H., 1960: Chemical steps involved in the production of mutations and chromosome aberrations by X-irradiation in *Drosophila*. I. The effect of post-treatment with cyanide in relation to dose-rate and oxygen tension. Int. J. Rad. Biol. **2**, 68–90.

SOBELS, F. H., 1961: Chemische Beeinflussung des röntgeninduzierten Mutationsprozesses bei *Drosophila*. Naturwissenschaften **48**, 146–155.

SOBELS, F. H., 1961: Recovery from mutational damage of X-irradiation in *Drosophila* spermatogenesis. Ninth Session, U. N. Scient. Com. on the Effects of Atomic Radiation, March 12–24, 1961.

SPARRMAN, B., L. EHRENBERG, and A. EHRENBERG, 1959: Scavenging of free radicals and radiation protection by nitric oxide in plant seeds. Acta chem. scand. **13**, 199–200.

SPARROW, A. H., 1951: Radiation sensitivity of cells during mitotic and meiotic cycles with emphasis on possible cytochemical changes. Ann. N. Y. Acad. Sc. **51**, 1500–1540.

SPENCER, W. P., and C. STERN, 1948: Experiments to test the validity of the linear r-dose mutation frequency relation in *Drosophila* at low dosage. Genetics **33**, 43–74.

STADLER, L. J., 1928: Genetic effects of X-rays in maize. Proc. nat. Acad. Sci. U. S. **14**, 69–75.

STAPLETON, G. E., and A. HOLLAENDER, 1952: Mechanisms of lethal and mutagenic action of ionizing radiations on *Aspergillus terreus*. II. Use of modifying agents and conditions. J. Cellul. Comp. Physiol. **39** (Suppl. 1), 101–113.

Stapleton, G. E., 1955: Variations in the sensitivity of *Escherichia coli* to ionizing radiations during the growth cycle. J. Bact. **70**, 357—362.

Stent, G. S., 1958: Mating in the reproduction of bacterial viruses. Advances in Virus Research **5**, 95—149.

Stone, W. S., M. L. Alexander, F. E. Clayton, and E. Dudgeon, 1954: The production of translocations in *Drosophila virilis* by fast neutrons from a nuclear detonation. Amer. Nat. **88**, 287—293.

Stubbe, H., 1930: Untersuchungen über die experimentelle Auslösung von Mutationen bei *Antirrhinum majus*. I. Versuche mit Röntgenstrahlen, ultraviolettem Licht, Temperaturschocks und Zentrifugierungen. Z. Abstammungslehre **56**, 1—38.

Taylor, J. H., 1957: The time and mode of duplication of chromosomes. Amer. Nat. **91**, 209—221.

Thoday, J. M., 1942: The effects of ionizing radiations on the chromosomes of *Tradescantia bracteata*. A comparison of neutrons and X-rays. J. Genet. **43**, 189—210.

Thoday, J. M., 1951: The effect of ionizing radiations on the broad bean root. Part IX: Chromosome breakage and the lethality of ionizing radiations on the root meristem. Brit. J. Radiol. **24**, 572—576; 622—628.

Thoday, J. M., and J. Read, 1947: Effect of oxygen on the frequency of chromosome aberrations produced by X-rays. Nature (London) **160**, 608—610.

Thoday, J. M., and J. Read, 1949: Effect of oxygen on the frequency of chromosome aberrations produced by alpha rays. Nature (London) **163**, 133—134.

Valencia, J. J., and H. J. Muller, 1949: The mutational potentialities of some individual loci in *Drosophila*. Hereditas, Suppl. Vol. 1949, 681—683.

Wacker, A., H. Dellweg und E. Lodemann, 1960: Strahlenchemische Veränderungen der Nucleinsäuren. Angew. Chemie **73**, 64—65.

Wacker, A., H. Dellweg und D. Weinblum, 1961: Strahlenchemische Veränderung der Bakterien-Desoxyribonucleinsäure in vivo. Naturwissenschaften **47**, 477.

Watson, J. D., and F. H. Crick, 1953: Genetic implications of the structure of desoxyribose nucleic acid. Nature (London) **171**, 964—967.

Westergaard, M., 1960: Chemical mutagenesis as a tool in macromolecular genetics. In H. Stubbe (Hrsgb.), Chemische Mutagenese, Erwin-Baur-Gedächtnisvorlesungen I, 1959, Abh. d. Dtsch. Akad. d. Wiss. Berlin, Kl. f. Medizin, **1**, 30—44.

Witkin, E. M., 1956: Time, temperature and protein synthesis: A study of ultra-violet-induced mutation in bacteria. Cold Spring Harbor Symposia on Quant. Biol. XXI, 123—140.

Witkin, E. M., 1958: Postirradiation metabolism and the timing of ultraviolet-induced mutations in bacteria. Proc. 10th Internat. Congress of Genetics **1**, 280—299.

Witkin, E. M., and E. C. Theil, 1960: The effect of posttreatment with chloramphenicol on various ultraviolet-induced mutations in *Escherichia coli*. Proc. nat. Acad. Sci. U. S. **46**, 226—232.

Wolff, S., 1957: Recent studies on chromosome breakage and rejoining. In: G. C. de Hevesy et al., Advances in radiobiology. Oliver & Boyd, Edinburgh, 463—470.

Wolff, S., 1960: Studies on the nature of chromosome breakage. Amer. Nat. **94**, 85—93.

Wolff, S., 1960: Problems of energy transfer in radiation-induced chromosome damage. Rad. Res. (Suppl. 2), 122—132.

Wolff, S., and K. C. Atwood, 1954: Independent X-ray effects on chromosome breakage and reunion. Proc. nat. Acad. Sci. U. S. **40**, 187—192.

Wolff, S., and R. C. v. Borstel, 1954: The effects of pre- and postirradiation centrifugation on the chromosomes of *Tradescantia* and *Vicia*. Proc. nat. Acad. Sci. U. S. **40**, 1138—1141.

Wolff, S., and H. E. Luippold, 1955: Metabolism and chromosome-break rejoining. Science (Washington) **122**, 231—232.

Wolff, S., and H. E. Luippold, 1956: The biochemical aspects of chromosome rejoining. In: Mitchell et al. (Ed.), Progress in radiobiology. Oliver & Boyd, Edinburgh, 217—221.

Wolff, S., and H. E. Luippold, 1956: The production of two chemically different types of chromosome breaks by ionizing radiation. Proc. nat. Acad. Sci. U. S. **42**, 510—514.

Zimmer, K. G., 1956: The development of quantum biology during the last decade. Acta radiol. (Stockholm) **46**, 595.

Zimmer, K. G., 1960: Studien zur quantitativen Strahlenbiologie. Abh. d. Math.-Naturwiss. Klasse, Akad. d. Wiss. u. Literatur, Mainz, Jahrg. 1960, Nr. 3.

Zirkle, R. E., 1954: The radiobiological importance of linear energy transfer. In: A. Hollaender (Ed.), Radiation biology. McGraw-Hill, New York, Vol. I, 315—350.

On the Nature of the Principal Target of Lethal and Mutagenic Radiation Effects

W. Szybalski and Z. Lorkiewicz

McArdle Memorial Laboratory, University of Wisconsin, Madison 6, Wis., U. S. A., and Institute of General Microbiology, University of M. Curie-Skłodowska, Lublin, Poland

A living cell responds in a variety of ways to the injurious effects of radiation, but only two types of responses will be discussed at this time: (1) a lethal radiation effect which results in loss of the cell's capacity for indefinite reproduction, and (2) a mutagenic radiation effect which produces cells with some permanently altered heritable traits but still viable and able to reproduce ad infinitum. Although far from conclusive, many recent experiments performed with whole cells, and on the level of polymeric and monomeric cell components, permit discussion of the lethal and mutagenic radiation effects in molecular terms and postulation as to the chemical mechanism of these phenomena.

The present discussion will not be all inclusive, and while admittedly biased toward the authors' present interest and firsthand experience, it will deal with two types of radiation (ultraviolet [UV] light and X-rays) directed toward bacterial and human cells and their isolated components, mainly biologically functional (genetically transforming) deoxyribonucleic acid in its native or chemically modified form.

Identification of the most radiosensitive cell component

The principal structural elements of which each cell is composed are proteins, lipids, polysaccharides, RNA and DNA. Which of these components is the most radiosensitive and thus principally responsible for lethal radiation effects? This question was difficult to answer for a variety of reasons. (1) Direct chemical or physical estimates of radiation-induced damage to isolated cell components may have little meaning in terms of biologically-significant injury. (2) Biological tests for the functionality of biologically-active molecules (enzymes, infective nucleic acids) usually measure only a small and undetermined fraction of the total functions. (3) Damage within the cell might be governed by the cell architecture and thus differ from injury to the same component extracellularly inflicted. (4) Moreover, some of the cell components (a) may be present as multiples of one particular indispensable unit or (b) if inactivated, might be regenerated under the guidance of other intact cell components. This fourth consideration makes the low molecular metabolic intermediates rather unlikely candidates for the principal target of lethal radiation effects. For the purpose of this discussion, we shall lump together the direct and indirect effects of radiation, considering the radiosensitivity of the structural cell components in terms of the end result, cell death or mutation.

A likely candidate for the most radiosensitive cell component is DNA since (1) it could be considered that no DNA molecule exists in duplicate in each cell (allowing the possibility that each complementary DNA strand has equivalent information, and taking into account di- and polyploidy), (2) damaged DNA cannot be independently resynthesized under the guidance of other cell components, and (3) DNA represents a very large target due to its immense molecular weight, particularly if each chromosome is considered as one molecule.

The logical way to detect the principal target of UV radiation is to determine the lethal action spectrum of UV light. Such data point clearly to nucleic acids, obviously DNA in the case of RNA-free bacteriophages, but do not readily distinguish between DNA and RNA in the case of cellular organisms (e. g., Setlow, 1957).

The same approach is not applicable for determination of the X-ray target. In the case of DNA phages in the cysteine-containing protective solution, X-ray damage is principally directed toward DNA, while under unprotected conditions the adsorption machinery residing in the tail structure of the phage (protein) is concurrently affected by X-rays (Hradecna et al., 1961). For cellular organisms the subject of the principal radiation target is at present in a highly controversial state.

A new method for determination whether DNA is or is not the most sensitive cell component became available after it was shown that specific halogenation of DNA molecules results in increased radiosensitivity of the whole cells and of the halogenated transforming DNA extracted from such cells (Opara-Kubinska, Lorkiewicz, and Szybalski, 1961). Thus instead of comparing the absolute sensitivities of whole cells and of their components, as was hitherto usually done, one could now compare the increments in sensitivity due to halogenation both for the whole cells and for DNA. This method is thus analogous to the "action spectra" method, but it is applicable not only to UV but also to other types of radiation, whenever radiosensitization occurs. It distinguishes between DNA and RNA, since under the experimental conditions halogenation is directed only toward DNA.

The principle of the method was based on: (1) determination of the survival curves (radiation dose versus logarithm of cell survival, in terms of colony-forming capacity) for two cell subcultures, one (A) grown in the absence and another (B) in the presence of a halogenated thymidine analog under conditions of effective DNA incorporation (Opara-Kubinska, Lorkiewicz and Szybalski, 1961; Djordjevic and Szybalski, 1960), (2) determination of the inactivation curves for the purified transforming DNA extracted from subcultures A and B (radiation dose versus logarithm of the number of colony-forming transformants under standard conditions), (3) comparison of the survival curves (slope ratio) for subcultures A and B and for the transforming DNAs extracted from these subcultures (Szybalski and Opara-Kubinska, 1961).

Quantitative comparison of the increments in radiation sensitivity for (a) whole viable cells and for (b) pure transforming DNA permits the following reasoning. Were DNA the most radiosensitive cell component, the incorporation of 5-halodeoxyuridines would be expected to result in identical increments in radiation lethality for both (a) intact cells and (b) the pure transforming DNA extracted from these cells. Experiments performed with UV-irradiated *Bacillus subtilis* cells and DNA, under conditions of 5-bromodeoxyuridine (BUdR) labeling, bear out this contention, and thus permit the conclusion that in this case DNA is the principal target of lethal UV effects. If survival of *B. subtilis* cells were dependent upon some more sensitive cell component, it would be difficult to understand why purified DNA, probably the only site of BUdR incorporation, should exhibit the same degree of radiosensitization as the intact cells from which it was extracted. The data and diagrams comparing the UV survival of normal and BUdR-labeled cells and of extracted transforming DNA were published earlier (OPARA-KUBINSKA, LORKIEWICZ and SZYBALSKI, 1961: p. 290, Fig. 1.; SZYBALSKI and OPARA-KUBINSKA, 1961).

The data obtained with X-rays point in the same direction, although as also in the case of UV there exist some technical difficulties, which will be briefly summarized here. Since bacterial DNA replicates in semiconservative manner (MESELSON and STAHL, 1958), after a single DNA replication in the presence of BUdR, the cells will contain only unifilarly BUdR-labeled DNA molecules. In the case of *B. subtilis* such cells seem to be viable and the unifilarly BUdR-labeled transforming DNA extracted from these cells is fully biologically active. Upon subsequent DNA replications in the presence of the analog, bifilarly BUdR-labeled DNA molecules are produced, which when extracted can still perform their transforming function (SZYBALSKI et al., 1960). However, the viability of the cells replicating in the BUdR-containing medium is progressively lost. Thus, while it is easy to compare the radiosensitivity of the unifilarly-labeled transforming DNA with the sensitivity of the corresponding donor cells, this comparison becomes more difficult to interpret upon subsequent cell replications in the BUdR-containing medium. Although it is still easy to separate non-labeled (NN) from unifilarly (NB) and bifilarly (BB) labeled DNAs by CsCl-gradient centrifugation, and to determine their radiosensitivity (Fig. 1, broken lines NB and BB), it becomes difficult to determine whether the viable cells among the total donor population represent a uniform sample of this population or only those particular cells which have undergone a minimal number of divisions in the presence of BUdR and thus have incorporated less BUdR than the average for the whole donor population containing both viable and "BUdR-killed" cells. The reason for the death of the BUdR-labeled bacterial cells, which seems to be associated with great cell fragility, is not quite clear; under conditions of massive BUdR incorporation human cells can be grown for an indefinite period of time (DJORDJEVIC and SZYBALSKI, 1960).

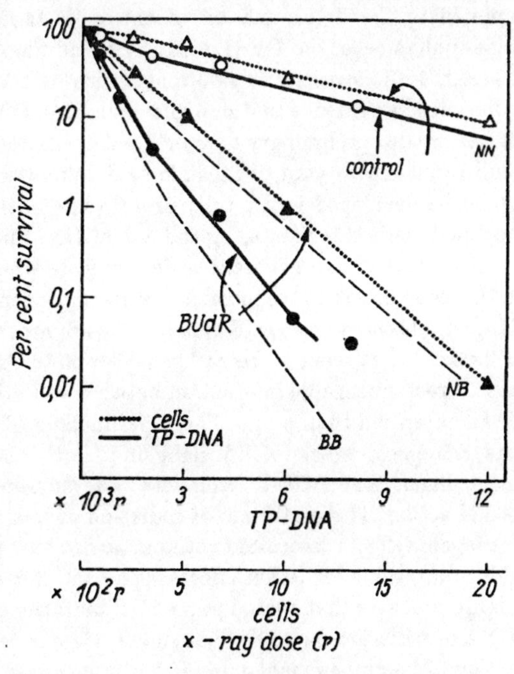

Fig. 1. X-ray survival (colony-forming ability) of *Bacillus subtilis* wild-type cells (dotted lines) grown for 4 hrs in the presence (BUdR) or in the absence (control) of BUdR (100 μg per ml) and FUdR (4 μg per ml), and of the non-fractionated transforming DNA (solid lines) isolated from the above described BUdR-free (control) and BUdR-labeled (BUdR)cells, irradiated in the presence of 0.4% AET and measured as ability to transform indole-requiring receptor cells to prototrophy. The X-ray survival of the transforming function of the CsCl-gradient fractionated unifilarly (NB) and bifilarly (BB) BUdR-labeled DNA fractions obtained from the above described non-fractionated BUdR labeled DNA is represented by the broken lines. — Curves based on the unpublished preliminary experiments of Z. OPARA-KUBINSKA.

Figure 1 illustrates the results of the comparison between the X-ray sensitivities of normal and BUdR-labeled cells and of isolated transforming DNA. It is evident that BUdR incorporation into cell DNA sensitizes the BUdR-labeled (viable) cells to the same degree ($3.5 x$) as the unifilarly labeled DNA (NB — broken line) isolated by fractionation in the preparative CsCl gradient. This is in agreement with the previous observation that under the present experimental conditions cells containing unifilarly labeled DNA are viable, while bifilarly labeled transforming DNA (BB), although more radiosensitive, would have no chance to contribute to the radiosensitivity of the total cell population since bacterial cells bearing such bifilarly-labeled DNA are very fragile or not viable and thus cannot be assayed by their colony forming property. The filled circles (solid lower line) indicate the intermediate sensitivity of non-fractionated DNA isolated from BUdR-labeled cells. These results (Szybalski and Opara-Kubinska, 1961, and unpublished data of Mrs. Z. Opara-Kubinska) point rather clearly toward DNA as the most X-ray sensitive cell component.

These experiments in their present form are applicable only for genetically transformable cells, but using the same principle of comparing the increments in radiosensitivity, the radiation effects on other biological or physico-chemical properties of DNA could be compared with cell survival (or mutagenicity) for both halogenated and non-halogenated cells. Specific labeling of other cell components for the purpose of assessing their contributions to cell radiosensitivity is also theoretically possible. Although this approach has not been applied to the determination of the most radiosensitive component of human and other mammalian cells, the UV (Djordjevic and Szybalski, 1960) and X-ray (Erikson and Szybalski, 1961) sensitization obtained with halodeoxyuridine-labeled cells can by analogy be construed as evidence that in higher cells also, DNA is the most radiosensitive cell component. If our present preliminary data on DNA-mediated transformation in human cells (transformation from loss to gain of inosinic acid pyrophosphorylase: Szybalski and Joannes, 1962) could be developed into a fully useful experimental system, direct application of the above outlined method for assessing the role of DNA in cell radiosensitivity will become possible.

Little data exist at present which indicate whether direct interaction of radiation with the DNA component is responsible for its mutagenic effects. With the exception of phage $\varkappa$ (kappa), discussed by Dr. Kaplan in this Symposium, and the study of Krieg (1959a), extracellular phages seem to be rather resistant to the mutagenic effects of radiation, indicating that direct mutagenic interaction between radiation and preformed DNA seems to be more of an exception than a rule. The experiments of Lorkiewicz and Szybalski (1960), assessing the mutagenic effects of UV light on BUdR-labeled *Escherichia coli* cells, indicated, on the other hand, that BUdR "enhances" the mutagenic effects of UV light in parallel with its lethal action. The multitude of radiation effects together with the assumed specificity of the mutagenic effects associated with particular base configurations will make it necessary to consider mutagenic radiation effects separately for each particular type of mutation. It is also highly probable that a large part of UV mutations are caused by incorporation into replicating DNA of radiation-modified deoxynucleotides in analogy with the mechanism of the mutagenic action of triethylene melamine (Lorkiewicz and Szybalski, 1961). The results of Kirchner (1961), indicating that many UV-produced mutations can be reverted by base analogs, point in the same direction.

Molecular mechanisms of radiation effects

Since DNA appears to be the principal target of lethal radiation effects, several questions arise and require experimental solutions: (1) which chemical changes arise in DNA, and in which order, as a result of radiation; (2) which of them are responsible for the lethal and mutagenic effects; and (3) which function of DNA is primarily affected by the radiation caused chemical modifications?

The thymine dimerization reaction under the influence of UV light, first demonstrated by BEUKERS and BERENDS (1961) and confirmed by several laboratories, will be discussed in this Symposium by Dr. SHUGAR and Dr. WACKER. This reaction can either cross-link two complementary polydeoxynucleotide strands, or link the adjoining thymines on the same strand. In both cases some degree of preliminary radiation-caused denaturation of the DNA molecule would help to bring the thymine bases into proper alignment. The studies reported by MARMUR and GROSSMAN (1961), and also those performed in our laboratory using a different experimental approach, indicate cross-link formation between the complementary strands of the DNA double-helix, but not between separate DNA molecules. On the other hand, the studies of Dr. WACKER reported here stress more the dimerization between adjoining thymine molecules. The biological importance of both of these reactions could be based on the facts that (1) this reaction is perceptible already at relatively low doses of radiation, and (2) it is photoreversible in the presence of a photoreactivating "enzyme", as discussed here by Dr. WACKER and published by MARMUR and GROSSMAN (1961). The results, however, obtained with BUdR-labeled DNA do not favor the all important role of thymine dimerization. On the one hand BUdR incorporation results in a 10- to 20-fold increase in the UV sensitivity of such modified DNA, while on the other hand, according to Dr. WACKER, the dimerization reaction between 5-bromouracil moieties or between 5-bromouracil and thymine appears to be strongly depressed. Obviously some other reaction could take over a predominant role in the halogenated DNA, but the authors prefer to believe that thymine dimerization is responsible only for a part of the UV radiation effects, mainly for the socalled "photoreactivable sector". The cross-linking of the complementary DNA strands may have a contributing, but unlikely a major role, since the double-stranded DNA configuration is not required for the lethal UV effects, as exemplified by phage ⌀X-174. Moreover, infective ⌀X-174 DNA seems to be 10 times more sensitive to UV light in the single-stranded configuration than when hydrogen bonded to a polymerase-synthesized complementary strand (SINSHEIMER, 1961).

X-ray effects do not seem to have any relationship to the thymine dimerization reaction, and several other mechanisms can be postulated for this and other radiation effects. These include: (a) breakage of the longitudinal DNA backbone, mainly the phosphate ester bonds, (b) breakage of glucoside linkages, (c) modification of purine and pyrimidine bases (in preformed DNA molecules or just prior to incorporation), (d) various cross-linking reactions in addition to thymine dimer formation.

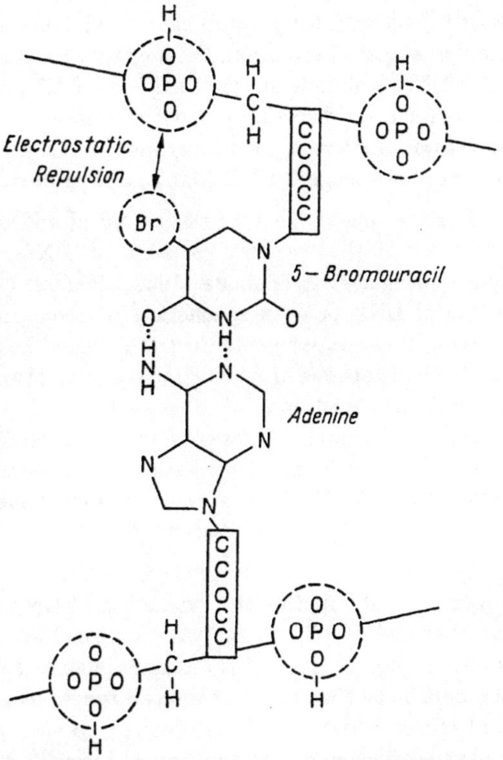

Fig. 2. Diagramatic representation of the steric relationship between the phosphate groups and the halogen (the latter replacing the methyl group of thymine) in the DNA molecule, represented by the adenine -5-bromouracil deoxynucleotide pair (SZYBALSKI, 1962)

Although at present it is difficult to assign particular biological roles to all of these possible types of damage, breakage of the phosphate ester bonds seems to us the most likely cause of most lethal radiation effects. The fact that replacement of the methyl groups in DNA thymine by halogens seems to affect the strength of the neighbouring phosphate ester bonds, as discussed by SZYBALSKI (1962) and as represented in Fig. 2, would easily explain the radiosensitizing effect of halogenated thymidine analogs, especially if one assumes that the phosphate ester bonds already are the most X-ray sensitive structures in the DNA molecule. There could be, however, more than one radiosensitizing mechanism in effect simultaneously, since as was discussed by Dr. WACKER during this symposium extensive dehalogenation takes place during irradiation of the 5-bromouracil nucleotides. One hopes that future systematic studies will determine the relative importance of all the individual reactions in radiation-caused damage to normal and modified DNA molecules.

Radiation damage to DNA functionality

The present discussion has pointed to DNA as the most radiosensitive cell component and briefly dealt with the possible structural damages inflicted on this molecule by the direct and indirect effects of radiation. On the other hand the present data do not pinpoint which of the vital DNA functions are primarily susceptible to radiation: (1) its role as a primer (KORNBERG, 1960) for directing the self replication of identical and functional DNA molecules, or (2) its function in governing and controlling the synthesis of proteins, probably via the route of "messenger" RNA (JACOB and MONOD, 1961).

(1) Exposure of DNA to physiological and even to massive doses of X-rays has hardly any effect on the "priming" capacity of DNA as measured in the in vitro enzymatic DNA-synthesizing system (BOLLUM, 1960; STACEY, 1961). This is not surprising, since small fragments of DNA or even synthetic oligodeoxy-nucleotides are still active as primers in this system. These experiments do not indicate, however, whether X-ray demaged DNA would prime the synthesis of biologically functional DNA molecules, or only of the inactive, similarly damaged ones.

The authors are not aware of any studies on the priming capacity of UV-irradiated double-stranded DNA in the enzymatic system described by KORNBERG and his collaborators (KORNBERG, 1960). In this case, thymine dimerization may cause steric inhibition of the priming action by UV irradiated DNA.[1])

(2) Several studies indicate that, both in vivo and in vitro, the induced synthesis of some enzymes is inhibited by the irradiation of DNA or by its inactivation by the decay of DNA-incorporated ^{32}P (NOVELLI, 1961; RUSHIZKY et al., 1960; MCFALL et al., 1958). Although the interpretation of some of these experiments seems to be still controversial (MCFALL, 1961), the possibility that many of the cell functions are inhibited as the result of irradiation may well reflect radiation induced damage to the structural or control regions of the DNA molecule. Since in many investigated cases the enzyme forming capacity of the cell, probably depending on "messenger" RNA, is very unstable and requires constant resynthesis of this information-transmitting polynucleotide, damage to DNA would very soon be reflected by

[1]) The recent data of BOLLUM and SETLOW (1962) on the "priming" activity of UV irratiated DNA corroborate this prediction.

the loss of the enzyme forming capacity of the cell. The radiation-caused inhibition of the mitotic activity of the cell is probably a secondary effect of the DNA damage, and reflects the inability of damaged DNA to direct the synthesis of some proteins necessary for the proper functioning of the mitotic apparatus. In his studies on the genetics of UV effects, KRIEG (1959b) was able to distinguish clearly between UV damage to the r_{II} gene function ("survival of phenotype") and to the recombinational and selfreplicating properties of phage genes ("cross reactivation").

Up to this point only the inactivation of DNA functionality has been discussed. Radiation could, however, inflict an operationally more serious deviation in DNA function: it might cause it to direct the synthesis of a "poisonous messenger" RNA, either by immediate damage to DNA or by appearance of modified "maverik" DNA during its subsequent replication. "Poisonous messenger" RNA could be defined as one which not only has lost its capacity to effect the synthesis of a given protein, but inhibits the functioning of other norma messenger RNA. This "poisonous messenger" RNA might perhaps be bound irreversibly either to the DNA molecules (including the cross-dimerization between the thymine and uracil moieties), or to the ribosomes, the sites of enzyme formation. The concept of the formation of a "poisonous messenger" RNA might explain the "single hit" curves obtained under special conditions with multinuclear cells, as discussed by Dr. ALPER during this Symposium. This type of damage, produced even in one out of several nuclei in the same cell, would result in an overall dominant lethal effect throughout the whole cell, if the molecules of the "poisonous messenger" RNA released by the damaged nucleus irreversibly blocked the limited number of ribosomes of a particular type, or in some other way interfered with protein synthesis or its control. In this connection, it might be interesting to note that recent studies by PERSON et al. (1961) are compatible with the idea that a lethal radiation effect produced in one cell could spread even beyond this cell, through the cluster of neighbouring cells, in a small glass-attached clone of cultured mammalian cells.

Since many other aspects of radiation effects have been so ably discussed by other speakers during this symposium, and since many aspects of our work were reviewed very recently elsewhere (SZYBALSKI, 1962), we shall close this discussion with one additional optimistic note. The study of the molecular mechanism of biological radiation effects has just entered a very profitable period of systematic chemical and physico-chemical analysis, and the sequence of chemical changes inflicted by radiation on the DNA molecule should soon be mapped out in detailed fashion.

Summary

Among the different cell components, DNA appears to be the most vulnerable to the lethal effects of radiation. This conclusion was based on experiments employing a new principle: comparison of the increments in radiation sensitivity of normal and halogenated transforming DNA, on the one hand, and of subcultures of the cells from which the two types of DNA were extracted, on the other hand. In addition, the various types of radiation-induced damage to the DNA molecules are outlined and their effect on the "priming" and protein synthesis-directing functions of DNA is discussed. With respect to the latter function, the concept of the dominant lethal role of "poisonous messenger" RNA synthesized under the influence of radiation-damaged DNA is proposed.

Acknowledgements

We are greatly indebted to Mrs. Z. Opara-Kubinska, for permitting us to quote her unpublished results on the X-ray inactivation of halogen-labeled transforming DNA and cells.

This study was supported by grants from the U. S. Public Health Service (CY-5415 and CY-3492).

References

Beukers, R., and W. Berends, 1961: The effects of U.V.-irradiation on nucleic acids and their components. Biochim. biophysica Acta **49**, 181—189.

Bollum, F. J., 1960: General discussion following session 3. In: J. S. Mitchell (Ed.), The Cell Nucleus. Academic Press, New York, p. 163.

Bollum, F. J., and R. B. Setlow, 1962: Effect of monochromatic UV radiation on "primer" DNA. Abstracts, Biophysical Soc. 6th Ann. Meeting, Washington, D. C., p. TA-11.

Djordjevic, B., and W. Szybalski, 1960: Genetics of human cell lines. III. Incorporation of 5-bromo-, and 5-iododeoxyuridine into the deoxyribonucleic acid of human cells and its effect on radiation sensitivity. J. Exper. Med. **112**, 509—531.

Erikson, R. L., and W. Szybalski, 1961: Molecular radiobiology of human cell lines. I. Comparative sensitivity to X-rays and ultraviolet light of cells containing halogen-substituted DNA. Biochem. Biophys. Res. Comm. **4**, 258—261.

Hradecna, Z., W. Arber, and G. Kellenberger, 1961: Personal communication.

Jacob, F., and J. Monod, 1961: Genetic regulatory mechanisms in the synthesis of proteins. J. Mol. Biol. **3**, 318—356.

Kirchner, C. E. J., 1961: The effects of the mutator gene on molecular changes and mutation in *Salmonella typhimurium*. J. Mol. Biol. **2**, 331—338.

Kornberg, A., 1960: Biologic synthesis of deoxyribonucleic acid. Science (Washington) **131**, 1503—1508.

Krieg, D. R., 1959a: Induced reversion of $T4_{rII}$ mutants by ultraviolet irradiation of extracellular phage, Virology, **9**, 215—227.

Krieg, D. R., 1959b: A study of gene action in ultraviolet-irradiated bacteriophage T4. Virology, **8**, 80—98.

Lorkiewicz, Z., and W., Szybalski, 1960: Genetic effects of halogenated thymidine analogs incorporated during thymidylate synthetase inhibition. Biochem. Biophys. Res. Comm. **2**, 413—418.

Lorkiewicz, Z., and W. Szybalski, 1961: Mechanism of chemical mutagenesis. IV. Reaction between triethylene melamine and nucleic acid components. J. Bact. **82**, 195—201.

McFall, E., 1961: Effects of P^{32} decay on enzyme synthesis. J. Mol. Biol. **3**, 219—224.

McFall, E., A. B. Pardee, and G. Stent, 1958: Effects of radiophosphorus decay on some synthetic capacities of bacteria. Biochim. biophysica Acta **27**, 282—297.

Marmur, J., and L. Grossman, 1961: Ultraviolet light induced linking of deoxyribonucleic acid strands and its reversal by photoreactivating enzyme. Proc. nat. Acad. Sci. U. S. **47**, 778—787.

Meselson, M., and F. W. Stahl, 1958: The replication of DNA in *Escherichia coli*. Proc. nat. Acad. Sci. U. S. **44**, 671—682.

Novelli, G. D., 1961: Mechanisms involved in the control of enzyme synthesis. Fifteenth Annual Symposium on Fundamental Cancer Research "The Molecular Basis of Neoplasia". M. D. Anderson Hospital, Houston, Tex. (in press).

Opara-Kubinska, Z., Z. Lorkiewicz, and W. Szybalski, 1961: Genetic transformation studies. II. Radiation sensitivity of halogen labeled DNA. Biochem. Biophys. Res. Comm. **4**, 288—291.

Person, S. R., L. J. Tolmach, and E. Sato, 1961: Interactions among mammalian cells irradiated in vitro. Radiation Research **14**, 491.

Rushizky, G. W., M. Riley, L. S. Prestidge, and A. B. Pardee, 1960: Inactivation of enzyme formation by ultraviolet light. I. Action spectra and size of the sensitive unit. Biochim. biophysica Acta **45**, 70—76.

Setlow, R., 1957: Action spectroscopy. Advances Biol. Med. Phys. **5**, 37—74.

SINSHEIMER, R., 1961: The replication of bacteriophage Øx 174. J. Chim. physique Physico-Chim. biol. 58, 986–993.

STACEY, K. A., 1961: The inhibition of DNA synthesis by calf thymus polymerase by X-irradiation of the primer DNA. Biochem. Biophys. Res. Comm. 5, 486–490.

SZYBALSKI, W., 1962: Properties and applications of halogenated deoxyribonucleic acid. In "The Molecular Basis of Neoplasia" (A collection of papers presented at the 15th Annual Symposium on Fundamental Cancer Research at the University of Texas M. D. Anderson Hospital and Tumor Institute). Pp. 147–171. Austin: University of Texas Press, Austin 1962, 147–171.

SZYBALSKI, W., and S. A. JOANNES, 1962: Genetic transformation in human cells. Microbial Genetics Bull., No. 18, 22–23.

SZYBALSKI, W., and Z. OPARA-KUBINSKA, 1961: DNA as principal determinant of cell radiosensitivity. Radiation Research, 14, 508–509.

SZYBALSKI, W., Z. OPARA-KUBINSKA, Z. LORKIEWICZ, E. EPHRATI-ELIZUR, and S. ZAMENHOF, 1960: Transforming activity of deoxyribonucleic acid labeled with 5-bromouracil. Nature (London) 188, 743–745.

Primary Effects of Ultraviolet Radiation on Nucleic Acids

D. Shugar

Institute of Biochemistry and Biophysics, Academy of Sciences and Dept. of Biochemistry, State Institute of Hygiene, Warsaw

With the increasing application of nuclear energy to scientific, industrial and medical purposes, and with space travel out of the stage of fantasy it is natural that so much attention is being devoted to the effects of all types of radiation on living cells and their constituents. Ultraviolet radiation possesses the distinct advantage in such studies in that the process of energy absorption is specific and may be correlated with the molecular structure of the components involved. It has even proven fruitful, in investigations on the effects of ionizing radiations, to conduct parallel studies with UV. Sufficient data has now been accumulated about the primary effects of UV on nucleic acids and their constituents to provide a basis for discussion between the physical chemist and photochemist on the one hand and the biologist and, in particular, the geneticist on the other. Recent developments in the field of radiation chemistry of nucleic acids suggest that we are approaching a similar situation with regard to the genetical effects of ionizing radiations, despite the greater complexities involved.

We shall concern ourselves, in what follows, principally with the effects of UV on nucleic acids and their constituents. Relatively little is as yet known about the final effects of irradiation, particularly when we are dealing with the intact cell. This is perhaps best exemplified by mutagenesis in microorganisms, where phenotypically mutant cells are formed only after a certain delay following irradiation. During this lag period, which precedes what may be regarded as the subcultivation period leading to the formation of stable mutants, the yield of the latter may be considerably modified by a variety of treatments, of which the best known is that commonly designated as photoreactivation or photorestoration (PR) (Kelner, 1949; Jagger, 1958). Cellular metabolism appears to be essential for mutant formation, which has been variously linked with DNA replication as the terminal event in the process, although there is some evidence that RNA and protein synthesis are involved (Witkin, 1956, 1958; Doudney and Haas, 1960; Weatherwax and Landman, 1960). Some UV-induced mutations have also been reported which are inert to postirradiation treatment. Mutagenesis by UV in extracellular RNA and DNA is unknown and is normally a rare event in bacteriophage (Krieg, 1959); a notable exception is the X-phage of *Serratia*, where up to 2% mutations to clearplaque types has been reported by irradiation with UV or X-rays of the free phage (Ellmauer and Kaplan, 1959).

Insofar as concerns the cellular components involved in the primary effects of irradiation, leading to such events as mutation, "killing", inactivation of enzyme induction, etc., evidence from action spectra studies implicate intact nucleic acids as the immediate receptors of the photochemically effective wavelengths. Action spectra alone do not distinguish between RNA and DNA (but cf. SETLOW and SETLOW, 1960) and in one instance nucleic acid precursors have been proposed as the initial receptors of radiation leading to mutations (HAAS and DOUDNEY, 1957). While this latter postulate lacks confirmation, enough is now known about the photoproducts of nucleic acid precursors to put it to a direct test. As regards reversal of the primary effects of radiation, there remains little doubt but that this is due to a direct reversal of injury to nucleic acid molecules and not to a metabolic by-pass mechanism.

From investigations on nucleic acid derivatives and related compounds, and on model oligo- and polynucleotides as well as natural nucleic acids, the following appear to be the primary initial effects resulting from irradiation of nucleic acids with UV doses of the order of magnitude of those producing biological effects (10^2–10^4 ergs/mm [JÄGGER, 1958]:

(a) Purine residues and carbohydrate moieties are unaffected; nor is there any rupture of internucleotide linkages.

(b) For uracil and cytosine residues, there may occur nucleophilic addition of water across the 5,6 double bonds, with resultant saturation of the latter and the formation of 5-hydro-6-hydroxy derivatives, as follows:

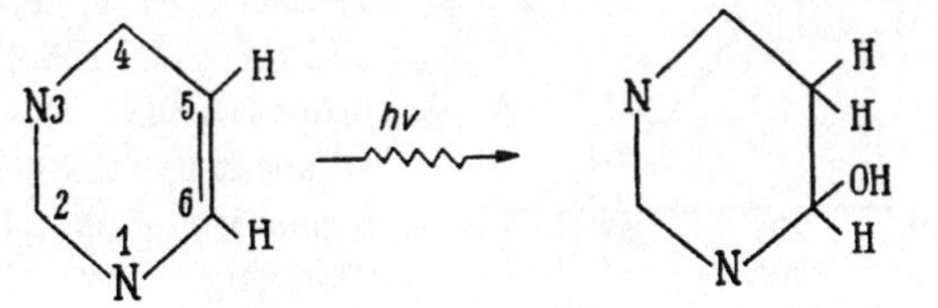

The water molecule may subsequently be eliminated in the dark, by what is essentially an acid-base catalyzed reaction, either by heating at neutral pH or, frequently, by acidification or alkalization at room temperature. Thymine residues do not undergo this photochemical addition of water (MOORE and THOMSON, 1957; WANG, 1958; SINGSHEIMER, 1957; SHUGAR and WIERZCHOWSKI, 1958; SHUGAR, 1960).

At neutral pH the hydrated photoproducts of uracil derivatives are relatively stable in the dark at room temperatures and rather elevated temperatures (70–90°C) must be employed to eliminate the water molecule at an appreciable rate. The photoproducts of cytosine derivatives are considerably less stable and, for some analogues, the reverse reaction (i. e. elimination of the water molecule) proceeds at a measurable rate in the temperature range 20–40°C (WIERZCHOWSKI and SHUGAR, 1961).

By analogy with dihydropyrimidines, the photoproducts of uracil derivatives should exhibit only end absorption in the quartz ultraviolet, and such is indeed the case. However, for cytosine photoproducts, suturation of the 5,6 double bond still leaves a pair of conjugated double bonds (the number 2 carbonyl and the 3,4 double bond), and these should exhibit a maximum in the neighbourhood of 2300 Å as for dihydrocytosine analogues (JANION and SHUGAR, 1960). This is indeed the case, as can be seen from Fig. 1, for cytidylic acid. It is necessary to emphasize this observation because of the postulate of WANG (1959) that the photoproducts of cytosine and its glycosides are not the 5-hydro-6-hydroxy derivatives, but rather tautomers in which the proton on the N_1 nitrogen has been shifted to position 5 with the accompanying displacement of the 5,6 double bond to N_1, C_6. This conclusion is based on a number of erroneous assumptions which we have dealt with elsewhere (WIERZCHOWSKI and SHUGAR, 1961).

(c) In addition to hydration, there is another photochemical reaction which occurs between adjacent pairs of uracil and/or thymine rings, likewise involving saturation of the 5,6

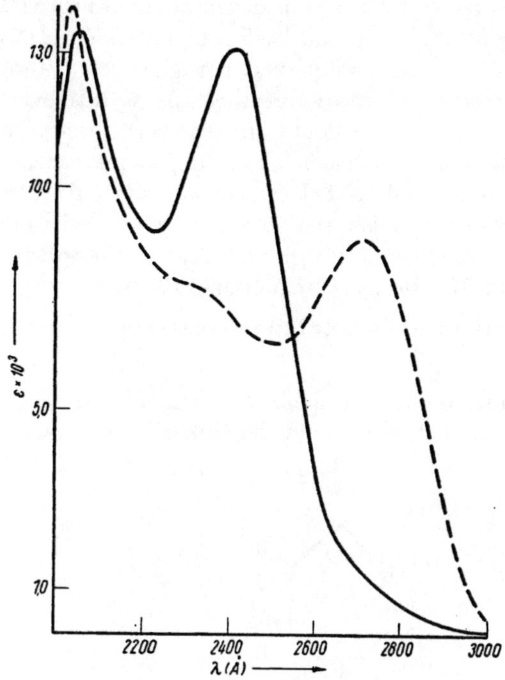

Fig. 1. Absorption spectrum at neutral pH of cytidine-3'-phosphate, Cp, (– – – – –) and its photoproduct, Cp · H_2O (———)

double bonds. The quantum yield for this latter reaction in a single-stranded polynucleotide chain is most likely greater than for hydration and, unlike the hydration reaction, is not heat reversible in the dark (WIERZCHOWSKI and SHUGAR, 1959; SHUGAR and WIERZCHOWSKI, 1960, 1961; SHUGAR, 1960, 1961). In polymers containing the thymine ring, this is the only reaction which takes place initially (WIERZCHOWSKI and SHUGAR, 1960) since, as mentioned above, thymine does not undergo hydration. The foregoing findings, plus parallel studies on the photochemical transformation of airdried films of natural and synthetic polynucleotides (BARANOWSKA and SHUGAR, 1960), led to the tentative proposal that this interaction between adjacent pyrimidine residues involved the formation of crosslinked products of the following form (SHUGAR and WIERZCHOWSKI, 1960, 1961; SHUGAR, 1960, 1961; BARANOWSKA and SHUGAR, 1960):

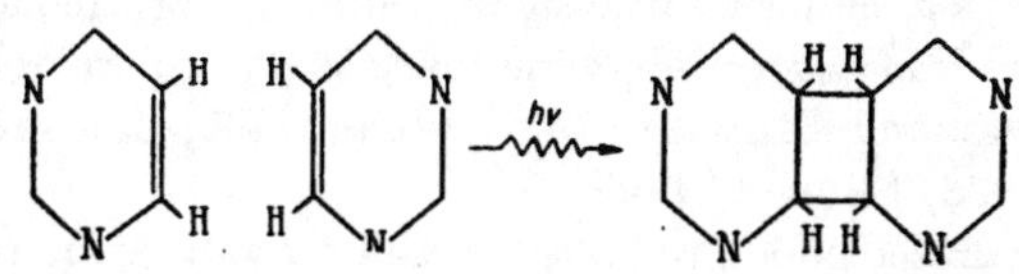

with formation of a cyclobutane ring between the 5,6 bonds of the two rings.

The validity of this assumption was established almost simultaneously by the experimental isolation of such a photodimer of thymine by BEUKERS and BERENDS (1960, 1961) from an ultraviolet irradiated frozen aqueous solution of thymine. This has been further confirmed by WULFF and FRAENKEL (1961) and such dimers of thymine have been reported as having been isolated chromatographically from UV irradiated DNA (WACKER, DELLWEG and WEINBLUM, 1960). These findings not only lend support to the previous interpretation of the kinetics of photochemical transformation of oligo- and polynucleotides of Up (WIERZCHOWSKI and SHUGAR, 1959); but also provide the basis for a more satisfactory interpretation of the concentration dependance of the kinetics of photolysis of simple monomers such as dimethyluracil, uridylic acid, etc. (WIERZCHOWSKI and SHUGAR, 1959; SMIETANOWSKA and SHUGAR, 1961).

We have found that a variety of 2,4-diketopyrimidines will undergo photodimerization in aqueous, frozen, medium and have isolated the dimer of uracil in crystalline form (SMIETANOWSKA and SHUGAR, 1961). Since the photodimers pressess 5,6 saturated bonds, they should exhibits only end absorption at neutral pH in the quartz ultraviolet; that this is so can be seen from Fig. 2. The photodimers are remarkably stable in acid but undergo degradation in hot alkali. At room temperature, however, they are reasonably stable in alkali and exhibit an absorption maximum at 2350 Å due to dissociation of the number 4 carbonyl (Fig. 2) with a pK greater than 12.3 for both thymine and uracil dimers. There are actually 4 dimer isomers possible, two cis and two trans (WULFF and FRAENKEL, 1961), and the pK value for the thymine dimer shown in Fig. 2 has been regarded as vidence for its being one of the cis forms.

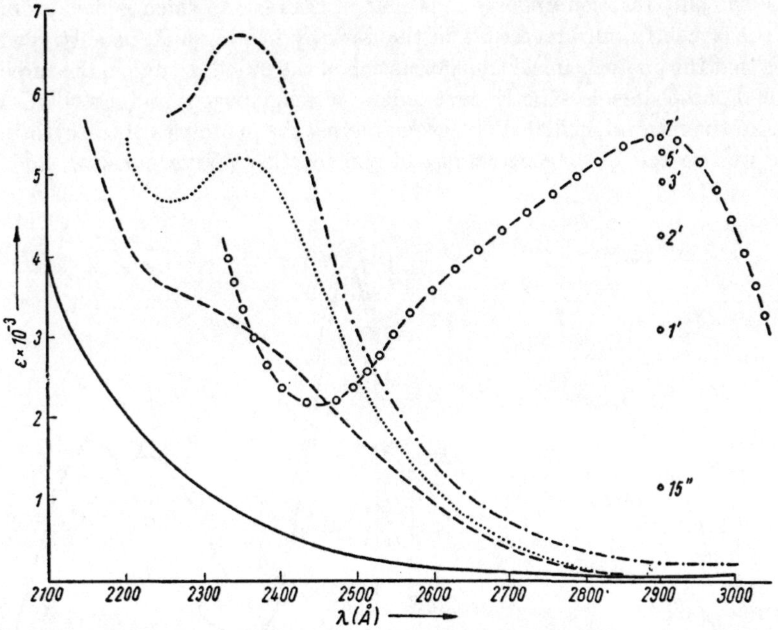

Fig. 2. Absorption spectrum of thymine dimer at neutral and alkaline pH values, and photochemical conversion of dimer to thymine at pH 12:
—·—·— dimer at pH 7; — — — — dimer in 0.01 N NaOH; . . . . . . dimer in 0.1 N NaOH; o — o — absorption spectrum resulting from irradiation of dimer at pH 12 for 7 mins. at incident intensity of 0.9×10^{17} quanta/cm²/min. at 2537 Å. Spectrum corresponds accurately to that of thymine at pH 12 and figures under maximum of curve show amount of thymine regenerated at various times of irradiation

BEUKERS and BERENDS (1960) showed that irradiation of thymine dimer in aqueous, liquid, medium resulted in dissociation of the cyclobutane ring with reformation of thymine, and the quantum yield for this reaction was estimated by SETLOW (1961) to be about 0.6. Measurement of the quantum yield for dimer dissociation at neutral pH at 2537 Å is rather difficult because of the low absorption at this wavelength. On the other hand the dimers possess appreciable absorption at this wavelength at alkaline pH; furthermore uracil and thymine themselves are relatively radiation resistant in alkaline medium. We have therefore measured the quantum yields at 2537 Å for photochemical dissociation of uracil and thymine dimers in the pH range 12–13 (see Fig. 2) and find that the values are 0.52 and 0.73 for uracil and thymine, respectively.

Photochemically induced dimer formation in polynucleotide chains procedes with a quantum yield of 0.05 for uracil residues (WIERZCHOWSKI and SHUGAR, 1959) and at least 0.01 for thymine (WIERZCHOWSKI and SHUGAR, 1960). In polymer chains photochemical dissociation of dimers, once formed, apparently does not readily occur due, most, likely, to a "cage effect" imposed by the internucleotide linkages. This problem requires further study since it is probable that an equilibrium state is established under continuous irradiation, with formation, dissociation and reformation of dimers (SZTUMPF and SHUGAR, in preparation).

Finally it should be emphasized that no positive evidence exists for such dimer formation between cytosine residues and the evidence from reaction kinetics rather argues against it.

(d) Several typical examples will now be cited for purposes of illustration:

(1) When a uracil or cytosine residue is flanked only by a purine or is sandwiched between two purines, the only reaction undergone is that of hydration, which is first order and the reaction is subsequently fully reversible in the dark by heating and, usually, acidification. Fig. 3 shows that the photochemical transformation of the uracil residue in the three isomeric trinucleoside-diphosphates is strictly first-order (WIERZCHOWSKI and SHUGAR, 1961). In Fig. 4 is shown the spectral modification accompanying the photolysis of the cytosine residue in ApCp as well as the 100% reversibility of the reaction (WIERZCHOWSKI and SHUGAR, (1960).

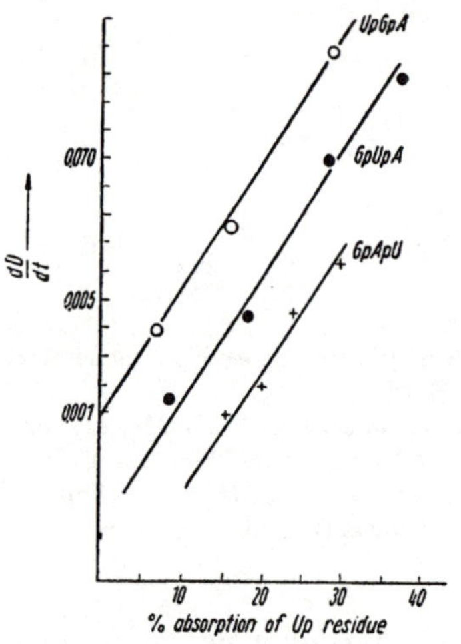

Fig. 3. Kinetics of photochemical transformation of Up residue in three isomeric trinucleotides at neutral pH; dD/dt is the rate of change of optical density (hence of concentration) with time. Incident light intensity 6.5 × 10^{-7} einsteins/cm^2/min. Photochemical reaction fully reversible in dark, by heating, for all three isomers

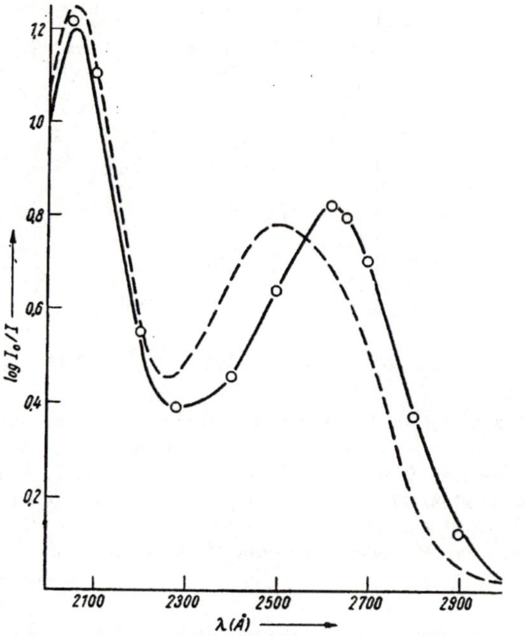

Fig. 4. Photochemical transformation of ApCp and subsequent thermal reactivation (TR) of photoproduct in 0.02 M-phosphate buffer pH 7: (———) prior to irradiation; (– – – – –) following irradiation for 150 mins. at an intensity of 2.6 × 10^{-1} einsteins/cm^2/min. and at a temperature of 6 °C to inhibit the reverse, dark reaction; (o) after removal of irradiation source and warming of solutions 180 mins. at 30 °C

(2) In a dinucleotide such as CpU, spectral observations and the kinetics of the reverse reaction shows that each residue undergoes only hydration. The reverse reaction takes place in two stages, the Cp residue first eliminating its water molecule at 30–40°C and the Up at a much higher temperature. If dimerization had taken place, even partially, this would have shown up in the reverse reaction (WIERZCHOWSKI and SHUGAR, 1960).

(3) In homopolymers of U, dimerization varies from 15–20% for a dinucleotide to over 35% for highly polymerized poly-U. The remaining residues undergo hydration. Dimerization occurs with a quantum yield somewhat greater than that for hydration, hence largely during

the initial stages of the reaction, a fact which is also made clear by the kinetics of the reaction (WIERZCHOWSKI and SHUGAR, 1959). In a compound such as TpT where hydration cannot occur, 34% dimerization has been reported (WACKER, DELLWEG, and LODEMANN, 1961). From 9 to 18% of the total thymine of the DNA of various bacterial strains have been isolated as dimers following irradiation (WACKER, DELLWEG, and LODEMANN, 1961) but it is as yet premature to link these results with the DNA thymine content and sequences. The quantum yield for thymine dimer formation in polymer chains is fairly high. Fig. 5 exhibits the decrease in absorption of an oligonucleotide of Tp as a result of irradiation at 2537 Å. The initial steep portion of the curve is due to dimer formation and the quantum yield for this reaction is 0.012, or 0.024 if excitation of one thymine ring is required.

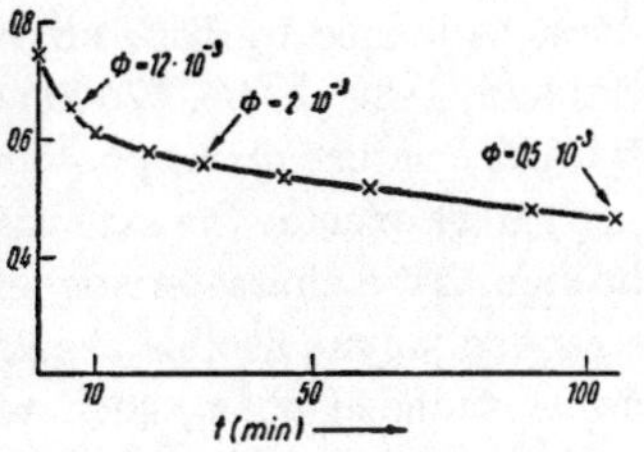

Fig. 5. Course of photodecomposition of poly-T in 0.02 M-phosphate buffer pH 7, followed by decrease in optical density of principal absorption maximum at 2670 Å with time of irradiation. Incident radiation intensity 2.6×10^{-7} einsteins/cm²/min. The calculated quantum yield varies during the course of the reaction as indicated by the arrows and drops to a limiting value of 0.5×10^{-3} equal to that for free Tp. Since the reaction initially proceeds uniquely via dimerization, the quantum yield shown in the figure as 12×10^{-3} should be 24×10^{-3}

(4) When dried films of nucleic acids are irradiated with UV, cross-linking takes place with a very high quantum yield (BARANOWSKA and SHUGAR, 1960; SETLOW and DOYLE, 1954; SHUGAR and BARANOWSKA, 1960). The resulting products are extremely insoluble in aqueous or salt solution and consist of fibres which are visible to the naked eye. The course of the reaction is best followed by the increased affinity of the irradiated films for methyl green, using a ^{14}C-labelled dye and window counting (Fig. 6).

Studies with model synthetic polymers indicated that the reaction involves exclusively the pyrimidine residues with formation of linkages between the 5,6 double bonds of the pyrimidine residues (BARANOWSKA and SHUGAR, 1960; SHUGAR and BARANOWSKA, 1960). Since this reaction takes place readily in air-dried films, which still contain considerable moisture, it is quite conceivable that it will occur to some extent in nuclear DNA. It appears somewhat unlikely that this reaction is reversible and this may be one of the reasons why photoreactivation (PR) is very rarely a 100% phenomenon. This is almost certainly the explanation for the failure to observe PR in UV-inactivated dried

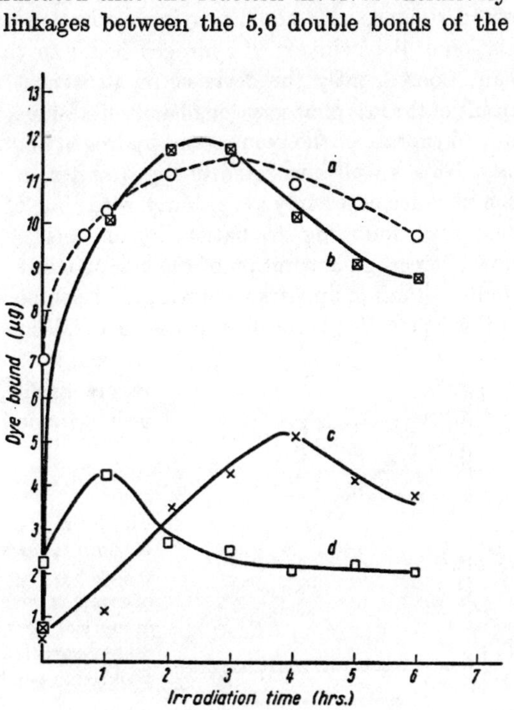

Fig. 6. Effect of ultraviolet irradiation on affinity for methyl green of irradiated, air-dried, films of (a) DNA, (b) apurinic acid, (c) reduced apurinic acid, and (d) RNA. The films were stained with ^{14}C-labelled methyl green following irradiation and then placed under an end-window counter for measurement

preparations of transforming DNA (T-DNA) and bacteriophages. Attention should be drawn to the fact that in vivo crosslinking of DNA has been reported to result from electron irradiation (Alexander and Stacey, 1959).

(e) For 5-substituted cytosine residues which, with a few exceptions (e. g. wheat germ, the DNA of which contains 5% 5-methylcytosine), are confined to the T-even bacteriphages, the photochemical transformation most likely involves ring opening and/or rearrangement. The precise mechanism of the reaction has not yet been elucidated. But experiments with other analogues of 4-aminopyrimidines, which exhibit similar photochemical behaviour, have demonstrated that this type of reaction is subject to heat reversibility in the dark. For example, the photoproduct in slightly alkaline medium of the pyrimidine component of vitamin B_1, 2-methyl-5-ethoxymethyl-4-aminopyrimidine, exhibits full thermal reversibility (TR) in the dark, as judged by both physico-chemical and microbiological criteria (Wierzchowski and Shugar, 1960, 1961). We have extensive experiments now under way with a view to clarifying the nature of the photoproducts of the 5-substituted cytosine derivatives.

(f) If we exclude the extensive cross-linking reaction which takes place in dried films, it will be seen that a characteristic feature of the foregoing photochemical transformations in single-stranded polynucleotide chains is that they are all potentially reversible. Even in a model twin- stranded chain, such as poly-A-poly-U, the initial photochemical reaction, involving only the uracil residues, is fully reversible by heat (but only following dissociation of the irradiated complex), suggesting water addition as the sole initial reaction and that dimerization is initially inhibited because of steric factors. It would be extremely useful to check this by an examination of the behaviour of the complex formed by poly A with poly-ribosethyminephosphate (Griffin, Todd and Rich, 1958), since excited thymine molecules will not undergo hydration. Even in poly-U itself the dimerization reaction is inhibited by more than 50% if the uracil residues are N-methylated (Wierzchowski and Shugar, 1960).

(g) In a twin-stranded chain such as poly-A-poly-U the photochemical transformation of a uracil residue, with accompanying saturation of the 5,6 double bond, results in the dissociation of the twin-strand hydrogen bonds to the adenine residue of the corresponding base pair. Consequently the decrease in ultraviolet absorption, which would be expected as a result of the loss of absorption of a uracil residue, is partically compensated for by the increase due to rupture of the twin-strand hydrogen bonds, i. e. by the resulting loss of hyperchromicity. This is well illustrated in Fig. 7, where it will be seen that the optical density of the complex decreases only very slowly with time of irradiation. If, however, hyperchromicity is destroyed, following irradiation, by dissociation of the complex, it becomes clear that the slow decrease in absorption of the complex was only apparent, and not real. A qualitatively similar situation appears to prevail for a complex such as poly-C-poly-I. Of some significance in this connection is the observation that UV inactivation of pneumococcal T-DNA is accompanied, by a small, but measureable, shift in the temperature profile, suggestive of the rupture of some of the twin-stranded hydrogen bonds (Marmur and Doty, 1959).

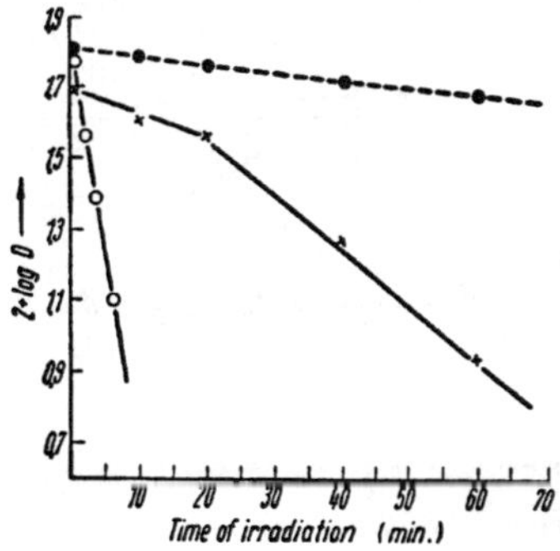

Fig. 7. Kinetics of photochemical transformation of poly-A-U complex, as followed by decrease in optical density at 2570 Å: ○ ○ ○ ○ ○ experimentally observed decrease in optical denity of complex with time of irradiation; × × × × × actual decrease in optical density obtained by enzymatic hydrolysis of irradiated complex with resultant destruction of twin-strand hypochromicity; ○ ○ ○ ○ ○ results for poly-U alone under same conditions

This latter observation is, in fact, probably the most sensitive criterion to date for detecting early physico-chemical modifications in irradiated DNA and merits further study.

(h) No oxygen effect has been observed in the physico-chemical studies of the above reactions in polymer chains, in agreement with the recognized general absence of an oxygen effect in photobiological processes such as mutation, etc. It is necessary to emphasize this because of the fact that an oxygen effect has been encountered with some pyrimidines, e. g. orotic acid (BEUKERS, IJLSTRA, and BERENDS 1959), and most likely due to some quenching of the excited state of the latter.

(i) Measurements of quantum yields for uracil and cytosine residues in synthetic single and twin-stranded polynucleotides lead to the conclusion that energy transfer is not an important factor in the above reactions. In all instances the quantum yields of incorporated residues are either unaltered or reduced, in some cases by a considerable factor. While there is very little doubt but that energy transfer does take place in nucleic acid chains, it does not appear to be photochemically effective in so far as the aromatic residues are concerned and it is consequently questionable whether one may invoke it to explain, for example, the complex nature of the UV survival curves for T-DNA or bacteriophages.

(j) Table I exhibits some quantitative data for the photochemical transformation of uracil and cytosine residues in various model polymer chains. The table also presents some data for the thermal, reverse reaction corresponding to elimination of the water molecules from the 5,6 double bonds of the pyrimidine rings. The extent of non-reversibility (by heat) is that due to dimerization for those instances where the polymers contain tracts of two or more uracil residues. The dimerization reaction can be placed in evidence, and examined quantitatively, by means of spectral (SMIETANOWSKA and SHUGAR, 1961) and other (SZTUMPF and SHUGAR, in preparation) methods.

Table 1

Reaction constants for reversible photolysis of Up and Cp residues in model oligo- and polynucleotides (Ø = quantum yields; k_{20} = 1st order rate constant for reverse reaction at 20° C, except where otherwise indicated; E = activation energy for reverse reaction; % TR = % thermal reversibility)

	Up residue in						Cp residue in					
	Up Up	poly-U	Ap Up	Cp U	Gp Up A Gp Ap U Up Gp A	poly-A poly-U complex	Cp Cp	poly-C	Ap Cp Ap Ap Cp	Cp U	Gp Cp Gp Cp Gp Cp	poly-I poly-C complex
Ø × 10^3	70**) 8.0	55**) 8	7.5	10	7.0 4.5 10.5	2.5**) 3.5	7.5	7.5	8.2	8.0	5.5	~ 4
$k_0^2 \times 10^4$ (sec^{-1})	0.01	0.01	0.015	0.015	–	–	0.2* 1.3	$\begin{bmatrix}1.7\\9.5\end{bmatrix}$ *) 30°	0.6	[1.8] 30°	–	–
E (kcal/mol	–	–	26.2	–	–	–	14*) 26	– –	18	–	–	–
% RT	85	65	100	90	> 90	100* 80	90	85	100	–	> 90	> 80

*) Reaction involves two steps.

**) Quantum yield changes from upper to lower values as reaction progresses.

Quantum yield measurements, and modifications in spectra for single-stranded chains, may be employed to make rough estimates of the size of a radiation-sensitive biological unit. For T-DNA such calculations range from 1 to about 5×10^5 in molecular weight for the size of a marker (SHUGAR, 1961). The functional unit responsible for tryptophanase induction in *E. coli* has been estimated in this way to possess a molecular weight of about 5×10^5, hence excluding s-RNA (RUSHIZKY, RILEY, PRESTIDGE, and PARDEE, 1961).

(k) Leaving aside the question of 5-substituted cytosine analogues, which is relevant mainly to the T-even bacteriophages, we see that hydration and dimerization of pyrimidine residues with the accompanying dissociation of hydrogen bonds are the principal inital effects of UV. One can as yet by no means exclude other reactions. Furthermore we do not as yet know what happens to the 11% 5-ribosyluracil in s-RNA, or to the RNA in intact TMV; it should be recalled that TMV itself is apparently not photoreactiveable, but infectious RNA is (BAWDEN and KLECZKOWSKI, 1959), suggesting some involvement of the TMV protein, most likely in the reverse reaction. It would obviously be desirable to conduct studies on additional model polynucleotides. Particularly necessary are some data on the photochemical behaviour of homopolymers of various chain lengths of thymine residues (cf. ref. WIERZCHOWSKI and SHUGAR, 1960).

Finally it will be necessary to obtain some data on the overall modifications produced in natural nucleic acid chains as a result of irradiation. The fact that we already know a good deal about the two principal photochemical reactions involved, plus the fact the that overall photochemically induced modifications are largely reversible, lends hope to an early resolution if this problem.

(l) Reversal of hydration and/or dimerization could be the function of the photoreactivating enzyme of RUPERT et al. (1958). It would be difficult to test this directly for hydration, bearing in mind the instability of the hydrated pyrimidines under hydrolysis conditions. But such a test should be possible for dimerization since the dimer photoproducts are sufficiently stable under acid hydrolysis conditions as to permit of their isolation. A logical experiment would consequently involve treatment of irradiated T-DNA with the PR enzyme system and estimations of biological activity and dimer content prior to and after photoreactivations. Such an experiment is actually under way at the present time by RUPERT and WULFF. As will be reported upon later in this symposium, Dr. WACKER (1961) is in possession of evidence to the effect that the PR enzyme will dissociate thymidine dimer.

PR of T2 phage by means of a light-activated extracellular enzyme system has likewise been reported (PICHL, 1960), but the degree of PR in relation to a control is rather low; and PAKULA (unpublished) has been unable to observe PR for T-DNA with the same system. Nonetheless it is likely that in living organisms the component which exhibits PR is not nucleoprotein but free nucleic acid.

It must, on the other hand, is recalled that other means of reversal of the effects of UV irradiation exist and that, of these, the most interesting and simultaneously the most effective is thermal reversal (TR). This was first demonstrated for UV irradiated bacteria and its magnitude found to be of the same order as PR. It is claimed to be not as general a phenomenon as PR but this is due simply to the fact that it is less spectacular and therefore not as extensively investigated.

BRESCH (1950) first reported TR for Tl phage adsorbed to the host cells; % TR was not high. Miss LUKIANIEC in our laboratory has been examining a number of phages for TR and the results show that it is not uncommon; in some instances the % TR is not much different from that obtained by PR. The main difficulty in establishing precise numerical values is the inactivation of the controls at the TR temperature employed, 45° C. In all instances adsorption to the host cell was a necessary prerequisite and no TR could be observed for the free phages, although heat denaturation interferes at temperatures above 45°C. It therefore appears, although this is not conclusive, that some cellular factor is necessary as for PR. Confirmation of this would tend to exclude hydration as a direct factor in reversibility of photobiological effects.

It is therefore of significance that both PR and TR in the absence any cellular factor has been reported for c-mutations in x-phage of *Serratia* (KAPLAN, WINKLER, and WOLF-ELL-

MAUER, 1960). Aside from the desirability of independent confirmation of this, it is essential that the dependance of the reported PR on wavelength be established, since, while TR of extracellular phages may be theoretically explicable by a pyrimidine photoproduct dehydration mechanism, that of PR is not so obvious. It should be recalled (see above) that this is the only extracellular DNA-containing particle for which direct mutation induction by irradiation has been reported, but as yet unconfirmed.

The question arises as to whether the various reversal mechanism and, in particular, TR and PR perform the same function. In the intracellular state this may be so since PR involves a dark reaction with an activation energy approximately the same as that for TR. For at least two microorganisms a combination of TR and PR, or conversely, gave the same final survival values as PR or TR alone, respectively.

Why is PR, or TR, not a 100% phenomenon? Is it because of the inefficiency of the process, or some cross-linking such as mentioned above, or because of some other UV induced reactions as yet unknown? What is the function of the PR enzyme? Is it a relic of the past which operated to protect primitive organisms at a time when shorter wavelengths from the sun penetrated to the earth's surface. Or is it an adaptive enzyme; or an enzyme which normally performs some other function, PR being merely incidental to its normal operation in the cellular environment.

From the above it will be seen that a good deal is already known about the primary effects of UV on nucleic acids. More important, it is now possible to plan physico-chemical experiments to cope with new biological findings. For example, the increased sensitivity of bromine-substituted DNA to UV and ionizing radiations invites an examination of the radiation chemistry of polymers of bromouridine, now in progress in our laboratory.

Two important fields of work remain relatively unexplored. One is that dealing with the effects of different wavelengths on photochemical transformations in NA derivatives and model oligo- and polynucleotides. In particular at wavelengths below about 2350 Å the photochemical reactions described above are accompanied by other, usually irreversible, reactions. The second is that dealing with the nature of the photochemically excited states of pyrimidine derivatives, both in the free state and when incorporated in polymer chains, and a knowledge of which would considerably extend our understanding of the reaction mechanisms involved in NA photochemistry, as well as facilitating the interpretation of at least some of the effects of ionizing radiations, particularly with respect to formation of pyrimidine hydroperoxides.

References

ALEXANDER, P., and K. A. STACEY, 1959: Cross-linking of deoxyribonucleic acid in sperm heads by ionizing radiation. Nature (London) **184**, 958—960.

BARANOWSKA, J., and D. SHUGAR, 1960: Photochemistry of model oligo- and poly-nucleotides. III. Cross-linking and staining properties of ultraviolet irradiated films of nucleic acids and oligonucleotides. Acta Biochim. Polon. **7**, 505—520.

BAWDEN, F. C., and A. KLECZKOWSKI, 1959: Photoreactivation of nucleic acid from tobacco mosaic virus. Nature (London) **183**, 503—504.

BEUKERS, R., and W. BERENDS, 1961: The effect of UV-irradiation on nucleic acids and their components. Biochim. biophysica Acta **49**, 181—189.

BEUKERS, R., and W. BERENDS, 1960: Isolation and identification of the irradiation product of thymine. Biochim. biophysica Acta **41**, 550—551.

BEUKERS, R., J. IJLSTRA, and W. BERENDS, 1959: The effect of UV-light on some components of the nucleic acids. IV. The influence of oxygen on the extent of the first irreversible reaction, shown by uracil and orotic acid. Recueil Trav. chim. Pays-Bas **78**, 879—882.

Bresch, C., 1950: Zur Reaktivierung von Bakteriophagen. Z. Naturforsch. **5b**, 420—422.

Doudney, C. O., and F. L. Haas, 1960: Some biochemical aspects of the postirradiation modification of ultraviolet-induced frequency in bacteria. Genetics **45**, 1481—1502.

Ellmauer, H., and R. W. Kaplan, 1959: Auslösung von Klarplaque-Mutationen durch UV-Bestrahlung des freien Phagen von *Serratia marcescens*. Naturwissenschaften **46**, 150.

Griffin, B. E., A. R. Todd, and A. Rich, 1958: A synthesis of ribothymidine-5'-pyrophosphate and its conversion to polyribothymidylic acid. Proc. nat. Acad. Sci. U. S. **44**, 1123—1128.

Haas, F. L., and C. O. Doudney, 1957: A relation of nucleic acid synthesis to radiation-induced mutation frequency in bacteria. Proc. nat. Acad. Sci. U. S. **43**, 871—883.

Jagger, J., 1958: Photoreactivation. Bact. Rev. **22**, 99—142.

Janion, C., and D. Shugar, 1960: Absorption spectra, structure and behaviour towards some enzymes of dihydropyrimidines and dihydro-oligonucleotides. Acta Biochim. Polon. **7**, 309—329.

Kaplan, R. W., U. Winkler, and H. Wolf-Ellmauer, 1960: Induction and reversion of c-mutations by irradiation of the extracellular phage of *Serratia*. Nature (London) **186**, 330—331.

Kelner, A., 1949: Effect of visible light on the recovery of *Streptomyces griseus* conidia from ultraviolet irradiation injury. Proc. nat. Acad. Sci. U. S. **35**, 73—79.

Krieg, D. R., 1959: Induced reversion of T4 mutants by ultraviolet irradiation of extracellular phage. Virology **9**, 215—227.

Marmur, J., and P. Doty, 1959: Heterogeneity in deoxyribonucleic acids. I. Dependence on composition of the configurational stability of deoxyribonucleic acids. Nature (London) **183**, 1427—1429.

Moore, A. M., and C. H. Thomson, 1957: Ultraviolet irradiation of pyrimidine derivatives. I. 1,3-dimethyluracil. Canad. J. Chem. **35**, 163—169.

Pichl, H., 1960: Wirtszellenfreie Photoreaktivierung UV-inaktivierter T2-Phagen. Naturwissenschaften **47**, 283—284.

Rupert, C. S., S. H. Goodgal, and R. M. Herriott, 1958: Photoreactivation in vitro of ultraviolet inactivated *Hemophilus influenzae* transforming factor. J. Gen. Physiol. **41**, 451—471.

Rushizky, G. W., M. Riley, L. S. Prestidge, and A. B. Pardee, 1961: Inactivation of enzyme formation by ultraviolet light. I. Action spectra and size of the sensitive unit. Biochim. biophysica Acta (in press).

Setlow, R., 1961: The action spectrum for the reversal of the dimerization of thymine induced by ultraviolet light. Biochim. biophysica Acta **49**, 237—238.

Setlow, R., and B. Doyle, 1954: The action of radiation on dry desoxyribonucleic acid. Biochim. biophysica Acta **15**, 117—125.

Setlow, J. K., and R. Setlow, 1960: Evidence for the existence of a single-stranded stage of T2 bacteriophage during replication. Proc. nat. Acad. Sci. U. S. **46**, 791—798.

Shugar, D., 1960: Photochemistry of nucleic acids and their constituents. In: E. Chargaff and J. N. Davidson (Editors), The nucleic acids, Vol. III, Chapt. 30. Academic Press, New York.

Shugar, D., 1960/61: Photoreactivation and reversible photolysis of nucleic acids. 3rd Internat. Congr. Photobiol., Copenhagen, 1960, Comm. No. **111**; Elsevier, Amsterdam, 1961.

Shugar, D., and J. Baranowska, 1960: Aggregation and staining behaviour of ultra-violet irradiated films of nucleic acids. Nature (London) **185**, 33—34.

Shugar, D., and K. L. Wierzchowski, 1958: Photochemistry of nucleic acids, nucleic acid derivatives and related compounds. Postepy Biochemii 4, (Suppl.), 243—296.

Shugar, D., and K. L. Wierzchowski, 1960/61: Photochemistry of pyrimidine residues in oligonucleotides and in single and double stranded polynucleotides. 3rd Internat. Congr. Photobiol., Copenhagen, 1960, Comm. No. **157**; Elsevier, Amsterdam, 1961.

Sinsheimer, R. L., 1954: The photochemistry of ridylic acid. Radiation Research **1**, 505—513.

Sinsheimer, R. L., 1957: The photochemistry of cytidylic acid. Radiation Research **6**, 121—125.

Smietanowska, A., and D. Shugar, 1961: Photochemistry of model oligo- and polynucleotides. V. Properties of photodimers of uracil and thymine. Bull. Acad. Polons Sci., Cl. II **9**, 375—380.

Wacker, A., H. Dellweg und E. Lodemann, 1961: Strahlenchemische Veränderungen der Nukleinsäuren in vivo und in vitro. 2 Mitteilungen. Angew. Chem. **73**, 64—65.

Wacker, A., H. Dellweg und D. Weinblum, 1960: Strahlenchemische Veränderung der Bakterien-Desoxyribonucleinsäure in vivo. Naturwissenschaften **47**, 477.

Wang, S. Y., 1958: Photochemistry of nucleic acids and related compounds. I. The first step in the ultraviolet irradiation of 1,3-dimethyluracil. J. Amer. Chem. Soc. **80**, 6196—6201.

Wang, S. Y., 1959: Phototautomerization of cytosine derivatives on ultra-violet irradiation. Nature (London) **184**, **4**, 184—185.

WEATHERWAX, R. S., and O. E. LANDMAN, 1960: Ultraviolet light-induced mutation and deoxyribonucleic acid synthesis in *Escherichia coli*. J. Bact. **80**, 528—535.

WIERZCHOWSKI, K. L., and D. SHUGAR, 1959: Studies of reversible photolysis in oligo- and polyuridylic acid. Acta Biochim. Polon. **6**, 313—334.

WIERZCHOWSKI, K. L., and D. SHUGAR, 1960: Photochemistry of model oligo- and poly-nucleotides. II. Homopolymers of adenylic, guanylic and cytidylic acids and several heteropolymers. Acta Biochim. Polon. **7**, 377—399.

WIERZCHOWSKI, K. L., and D. SHUGAR, 1960/61: Photochemistry of 4-aminopyrimidines. 3rd Internat. Congr. Photobiol., Copenhagen, 1960, Comm. No. **156**, Elsevier; Amsterdam, 1961.

WIERZCHOWSKI, K. L., and D. SHUGAR, 1961: Photochemistry of cytosine nucleosides and nucleosides II. Acta Biochem. Polon. **8**, 219—234.

WIERZCHOWSKI, K. L., and D. SUGAR, 1962: Photochemistry of model oligo- and poly-nucleotides IV J. Photochem. Photobiol. **1**, 21–36.

WITKIN, E. M., 1956: Time, temperature and protein synthesis: A study of ultraviolet-induced mutation in bacteria. Sympos. Quant. Biol. **21**, 123—140.

WITKIN, E. M., 1958: Post-irradiation metabolism and the timing of ultraviolet-induced mutations in bacteria. Proc. 10th Internat. Congr. Genet. **1**, 280—299.

WULFF, D. L., and G. FRAENKEL, 1961: On the nature of thymine photoproduct. Biochim. biophysica Acta **51**, 332–339.

Strahlenchemische Veränderung der Nucleinsäuren in vivo und in vitro

A. Wacker

Institut für Therapeutische Biochemie der Universität Frankfurt am Main

Im Verlauf unserer Arbeiten über die chemische Veränderung der Bakteriennucleinsäure in vivo durch den Einbau Pyrimidin- und Purin-analoger Verbindungen (Weygand, Wacker und Dellweg, 1952; Wacker, Kirschfeld und Weinblum, 1960; Wacker, Kirschfeld, Hartmann und Weinblum, 1960; Wacker, Kirschfeld und Träger, 1960) untersuchten wir ebenfalls die strahlenchemische Veränderung der Basenbausteine der Desoxyribonucleinsäure (DNS) und Ribonucleinsäure in vivo und in vitro.

Die Wirkung von UV-Licht

Die strahlenchemische Veränderung des Thymins

Bestrahlt man eine Bakteriensuspension mit UV-Licht der Wellenlänge 254 mμ, so findet man nach Perchlorsäurespaltung der gesamten Bakterien oder der aus ihnen isolierten DNS eine zweite radioaktive Verbindung, die mit dem Thymin-Dimeren identisch ist (Beukers und Berends, 1960; Wacker, Dellweg und Weinblum, 1960; Wacker, Dellweg und Lodemann, 1961) (Abb. 1).

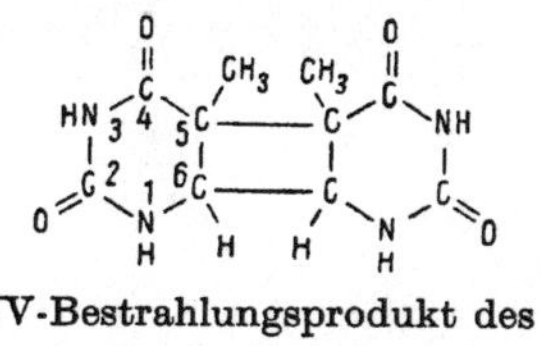

Abb. 1. UV-Bestrahlungsprodukt des Thymins

Wie aus Tabelle 1 hervorgeht, hängen die Überlebensrate der Bakterien und die Menge des Thymin-Dimeren (T_D) von der eingestrahlten UV-Dosis ab. Die maximal gebildete Menge T_D ist anscheinend je nach Bakteriengattung verschieden. So fanden wir bei drei verschiedenen *E.-coli*-Stämmen (*E. coli* B, *E. coli* B_r und *E. coli* 15 T−) bei Bestrahlung mit $5{,}8 \times 10^5$ erg/mm^2 im Durchschnitt 9,5% T_D, bei Bestrahlung von *Enterococcus* Stei dagegen 17% T_D. Es sei hinzugefügt, daß das Basenverhältnis (G + C)/(A + T) für *E. coli* 1,0 und für *E.* Stei 0,69 beträgt. Entfernt man aus der DNS von *E.* Stei durch Hydrolyse die Purine, so steigt die Menge T_D von 17 auf 29% an.

Tabelle 1

Umwandlung von Thymin in der Bakterien-DNS von *E. coli* durch UV-Strahlen

UV-Bestrahlung $[erg/mm^2] \times 10^{-4}$	Überlebende [%]	Thymindimer [%]
0,019	18 ± 2	0,059
0,057	0,5 ± 0,4	0,094
0,54	> 0,01	1,4
1,08	> 0,0001	1,9
2,16		3,4
32,4		9,7

Mit einer Reihe von Modellsubstanzen haben wir die Dimerisierung des Thymins durch UV-Licht näher untersucht. Die Erklärung dafür, daß nicht das gesamte Thymin in der DNS in T_D umgewandelt wird, sehen wir darin, daß, abgesehen von der Einstellung eines Gleichgewichtszustandes, nicht alle Thyminmoleküle ein zweites Thymin in Nachbarschaft haben. In der Apurinsäure setzt sich mehr Thymin zu T_D um, da durch die Herausnahme der Purine mehr Thyminmoleküle in Nachbarschaft zueinander stehen können und außerdem die Wasserstoffbrücken zwischen den beiden DNS-Strängen nicht mehr vorhanden sind. Das gleiche gilt für hitzedenaturierte DNS; hier beobachten wir einen Anstieg um etwa 20% an T_D. Bestrahlt man eine wäßrige Lösung von Thymin (0,0001 m), Thymindesoxyribosid oder Thymindesoxyribosid-5 -monophosphat mit 5×10^5 erg/mm², so bildet sich die dimere Verbindung nur zu etwa 3%, dagegen bei Bestrahlung einer Lösung von Thymidylyl-(3' → 5')-thymidin (TpT) zu etwa 34%. Ersetzt man in dem Dinucleotid ein Thymin durchAdenin oder vergrößert den Abstand der beiden Thyminmoleküle durch eine Pyrophosphatbrücke, so sinkt die Ausbeute an Bestrahlungsprodukt auf 6%. Im Hinblick darauf, daß die Bildung des T_D durch UV-Licht die Nachbarschaft zweier Thyminmoleküle in der DNS zur Voraussetzung hat, weiterhin ein Drittel der Thyminmoleküle maximal dimerisieren kann (wie dies die Untersuchungen mit TpT zeigen), errechnet sich für die in der *E.-coli*-DNS benachbarten Thyminmoleküle $3 \times 9{,}5\% = 28{,}5\%$ und für *E.* Stei $3 \times 17\% = 51\%$ Thyminmoleküle in Nachbarschaft. Dieser Wert steht in guter Übereinstimmung mit einer Sequenzanalyse der Basenbausteine in der *E.-coli*-DNS, über die kürzlich A. KORNBERG und Mitarbeiter (JOSSE, KAISER und KORNBERG, 1961) berichtet haben.

Außerdem haben wir die Dimerisierung von TpT und die Spaltung des T_D (beides in wässeriger Lösung) in Abhängigkeit von der Wellenlänge untersucht (Tabelle 2 und 3).

Tabelle 2

Bildung und Spaltung der Thymindimeren durch UV-Licht verschiedener Wellenlänge. Strahlenquelle: HBO 500 + Zeissfilter; Abstand der Strahlenquelle: 23 cm; Bestrahlungszeit: 20 min; Konzentration der Dimeren: 200 γ/ml; Konzentration der TpT-Lösung: 12 γ/ml

Filter	Durchlässigkeitsbereich [mμ]	Dimerenbildung aus TpT [%]	Dimerenspaltung [%]
WG 1	340 – > 3000	0	0
WG 3	310 – > 3000	0	0
WG 5	280 – > 3000	1,5	0,1
WG 6	260 – > 3000	4	0,2
WG 7	250 – > 3000	5	0,5
UG 5	220 – 450	9	7,7
ohne		9	

Tabelle 3
Spaltung der Thymindimeren mit UV-Licht verschiedener Wellenlänge. Strahlenquelle: XBO 501 + Doppelmonochromator; eingestrahlte Energie: $1{,}5 \times 10^3$ erg/mm²

Wellenlänge [mμ]	Dimerenspaltung [%]	Wellenlänge [mμ]	Dimerenspaltung [%]
253	4,6	290	0
271	4,2	315	0
285	1,3	325	0

Versuche zur Photoreaktivierung

Bestrahlt man Thymidesoxyribosid-Dimeres oder eine mit Thymin-[2-^{14}C] markierte, UV-bestrahlte DNS, in der etwa 10% der Thyminmoleküle dimerisiert sind, mit sichtbarem Licht einer Philips-Tageslichtlampe HRP 125 W zwei Stunden in 10 cm Abstand in Gegenwart eines zellfreien Extraktes aus Bäckerhefe, so werden etwa 10% der dimerisierten Thyminmoleküle wieder zu der monomeren Verbindung aufgespalten. T_D wird unter diesen Bedingungen nicht beeinflußt. Dagegen wird diese Verbindung mit Uranylacetat (10^{-3} m) und sichtbarem Licht zu etwa 10% zu Thymin aufgespalten.

Elektronenverteilung im Thymindimeren

Die Elektronenverteilung im Pyrimidinring in Abhängigkeit von den Substituenten zeigt sich auch in der Bindefestigkeit der Desoxyribose. Hydrolysiert man unter gleichen Bedingungen (Webb und Levy, 1955) verschiedene Desoxyriboside und setzt dabei die hydrolisierte Menge Desoxyribose aus Thymindesoxyribosid (TDR) gleich 1, so ergibt sich dabei das folgende Bild: TDR 1; 5-Bromuracildesoxyribosid (BUDR) 7,7; Cytosindesoxyribosid (CDR) 19,5; TDR-Dimer 93. Demnach steht BUDR in seinem Verhalten zwischen TRD und CDR. Dies könnte eine Erklärung für sein mutagenes Verhalten sein, d. h. seine Paarungstendenz mit Guanin an Stelle mit Adenin bei der Verdopplung der DNS (vgl. hierzu Freese, 1959), ebenso wie die extrem veränderten Verhältnisse beim TDR-Dimeren einen Hinweis für die mutagene Wirkung des UV-Lichtes geben könnten.

Photochemische Reaktionen von Uracil und Uridin

Durch spektralphotometrische Untersuchungen entdeckten Sinsheimer und Hastings (1949) die reversible photochemische Umwandlung des Uracils und Uridins. Mit Hilfe der Papierchromatographie bestätigten Moore (1958) die Wasseranlagerung an die 5,6-Doppelbindung des Uracilmoleküls (Shugar und Wierzchowski, 1957). Durch Verwendung ^{14}C-markierten Uracils und Uridins konnten wir kürzlich zeigen (Wacker, Träger und Weinblum, 1961), daß die beiden Verbindungen bei Bestrahlung mit UV-Licht der Wellenlänge 254 mμ dimerisieren, wie dies erstmals Beukers und Berends (1960) für Thymin gefunden haben. Nachfolgend berichten wir ausführlicher über die photochemischen Veränderungen des Uracils und Uridins.

Bestrahlt man wäßrige Lösungen von Uracil-[2-^{14}C] flüssig oder eingefroren, so bilden sich in Abhängigkeit von der eingestrahlten Dosis UV-Licht (und von der Konzentration der Lösung) zwei Verbindungen A und B, die sich papierchromatographisch (s. Abb. 2) im Lösungsmittelsystem n-Butanol/Wasser (86: 14) trennen lassen (R_F-Werte: Verbindung A 0,04; B 0,18; Uracil 0,31). Wie aus Abb. 2 hervorgeht, bildet sich bei einer kleineren UV-Dosis mehr A, bei einer größeren mehr B. Höhere Uracilkonzentrationen begünstigen die Bildung von A, niedrigere die von B. Das Absorptionsspektrum der Verbindungen A und B zeigt kein Maximum bei 258 mμ, woraus man schließen kann, daß das Uracil unter Absättigung der 5,6-Doppelbindung reagiert hat. Durch Hitze (100°, 15 min) wird A sehr wenig, B zum größten Teil in Uracil verwandelt. Durch UV-Bestrahlung der wäßrigen Lösungen wird A in Uracil überführt, B dagegen nicht verändert. Gegenüber Säuren und Alkalien sind beide Verbindungen wenig beständig. Nach ihrem R_F-Wert in n-Butanol/Wasser (86: 14) (MOORE (1958) und ihren physikalisch-chemischen Eigenschaften ist die Verbindung B mit dem Wasseradditionsprodukt des Uracils identisch. Die Verbindung A zeigt all die Eigenschaften einer dimeren Verbindung analog der des Thymins (BEUKERS und BERENDS, 1960; WACKER, DELLWEG und WEINBLUM, 1960; WACKER, DELLWEG und LODEMANN, 1961). Offenbar ist das Uracildimere mit dem von RÖRSCH et al. (RÖRSCH, BEUKERS, IJLSTRA und BERENDS, 1958) beschriebenen 1. irreversiblen Photoprodukt des Uracils identisch.

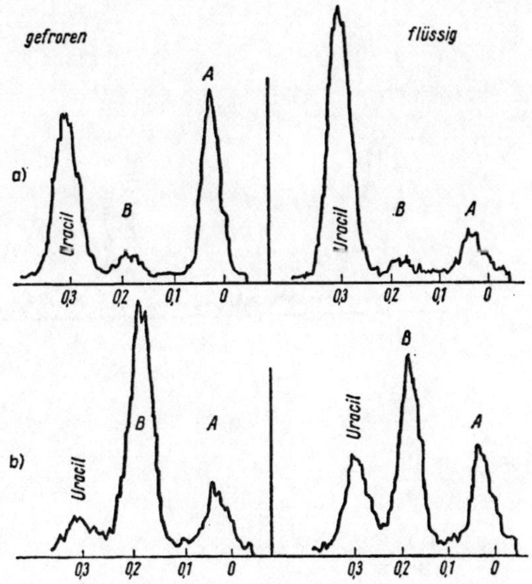

Abb. 2. Bildung von Uracildimerem (A) und Uracil-Wasseradditionsprodukt (B) in Abhängkeit von der UV-Dosis und vom Zustand der bestrahlten Lösung. Chromatogramme von UV-bestrahltem Uracil-[2-^{14}C] (spezif. Aktivität 2 mC/mM; Konzentration 10^{-3} m). a) UV-Dosis $5,4 \times 10^4$ erg/mm^2; b) UV-Dosis $8,6 \times 10^5$ erg/mm^2. Abszisse: R_F-Wert im Lösungsmittelsystem n-Butanol/Wasser (86:14); Ordinate: Radioaktivität Imp/min

Das gleiche Verhalten wie Uracil zeigt Uridin. Bestrahlt man eine wäßrige Lösung von Uridin-[2-^{14}C] (spezif. Aktivität 0,28 mC/mM; Konz. 5×10^{-4} m) mit verschiedenen Dosen UV-Licht ($7,2 \times 10^4$ erg/mm^2; $5,8 \times 10^5$ erg/mm^2), so erhält man mit der kleinen Dosis 27% Dimeres und 27% Wasseranlagerungsprodukt, mit der großen Dosis 19% Dimeres und 58% Wasseranlagerungsprodukt. Wird die Uridinlösung in gefrorenem Zustand bestrahlt, so ist die Dimerisierung im Vergleich zur Wasseraddition begünstigt. R_F-Werte im Lösungsmittelsystem n-Butanol/Wasser (86:14): Uridin 0,17; Wasseradditionsprodukt des Uridins 0,09; Uridindimeres 0,02. Die Stabilität des Wasseranlagerungsproduktes und der dimeren Verbindung gegenüber Hitze, UV-Licht, Säure und Alkali ist ähnlich wie beim Uracil.

Bestrahlt man in Eis eine 5×10^{-4}molare Lösung von Thymin-[2-^{14}C] und Uracil-[2-^{14}C] mit einer UV-Dosis von $5,4 \times 10^4$ erg/mm^2 und chromatographiert anschließend die Reaktionsmischung, so erhält man das in Abb. 3c wiedergegebene Bild. Eluiert man die photochemisch entstandenen Verbindungen und bestrahlt ihre wäßrigen Lösungen mit UV-Licht (Dosis $5,4 \times 10^4$ erg/mm^2), so erhält man bei der Verbindung mit dem kleinsten R_F-Wert nur Uracil zurück, bei der Verbindung mit dem nächsthöheren R_F-Wert in äquimolaren Mengen Thymin und Uracil. Die Verbindung mit dem R_F-Wert 0,13 ist mit dem Thymin-

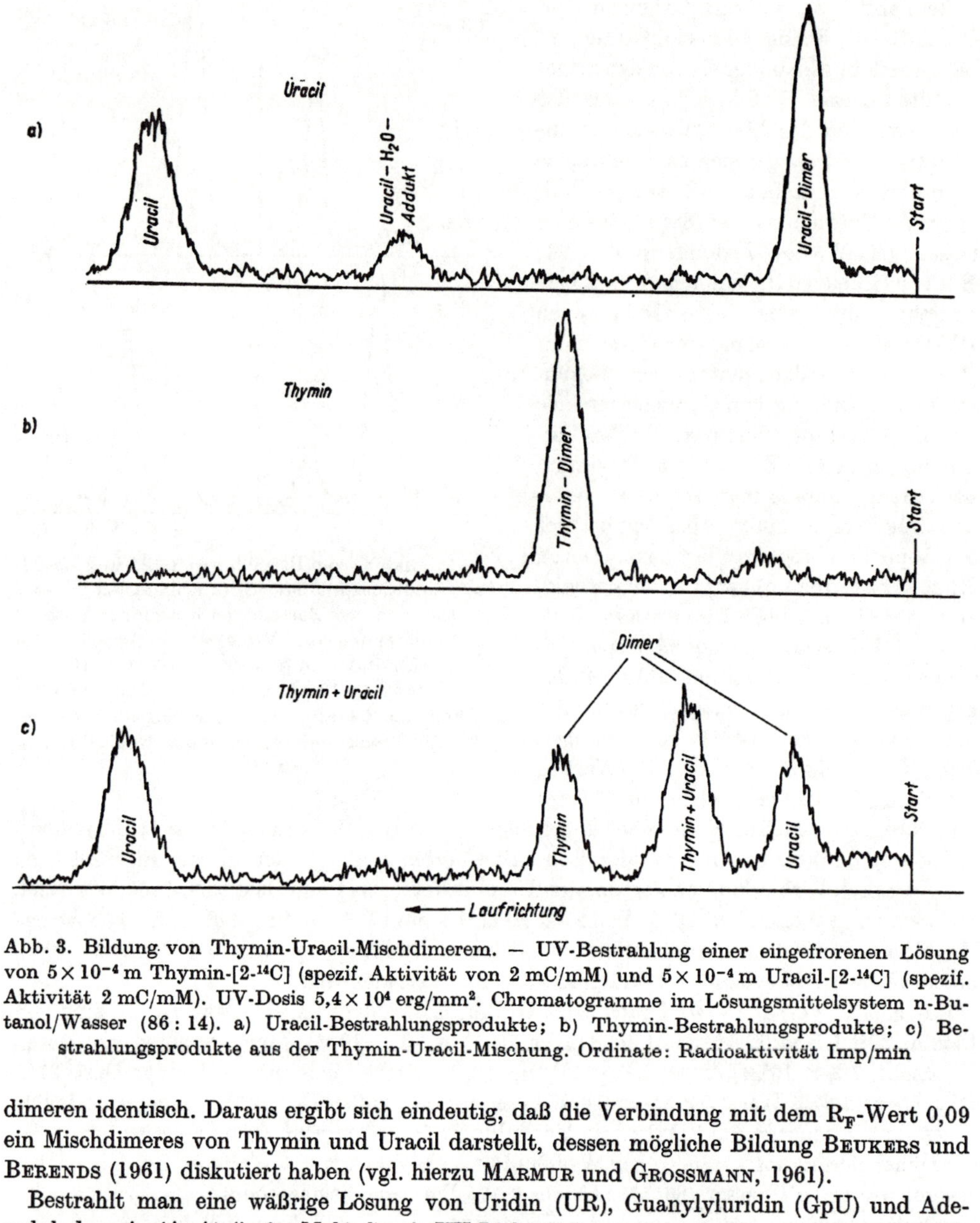

Abb. 3. Bildung von Thymin-Uracil-Mischdimerem. — UV-Bestrahlung einer eingefrorenen Lösung von 5×10^{-4} m Thymin-[2-^{14}C] (spezif. Aktivität von 2 mC/mM) und 5×10^{-4} m Uracil-[2-^{14}C] (spezif. Aktivität 2 mC/mM). UV-Dosis $5{,}4 \times 10^4$ erg/mm². Chromatogramme im Lösungsmittelsystem n-Butanol/Wasser (86 : 14). a) Uracil-Bestrahlungsprodukte; b) Thymin-Bestrahlungsprodukte; c) Bestrahlungsprodukte aus der Thymin-Uracil-Mischung. Ordinate: Radioaktivität Imp/min

dimeren identisch. Daraus ergibt sich eindeutig, daß die Verbindung mit dem R_F-Wert 0,09 ein Mischdimeres von Thymin und Uracil darstellt, dessen mögliche Bildung Beukers und Berends (1961) diskutiert haben (vgl. hierzu Marmur und Grossmann, 1961).

Bestrahlt man eine wäßrige Lösung von Uridin (UR), Guanylyluridin (GpU) und Adenylyladenosin (ApA) (je 0,4 Mol/ml) mit UV-Licht ($5{,}8 \times 10^5$ erg/mm²), so beträgt der Abfall von der Ausgangsextinktion am jeweiligen Maximum bei UR 86%, bei GpU 15% und bei ApA 0%. Erhitzt man die bestrahlten Lösungen (100°, 15 min), so steigt die Extinktion der Lösungen bei UR um 33% und bei GpU um 13% an. Vergleicht man diese Ergebnisse mit den mitgeteilten Beständigkeiten der dimeren Verbindung des Uridins und des Wasseradditionsproduktes des Uridins, so ergibt sich hieraus, daß im Dinucleotid GpU das Uracilmolekül bei der UV-Bestrahlung lediglich Wasser anlagert, jedoch nicht dimerisiert. Zu dem gleichen Ergebnis kam auch D. Shugar (persönliche Mitteilung) bei der UV-Bestrahlung anderer uracilhaltiger Nucleotide.

Wuchsstoffeigenschaften der Photoprodukte des Thymins und Uracils

Die Wuchsstoffeigenschaften des Thymins und Thymindesoxyribosids sind bei den dimeren Verbindungen nicht mehr vorhanden. Dies gilt für *Sc. faecalis* R, *Lb. leichmannii* 313 und *Lb. acidophilus* R 26. Auch läßt sich die Sulfonamidhemmung mit Thymindimeren statt Thymin bei einer Reihe von Bakterien nicht mehr aufheben. Die Wasseradditionsprodukte von Uracil und Uridin und die dimeren Verbindungen des Uracils und Uridins sind für die Uracil-Mangelmutante *E. coli* 63–86, die entweder Uracil oder Uridin zum Wachstum benötigt, keine Wuchsstoffe mehr.

Strahlensensibilisierende Wirkung des 5-Bromuracils

Es war von Interesse, festzustellen, wie in der DNS an Stelle von Thymin eingebautes 5-Bromuracil (BU) strahlenchemisch verändert wird, und weiterhin, wie sich der teilweise Ersatz von Thymin durch BU in der DNS auf die Dimerisierung des Thymins auswirkt.

Bestrahlt man eine wäßrige Lösung von BU (10^{-4} m) mit $1{,}5 \times 10^5$ erg/mm^2 UV-Licht der Wellenlänge 254 mμ, so wird dabei das BU nicht nachweisbar verändert. 5-Bromuracil-desoxybribosid wird mit der gleichen Dosis zu etwa 3% in Uracildesoxyribosid umgewandelt. Bei Bestrahlung der eingefrorenen Lösungen erhält man das gleiche Ergebnis. Daraus ergibt sich, daß BU im Gegensatz zu Thymin und Uracil unter der Einwirkung von UV-Licht nicht dimerisiert. Eingebaut in die DNS wird jedoch überraschenderweise BU bei Bestrahlung entweder der Bakterien oder der aus ihnen isolierten DNS mit der gleichen UV-Dosis zu 56% (Bakterien) bzw. 72% (DNS) (s. Abb. 4) verändert. Die Menge des in der DNS umgewandelten BU hängt von der eingestrahlten Dosis ab. Wird sie auf $5{,}8 \times 10^5$ erg/mm^2 erhöht, so werden 83% (Bakterien) BU verändert. – Es ist anzunehmen, daß ein großer Teil des isolierten, aus BU entstandenen radioaktiven Uracils in der Bakterien-DNS nach der UV-Bestrahlung in einer anderen Form vorgelegen hat (z. B. dimerisiert, in Form eines Wasseradduktes oder auch in Form eines Mischdimeren, wenn sich in Nachbarschaft zu Uracil Thymin befand), da die Uracilphotoprodukte unter den verwendeten Hydrolysebedingungen nicht stabil sind.

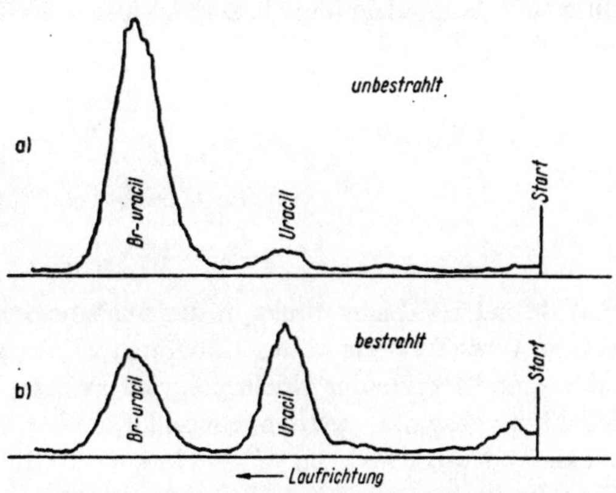

Abb. 4. Dehalogenierung von 5-Bromuracil in der Bakterien-DNS durch UV-Bestrahlung. – Papierchromatographie des Perchlorsäurehydrolysates von *Enterococcus* Stei in n-Butanol/Wasser (86 : 14). 2 γ 5-Bromuracil-[2-^{14}C]/ml Nährmedium, spezif. Aktivität 10 mC/mM; UV-Dosis $1{,}5 \times 10^5$ erg/mm^2. (Wie aus der Papierchromatographie der unbestrahlten Bakterien hervorgeht, wird BU zu ca. 5% von den Bakterien dehalogeniert und in Uracil überführt.) Ordinate: Radioaktivität in Imp/min

Bei Bestrahlung von *Enterococcus* Stei mit $1{,}7 \times 10^5$ erg/mm^2 UV-Licht (λ 254 mμ) dimerisieren 10,5% der Thyminmoleküle in der Bakterien-DNS, was man dadurch einfach nachweisen kann, indem man die Bakterien mit 2 γ Thymin-[2-^{14}C]/ml Nährmedium (spezif. Aktivität 10 mC/mM) kultiviert. Befinden sich im Nährmedium noch zusätzlich 16 γ BU/ml, wodurch etwa 20% Thymin in der DNS durch BU ersetzt werden, so werden bei Bestrahlung mit der gleichen Dosis nur 6,8% Thymin dimerisiert und bei einem weiteren Ersatz des Thymins durch BU von 35% (2 γ Thymin-[2-^{14}C] und 64 γ BU/ml) sinkt die Ausbeute an Thymindimeren auf 4,7%. Bei *E. coli* 15 T$^-$ lassen sich mit 1,7 × 10^5 erg/mm^2 UV-Licht 7,7% Thymin dimerisieren. In Anwesenheit von 2 γ Thymin-[2-^{14}C]/ml und 16 γ BU/ml, wodurch etwa 35% der Thyminmoleküle durch BU ersetzt werden, sinkt die Menge Thymindimeres bei gleicher UV-Dosis auf 6% ab.

Wir konnten nun weiter oben zeigen, daß ein Zusammenhang zwischen der Menge Thymindimeres und der Überlebensrate von Bakterien nach UV-Bestrahlung besteht (vgl. Tab. 1). Bei 18% Überlebenden bei *E. coli* 15 T$^-$ finden wir 0,059% Thymindimeres, bei 0,5% Überlebenden 0,094% Thymindimeres. Diese Befunde weisen darauf hin, daß die letale Wirkung des UV-Lichtes mit der Dimerisierung des Thymins zusammenhängen kann, wobei aber noch unklar ist, ob hierbei allein die Menge Thymindimeres unabhängig von dem Ort, an dem die Dimerisierung erfolgt, entscheidend ist, oder ob eine Dimerisierung zwischen den beiden DNS-Ketten ausschlaggebend ist. In scheinbarem Gegensatz zu dieser Annahme steht die Tatsache, daß nach Einbau von BU in die DNS die Menge Thymindimeres zurückgeht, gleichzeitig aber die UV-Empfindlichkeit der Zelle wesentlich zunimmt (Szybalski, 1959; Greer und Zamenhof, 1959). Berücksichtigt man aber, daß BU in der DNS sehr leicht durch UV-Licht dehalogeniert wird, Uracil aber kein Baustein der DNS ist, ferner, daß dabei auch einzelständig entstehende Uracilmoleküle durch UV-Licht unter Wasseraddition weiterreagieren können (was bei einem einzelständigen Thymin nicht der Fall ist), so sieht man, daß das Ausmaß der Störungen in einer BU-haltigen DNS-Kette (falscher Baustein, Dimerisierung, Mischdimerisierung, Wasseraddition) weit größer ist als in einer DNS, die kein BU enthält.

Die Wirkung von Röntgenstrahlen

Während UV-Licht direkt in der oben beschriebenen Weise die Pyrimidine der Nucleinsäuren verändert, zerstören Röntgenstrahlen auf primärem und sekundärem Wege Pyrimidine und Purine der Nucleinsäuren. Primär, indem die Moleküle durch die energiereiche Strahlung gespalten werden, sekundär, indem die Röntgenstrahlen im Milieu der Zelle aus Wasser oder organischen Verbindungen Radikale bzw. Peroxyde erzeugen, die mit den Basenbausteinen reagieren (Butler, 1959).

Tabelle 4
G-Werte zur Röntgenbestrahlung verschiedener Purine und Pyrimidine. Konzentration der Basen 10^{-4} mit Röntgenbestrahlung bei 60 kV

	G-Wert		G-Wert
Thymin	1,1	Bromuracil	1,5
Cytosin	1,2	Adenin	0,6
Uracil	1,1	Guanin	0,5

Nach der Einwirkung von Röntgenstrahlen auf eine wäßrige Lösung von Purinen und Pyrimidinen beobachtet man ein Absinken der UV-Extinktion. Wie aus Tab. 4 hervorgeht, nimmt die Extinktion[1]) der Pyrimidine Thymin, Cytosin und Uracil schneller ab als die der Purine Adenin und Guanin. Bromuracil reagiert am raschesten. Bestrahlt man die Basen in Gegenwart von Methionin, so fängt die Aminosäure einen Teil der primär gebildeten Radikale ab, und die Basen werden in geringerem Ausmaß angegriffen; das Methionin setzt sich dabei zu Methioninsulfoxyd um. Untersucht man papierchromatographisch röntgenbestrahlte, radioaktiv markierte Pyrimidine, z. B. Thymin-[2-^{14}C] (Abb. 5), so kann man unter anderen radioaktiven Verbindungen Harnstoff nachweisen, dessen Menge von der eingestrahlten Dosis

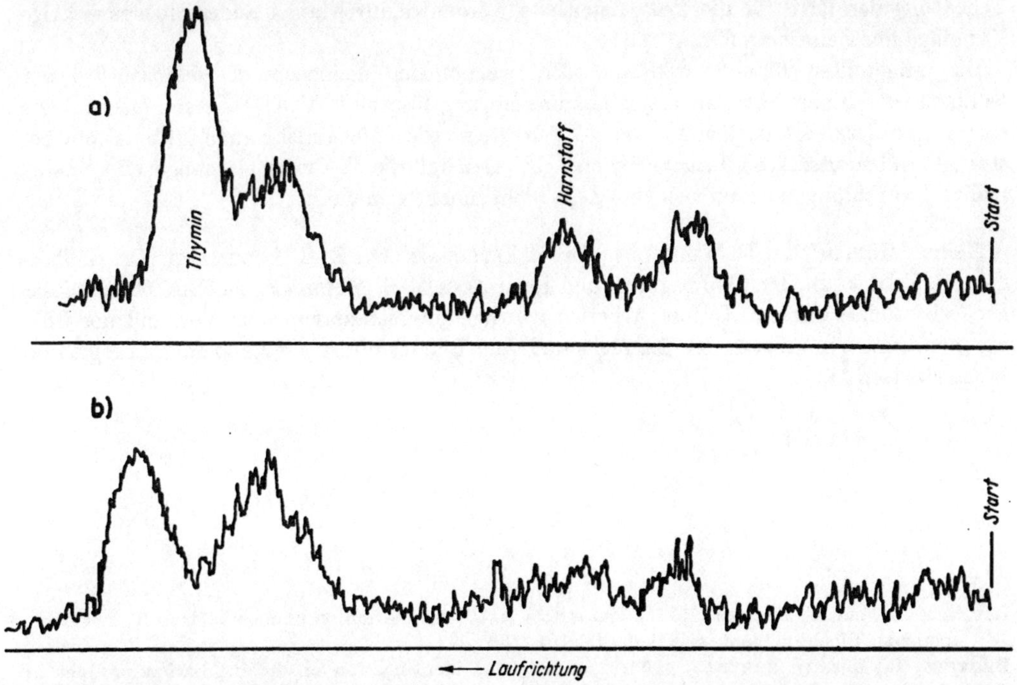

Abb. 5. Zersetzung von Thymin-[2-^{14}C] unter Einwirkung von Röntgenstrahlung oder Wasserstoffperoxyd. a) Thymin-[2-^{14}C], 10^{-4} m, Röntgendosis 60 kr, 60 kV; b) Thymin-[2-^{14}C], 10^{-4} m + 15% H_2O_2, 72 Std., 20° C; Chromatogramme in n-Butanol/Wasser (86:14). Ordinate: Radioaktivität in Imp./min

abhängt. Auch aus Bakterien, deren DNS mit Thymin-[2-^{14}C] markiert war, konnten wir nach Röntgenbestrahlung radioaktiven Harnstoff isolieren (Wacker und Lochmann, 1961). Im Gegensatz zu der durch UV-Licht bewirkten Dimerisierung des Thymins zerstören Röntgenstrahlen die Pyrimidine und Purine; hierdurch entstehen in der DNS „Löcher". Die primäre Wirkung der Röntgenstrahlen ist also im Vergleich zur spezifischen Wirkung des UV-Lichts unspezifisch und vielfältig.

Zur sekundären Wirkung der Röntgenstrahlen auf das genetische Material gehört u. a. die längst bekannte Bildung von Wasserstoffperoxyd. H_2O_2 ist das einfachste Peroxyd; interessanter dürften die im Milieu einer Zelle unter der Einwirkung von Röntgenstrahlen entstandenen organischen Peroxyde sein, wie z. B. Bernsteinsäuremonoperoxyd. Wie Untersuchungen mit markiertem Thymin zeigten, bewirkt Wasserstoffperoxyd einen ähnlichen Abbau des Thyminmoleküls wie Röntgenstrahlen (Abb. 5). Adenin zeichnet sich in diesem Zusammenhang besonders dadurch aus, daß es unter dem Einfluß von Wasserstoffperoxyd

[1]) An Stelle des prozentualen Abfalls der Extinktion ist es vorteilhafter, den G-Wert anzugeben.

bei Zimmertemperatur, also unter milden Bedingungen, sehr leicht Adenin-N-oxyd bildet (Stevens, Magrath, Smith und Brown, 1958). Diese Umwandlung von Adenin in Adenin-N-oxyd kann bei der Verdopplung der DNS eine Störung der Paarung mit Thymin bewirken. Die Bildung von Adenin-N-oxyd in der DNS nach Röntgenbestrahlung oder H_2O_2-Einwirkung konnten wir ebenfalls nachweisen (Wacker und Lochmann, 1961).

Da die Röntgenstrahlen ungerichtet auf die Moleküle einwirken, werden die Basenbausteine der Ribonucleinsäure (RNS) in dem gleichen Umfang wie die der DNS zerstört. Bei der Bedeutung der RNS für die Proteinsynthese werden dadurch auch andere lebenswichtige Vorgänge der Zelle betroffen.

Die mitgeteilten Befunde erklären auch, warum sich mehrfache Einzeldosen bei der Röntgenbestrahlung über längere Zeiträume hinweg bis zur letalen Dosis summieren. Die einmal gebildeten Leerstellen auf der DNS-Kette werden nicht mehr aufgefüllt. Ist ein bestimmter Prozentsatz an Leerstellen erreicht, gleichgültig ob durch einmalige oder mehrmalige Bestrahlung, so kann sich eine Zelle nicht mehr vermehren.

Meinen Mitarbeitern D. Weinblum, Dr. H. Dellweg, Dr. E.-R. Lochmann, Dr. L. Träger, E. Lodemann, Dr. D. Jacherts und Frau Zeinab H. Moustafa möchte ich an dieser Stelle besonders danken. Unsere Arbeiten wurden unterstützt von dem Verband der Chemischen Industrie – Fonds der Chemie – und dem Bundesminister für Atomkernenergie und Wasserwirtschaft.

Literatur

Beukers, R., and W. Berends, 1961: The effects of UV-irradiation on nucleic acids and their components. Biochim. biophysica Acta **49**, 181–189.

Beukers, R., and W. Berends, 1960: Isolation and identification of the irradiation product of thymine. Biochim. biophysica Acta **41**, 550–551.

Butler, J. A. V., 1959: Changes induced in nucleic acids by ionizing radiations and chemicals. Rad. Res. (Suppl. 1), 403–416.

Freese, E., 1959: The specific mutagenic effect of base analogues on phage T4. J. Mol. Biol. **1**, 87–105.

Greer, S., and S. Zamenhof, 1957: Effect of 5-bromouracil in desoxyribonucleic acid of *E. coli* on sensitivity to ultraviolet irradiation. Amer. Chem. Soc. Abstr. (131st meeting), p. 3c.

Josse, J., A. D. Kaiser, and A. Kornberg, 1961: Enzymatic synthesis of desoxyribonucleic acid. VIII. Frequencies of nearest neighbour base sequences in desoxyribonucleic acid. J. Biol. Chem. **236**, 864–875.

Marmur, J., and L. Grossmann, 1961: Ultraviolet light induced linking of desoxyribonucleic acid strands and its reversal by photoreactivating enzyme. Proc. nat. Acad. Sci. U. S. **47**, 778–787.

Moore, A. M., 1958: Ultraviolet irradiation of pyrimidine derivatives. Canad. J. Chem. **36**, 281–283.

Rörsch, A., R. Beukers, J. Ijlstra, and W. Berends, 1958: The effect of UV-light on some components of the nucleic acids. I. Uracil, thymine. Rec. trav. chim. **77**, 423–429.

Shugar, D., and K. Wierzchowski, 1957: Reversible photolysis of pyrimidine derivatives, including trials with nucleic acids. Biochim. biophysica Acta **23**, 657–658.

Sinsheimer, R. L., and R. Hastings, 1949: A reversible photochemical alternation of uracil and uridine. Science (New York) **110**, 525–526.

Stevens, M. A., D. I. Magrath, H. W. Smith, and G. B. Brown, 1958: Purine N-oxides. I. Monooxides of aminopurines. J. Amer. chem. Soc. **80**, 2755.

Szybalski, W., 1959: Effect of chemical modification of nucleic acids on the radiation sensitivity of human cell lines. Animal Cell inf. Serv. **1**, 5–7.

Wacker, A., H. Dellweg und E. Lodemann, 1961: Strahlenchemische Veränderung der Nukleinsäuren. Angew. Chem. **73**, 64–65.

WACKER, A., L. TRÄGER und D. WEINBLUM, 1961: Strahlenchemische Veränderung von Pyrimidinribosiden und Pyrimidinen. Angew. Chem. **73**, 65.

WACKER, A., H. DELLWEG und D. WEINBLUM, 1960: Strahlenchemische Veränderung der Bakterien-Desoxyribonukleinsäure in vivo. Naturwissenschaften **47**, 477.

WACKER, A., S. KIRSCHFELD, D. HARTMANN und D. WEINBLUM, 1960: Über den Einbau von 5-Nitrouracil, 5-Aminouracil und 2-Thiothymin in die Bakterien-Desoxyribonukleinsäure. J. Mol. Biol. **2**, 69–71.

WACKER, A., S. KIRSCHFELD und L. TRÄGER, 1960: Über den Einbau Purin-analoger Verbindungen in die Bakterien-Nukleinsäure. J. Mol. Biol. **2**, 241–242.

WACKER, A., S. KIRSCHFELD und D. WEINBLUM, 1960: Über den Einbau von 5-Bromcytosin und 5-Bromuracil in die Bakterien-Desoxyribonukleinsäure. J. Mol. Biol. **2**, 72–74.

WACKER, A., und R. LOCHMANN, 1961: Unveröffentlichte Ergebnisse.

WEBB, J. M., and H. B. LEVY, 1955: A sensitive method for the determination of desoxyribonucleic acid in tissues and microorganisms. J. Biol. Chem. **213**, 107–117.

WEYGAND, F., A. WACKER und H. DELLWEG, 1952: Stoffwechseluntersuchungen bei Mikroorganismen mit Hilfe radioaktiver Isotope. II. Kompetitive und nicht-kompetitive Enthemmung von 5-^{82}Br-Uracil. Z. Naturforsch. **7b**, 19–25.

Effects of Ionizing Radiations on Nucleic Acids and Nucleoproteins

G. SCHOLES and J. J. WEISS

University of Durham, King's College, Newcastle upon Tyne, 1

The degradation of nucleic acids and of nucleoproteins by ionizing radiations has attracted much interest because of the obvious radiobiological significance. The occurrence of chemical changes in polynucleotides on irradiation of these substances in aqueous solution was established as long ago as 1949, when effects on the constituent bases as well as on the sugar moieties were noted (SCHOLES, STEIN, and WEISS, 1949; SCHOLES and WEISS, 1952). More recently, it has been possible to obtain a quantitative appraisal of the relative extent of degradation of the nucleic acid components and particularly by experiments with purines, pyrimidines, nucleosides, and nucleotides, some further understanding of the nature of the radiation-induced reactions involved (SCHOLES, WARD, and WEISS, 1960). In this paper we wish to present a resume of recent data obtained in our Laboratory along these lines, together with some results obtained with calf thymus nucleoprotein solutions, a preliminary account of which has already been reported (EMMERSON, SCHOLES, THOMSON, WARD, and WEISS, 1960).

Results

All of the irradiations described below were carried out with X-rays (200 kV) and in the presence of oxygen (1 atm.). The radiation yields have been expressed as G-values, i. e. the number of molecules changed per 100 eV of absorbed radiation energy.

a) Pyrimidines and purines

Aqueous solutions (2×10^{-4} M, pH $\sim$ 5) of the individual bases were irradiated and the initial yields for the extent of destruction determined by (1) spectrophotometry at the appropriate wavelength (measurements of this kind are only meaningful if none of the radiation products absorb at the particular wavelength used), (2) paper chromatographic separation of the irradiated solutions, the individual spots of unchanged base being eluted and measured. The results are given in Table 1. Fairly good agreement was obtained using these methods. Cytosine, however, was unusual in that it gave irreproducible results and this behaviour was traced to a post-irradiation effect, involving a slow decrease of the optical density of the irradiated solution. In the case of adenine, very small amounts of 8-oxyadenine were present

Table 1

Initial yields of destruction of pyrimidine and purine bases in aqueous solution (2×10^{-4} M) on irradiation with X-rays (200 kV) in the presence of oxygen, pH 5.1—5.2

Base	Yield (G) UV method	Yield (G) Chromatography	Base	Yield (G) UV method	Yield (G) Chromatography
Thymine	1.89	1.91	Adenine	1.09	1.13
Cytosine	2.05—2.28	2.16	Xanthine	1.36	1.32
Uracil	1.93	—	Uric acid	1.39	1.48
Dimethyl uracil	1.85	1.91			

but they were formed with yields of less than $G \sim 0.1$; similarly very small amounts of uric acid were formed from xanthine. In these solutions, there is a striking difference in the extent of destruction of the purines and pyrimidines; the low yield for adenine is particularly noticeable. Experiments were therefore carried out in which the relative reactivities of the purine and pyrimidine bases were examined. Solutions of mixtures of various purines and pyrimidines were irradiated and the extents of radiolytic decomposition of the individual components determined by paper chromatographic separation. It was found, for example, that in a solution of adenine (10^{-4} M) + thymine (10^{-4} M), G(-adenine) = 0.58 and G(-thymine) = 1.02, and that in a solution of thymine (5×10^{-4} M) + cytosine (5×10^{-4} M), G(-thymine) = 0.92 and G(-cytosine) = 1.02. From these and other experiments, it has been concluded that the purines and pyrimidines examined show about equal reactivity towards the reactive species produced from the water.

Irradiation of a solution containing equimolar amounts of each of the four bases thymine, cytosine, adenine and guanine led to values for the extents of destruction of the individual bases given in Table 2.

Table 2

Yields of destruction of bases in a solution containing thymine, cytosine, adenine, and guanine on irradiation with X-rays (200 kV) in the presence of oxygen (1 atm.); pH = 5.3

Solution	Yields G (-thymine)	G (-cytosine)	G (-adenine)	G (-guanine)
Each base at a concentration of 3.3×10^{-5} M	0.40	0.38	0.22	0.17

It has been shown that hydroperoxides are formed from the pyrimidines, their nucleosides and nucleotides, when their solutions are irradiated in the presence of oxygen (Scholes, Weiss, and Wheeler, 1956). In the case of thymine, it has been proved that the radiation product is 4-hydroxy, 5-hydroperoxy thymine by chromatography and by comparing the rates of reaction with iodide ions of the radiation-produced hydroperoxide with that of the synthetic hydroperoxide (Ekert and Monier, 1959; Scholes and Weiss, 1960). The yields of hydroperoxide from several pyrimidines have been determined under various experimental conditions. Table 3 shows the influence of pH on the initial yields of thymine

Table 3

pH-Dependence of the initial yields of thymine hydroperoxide and hydrogen peroxide in aqueous solution (2×10^{-4} M) on irradiation with X-rays (200 kV) in the presence of oxygen (1 atm.)

pH	G (hydroperoxide)	G (H_2O_2)	pH	G (hydroperoxide)	G (H_2O_2)
8.00	1.02	1.81	1.95	0.33	2.35
5.08	1.05	1.73	1.09	0.14	2.69
3.05	0.58	1.79			

hydroperoxide and of hydrogen peroxide on irradiation of 2×10^{-4} M thymine solutions. Hydroperoxide formation depends markedly upon pH, the yield decreasing with decreasing pH. Cytosine hydroperoxide is unstable, particularly under neutral and alkaline solutions. The yields of hydroperoxide from neutral uracil solutions (G = 1.16) were comparable to those obtained from thymine; dimethyl uracil also yielded a hydroperoxide in somewhat similar amounts (G = 0.75). In contrast to the pyrimidines, the purines whether free or as nucleosides or nucleotides do not form hydroperoxides (or at least any stable hydroperoxides) on irradiation in oxygenated aqueous system.

Comparison of the data for the yields of destruction of the pyrimidines and those of hydroperoxide formation show that whilst the latter compounds are the main products of irradiation, other substances must also be formed. It was observed that treatment of irradiated solutions of uracil and dimethyl uracil with sulphuric acid led to an increase in absorption at 260 mμ. The products responsible for this increased absorption were still present after the hydroperoxide had been removed. In the case of dimethyl uracil the nature of the product could be studied more easily, since unchanged pyrimidine could be removed from the irradiated solution by extraction with chloroform. The spectrum in acid solutions of the separated radiation product agreed well with that obtained on acid treatment of solutions of dimethyl uracil glycol; it was also found that such treatment of both the radiation product and dimethyl uracil glycol gave dimethyl isobarbituric acid. Hence irradiation of demithyl uracil solutions in the presence of oxygen leads to the formation of dimethyl uracil glycol. The initial yield in 2×10^{-4} M dimethyl uracil solutions at pH 5.5 was found to be G(glycol) = 0.4. In the case of uracil, glycol is also an irradiation product, corresponding to a yield of G(glycol) = 0.45 at pH 5.5.

Irradiated solutions of thymine and cytosine did not show any increase in absorption at 260 mμ on acid treatment. However, these findings do not necessarily indicate the absence of glycol since it has been found that acid treatment of solutions of synthetic thymine glycol does not lead to spectral changes.

No lower aldehydes or dialdehydes could be detected. It appears, therefore, that the main products of irradiation of pyrimidines are those formed by saturation of the 4:5 double bond and that no ring-opening occurs.

b) Nucleosides and nucleotides

Table 4 records some values for the extent of base destruction, measured spectrophotometrically, on irradiation of oxygenated solutions of certain nucleosides and nucleotides. Some results for the free bases are included for comparison. In the compounds investigated the extent of base destruction decreases in the order: free base, nucleoside, nucleotide. Ion-exchange chromatography of irradiated solutions of adenylic acid showed that some ultra-

Table 4

Yields of destruction of base components of various purine and pyrimidine nucleosides and nucleotides on irradiation of their aqueous solutions (2×10^{-4} M) with X-rays (200 kV) in the presence of oxygen (1 atm.)

Compound	Yield (G (-base))	Compound	Yield(G (-base))
Uracil (pH 5.3)	1.95	Thymidylic acid (pH 7)	1.65
Uridine pH 5.3—7.0)	1.72	Adenine (pH 7)	1.09
Uridylic acid (pH 3.8—7.0)	1.65	Adenosine (pH 7)	0.92
Thymine (pH 7)	1.90	Adenylic acid (3′) (pH 7)	0.75
Thymidine (pH 7)	1.75		

violet absorbing products other than free adenine are produced, but the yields are very low ($G < 0.1$). We have confirmed in this Laboratory the findings of HEMS (1958) that the spectral changes presumed to be associated with the formation of a pyrimidine formamido-compound in irradiated solutions of guanylic acid are not found in solutions irradiated in the presence of oxygen.

c) Nucleic acids

Irradiated solutions of calf thymus deoxyribonucleic acid (DNA) were frozen dried and hydrolysed with formic acid or with perchloric acid. The liberated bases were separated by paper chromatography and the yields of destruction of the individual bases were obtained by UV absorption of the eluates. Fairly reproducible results were obtained with a given sample (Table 5).

Table 5

Initial yields of destruction of purine and pyrimidine bases on irradiation of aqueous solutions of DNA with X-rays (200 kV) in the presence of oxygen (1 atm.)

DNA concentration (wt. %)	G (-base)			
	Thymine	Cytosine	Adenine	Guanine
0.005	0.33	0.23	0.25	0.18
0.200	0.64	0.38	0.39	0.26

Irradiation of DNA solutions in the presence of oxygen has been shown to lead to the formation of hydroperoxides on the pyrimidine moieties of the nucleic acid molecule. Table 6 shows the yields of the radiation-produced hydroperoxide in DNA solutions of different concentrations.

Table 6

Concentration dependence of the initial yields of DNA hydroperoxide in aqueous solutions irradiated with X-rays (200 kV) in the presence of oxygen (1 atm.)

DNA (wt. %)	G (hydroperoxide)
0.005	0.17
0.025	0.30
0.10	0.55

Experiments with deoxyribonucleotides have shown that the deoxyribose component does not form a hydroperoxide and that in neutral solutions only thymidylic acid forms a stable hydroperoxide. We conclude, therefore, that the stable hydroperoxide found in irradiated DNA solutions is essentially associated with the thymine molecule. It also appears that the thymine hydroperoxide remains attached to the DNA since it is completely precipitated on treatment of the irradiated solution with trichloroacetic acid.

d) Nucleoproteins

Some experiments have been carried out on the extent of base destruction in irradiated solutions of calf thymus nucleoprotein (Emmerson et al., 1960). Nucleoprotein solutions were prepared from fresh calf thymus glands by the method of Doty and Zubay (1956). After irradiation the solutions were hydrolysed as above and the base composition determined chromatographically. The results showed that up to doses of about 1.3×10^5 rads the bases of the nucleic acid were not significantly attacked by the radiation; within the experimental error ($\pm$ 5%), no noticeable degradation was observed. In contrast to these findings, it was found that on irradiation of solutions of synthetic mixtures of DNA and histone, there was some degradation of the DNA component, but much less than in pure DNA solutions. In the synthetic mixtures, the protein protected the DNA to some extent, as would be expected from the fact that the reactive species produced from the water by irradiation can distribute themselves between the nucleic acid and the protein.

Discussion

The decomposition of water by ionizing radiations is conventionally represented by the equation (Weiss, 1944; Allen, 1952)

$$H_2O \longrightarrow H, OH, H_2, H_2O_2 \tag{1}$$

leading to the free radicals H and OH. Although currently there is a good deal of discussion about the exact nature of the reactive species produced, this does not affect the considerations below and for sake of simplicity we have designated the reactive entities as H and OH.

The quantitative results on the degradation of the pyrimidines can be accounted for if we consider that the OH radicals (but not HO_2) react with the solute in solutions irradiated in the presence of oxygen. The formation of hydroperoxide can be accounted for by the following reaction scheme (A = pyrimidine):

$$A + OH \longrightarrow A(OH). \tag{2}$$

$$A(OH). + O_2 \longrightarrow A(OH)O_2. \tag{3}$$

$$H + O_2 \longrightarrow HO_2 \rightleftarrows H^+ + O_2^- \tag{4}$$

$$A(OH)O_2. + O_2^{-.} \longrightarrow A(OH)O_2^- + O_2 \tag{5}$$

where reactions (2) and (3) represent the addition of the OH radical and the oxygen molecule respectively to the pyrimidine 4:5 double bond. The extent of scavenging (reaction 2) is G $\sim$ 2 and is what one would expect, in general, in solutions of the concentrations employed.

In this mechanism reactions (4) and (5) are the pH-dependent steps and would account for a decrease in hydroperoxide yield with increasing acidity.

Little is known about the detailed mechanism of the radical attack in the purine molecule, although it has been postulated that OH radicals attack the central ethylenic bond (SCHOLES and WEISS, 1952). The production of formamido-pyrimidines from purines irradiated in oxygen-free solutions (HEMS, 1958) imply an attack on the imidazole ring under these conditions. If the opening of the imidazole ring is due to OH radicals than the same site should be attacked in the presence of oxygen and the formation of some UV absorbing product would be anticipated. Since this is not so, it is probable that the formation of the formamido compound is specifically associated with initial reaction with the H atom species, which alone attacks the imidazole part of the molecule. It is interesting to speculate that addition of H to the –C=N-bond of the imidazole ring followed by the introduction of OH (either by reaction with an OH radical or by dismutation between two purine radicals) and subsequent breakage of the C–N linkage because of instability of the product, would lead to the formamido-derivative. Some support for the suggestion could come from the observation that dihydro-hydroxy derivatives are formed on irradiation of uracil and dimethyl uracil in oxygen-free solutions (SCHOLES, WARD, and WEISS, 1961).

It can be seen from Table 4 that the presence of sugar or sugar-phosphate residues does not drastically change the extent of chemical attack upon the base. In adenylic acid, about 30% of the radicals react with the pentose molecule whereas in uridylic acid only about 15% are thus diverted.

The conclusion that base attack predominates in DNA has been arrived at from studies of the chemical attack on the sugar phosphate components of the DNA. It has been possible to measure directly the number of breaks in the sugar phosphate backbone by treating the irradiated solutions with prostate phosphomono-esterase which liberated inorganic phosphate from phosphate end-groups, since at the site of each break a phosphate end-group will be formed. Experiments with this enzyme have shown that such end-groups are in fact formed (COLLYNS and OKADA, unpublished results). In irradiated 0.2% DNA solutions, for example, it has been found that the number of phosphate end-groups corresponds to a yield of G $\sim$ 0.6 whereas the yield of base destruction at this concentration is G $\sim$ 1.67 (table 5). Thus it can be seen that about 75% of the available radicals react with the bases.

Somewhat similar conclusions regarding the relative reactivity of the base and sugar moities in nucleosides, nucleotides and DNA have been arrived at by HEMS (1960 a, b), using 15 MeV electrons.

It should be mentioned that hydrogen bond rupture is also an important consequence of the chemical action of ionizing radiations on DNA in aqueous solutions (SCHOLES and WEISS, 1950; COX, OVEREND, PEACOCKE, and WILSON, 1955). It has been found that in 0.1% DNA solutions the number of hydrogen bond pairs broken per chain break amounts to a yield of about G $\sim$14 or 15 (SCHOLES, WARD, and WEISS, 1960). The fact that a relatively large number of hydrogen bond pairs can be broken per single chain break can be explained if one assumes that, following the chain break, water molecules can permeate the double helix of the DNA. In this way hydrogen bonding between the conjugate base pairs will be replaced to some extent by hydrogen bonding between the individual bases and the water molecules.

The experiments with nucleoprotein solutions described above indicate that the radiation-produced free radicals do not attack the purine or pyrimidine bases of the component DNA. This view is supported by the fact that no hydroperoxide could be detected in irradiated nucleoprotein solutions. On the other hand there is no doubt that, in these experiments, the nucleoprotein complex has been attacked. Even after relatively low doses ($\sim$ 30,000 rads) a general increase in absorption in the UV region has been noted and this can probably be attributed to scattering phenomena, since, if irradiation is continued long enough (to around 3×10^5 rads) a visible precipitate is formed. Furthermore, marked viscosity changes occur on irradiation (WARD, 1959). It can be concluded, therefore, that in the irradiation of intact nucleoprotein isolated from calf thymus glands, the free radicals react mainly with the protein, and this chemical evidence is in agreement with the observation of KAUFMANN, McDONALD, and BERNSTEIN (1955) that irradiation of nucleoprotein did not appreciably alter the viscosity of the DNA component. These findings may be of interest in regard to the high radiosensitivity of the nucleus during certain stages of the mitotic cycle, when the protein is shed from the DNA.

Summary

The effects of X-rays on aqueous oxygenated solutions of purines, pyrimidines, nucleosides and nucleotides have been discussed, particularly with regard to their relative reactivities and to the formation of hydroperoxides. Quantitative data on the extent of base destruction on irradiation of DNA in aqueous solutions have been presented. Most of the available OH radicals (at least 75%) attack the base moieties in the polynucleotides. It has been shown that in the irradiation of calf thymus nucleoprotein solutions the purines and pyrimidines of the DNA component are essentially unattacked.

Acknowledgements

We thank the Rockefeller Foundation and the Northern Council of the British Empire Cancer Campaign for financial support.

References

ALLEN, A. O., 1952: Radiation chemistry. Annual Rev. Phys. Chem. **3**, 57–80.

COX, R. A., W. G. OVEREND, A. R. PEACOCKE, and S. WILSON, 1955: Effects of γ-rays on solutions of sodium deoxyribonucleate—possible chemical mechanisms for their biological effects. Nature (London) **176**, 919–921.

DOTY, P., and G. ZUBAY, 1956: The preparation and properties of completely dispersed thymus nucleoprotein. J. Amer. chem. Soc. **78**, 6207–6208.

EKERT, B., and R. MONIER, 1959: Structure of thymine hydroperoxide produced by X-irradiation. Nature (London) **184**, 58–59.

EMMERSON, P., G. SCHOLES, D. H. THOMSON, J. F. WARD, and J. WEISS, 1960: Chemical effects of ionizing radiations on nucleic acids and nucleoproteins. Nature (London) **187**, 319–320.

HEMS, G., 1958: Effects of ionizing radiation on aqueous solutions of guanylic acid and guanosine. Nature (London) **181**, 1721–1722.

HEMS, G., 1960a: Effect of ionizing radiation on aqueous solutions of ionsine and adenosine. Radiation Research **13**, 777–787.

HEMS, G., 1960b: Chemical effects of ionizing radiation on deoxyribonucleic acid in dilute aqueous solution. Nature (London) **186**, 710–712.

KAUFMANN, B. P., M. R. McDONALD, and M. H. BERNSTEIN, 1955: Cytochemical studies of changes induced in cellular materials by ionizing radiations. Ann. N. Y. Acad. Sc. **59**, 553–566.

SCHOLES, G., G. STEIN, and J. WEISS, 1949: Action of X-rays on nucleic acids. Nature (London) **164**, 709–710.

SCHOLES, G., and J. WEISS, 1950: Chemical action of ionizing radiations on nucleic acids in aqueous systems. Nature (London) **166**, 640–642.

SCHOLES, G., and J. WEISS, 1952: Chemical action of X-rays on nucleic acids and related substances in aqueous systems Exper. Cell Res. (Suppl. 2), 219–241.

SCHOLES, G., J. WEISS, and C. M. WHEELER, 1956: Formation of hydroperoxides from nucleic acids by irradiation with X-rays in aqueous systems. Nature (London) **178**, 157.

SCHOLES, G., J., and J. WEISS, 1960: Organic hydroxy-hydroperoxides: A class of hydroperoxides formed under the influence of ionizing radiations. Nature (London) **185**, 305–306.

SCHOLES, G., J. F. WARD, and J. WEISS, 1960: Mechanism of the radiation-induced degradation of nucleic acids. J. Mol. Biol. **2**, 379–391.

SCHOLES, G., J. F. WARD, and J. WEISS, 1961: Action of gamma-irradiation on dimethyl uracil in aqueous solution in absence of oxygen. Science (Washington) **133**, 2016–2017.

WARD, J. F., 1959: Action of radiation produced radicals on nucleic acids and related compounds. Ph. D. Thesis, University of Durham.

WEISS, J., 1944: Radiochemistry of aqueous solutions. Nature (London) **153**, 748–750.

Die Erzeugung von organischen Peroxyden sehr hoher biologischer Wirksamkeit durch Bestrahlung von Agar mit hohen Dosisleistungen

K. Sommermeyer und H. Magnus

Radiologisches Institut der Universität Freiburg i. Br. und Städtisches Krankenhaus Karlsruhe, Strahlenabteilung

Häufig ist die Hypothese ausgesprochen worden, daß der Sauerstoffeffekt in der Strahlenbiologie und ebenso auch in der Strahlengenetik auf der Bildung von organischen Peroxyden mit hoher letaler oder mutagener Wirkung beruht. Wir haben durch eingehende Versuche über die Inaktivierung von Bakterien auf vorbestrahltem Agar die Bedingungen näher festzulegen versucht, unter denen solche organischen Peroxyde sehr hoher biologischer Wirksamkeit entstehen. Die Bestrahlung des Agars (2,5% Pronagar und 1% Pepton) wurde vor allem mit energiereichen Strahlen ausgeführt.

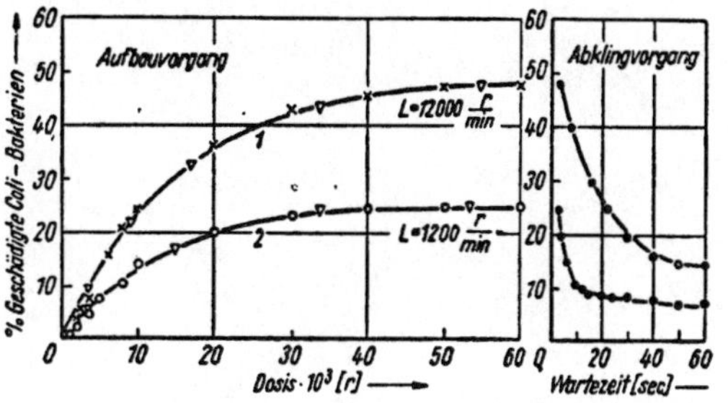

Abb. 1. Bildung des IS-Stoffes bei Anwendung von Röntgenstrahlen

Der Prozentsatz der inaktivierten *Coli*-Zellen, den man ermittelt, wenn man die Bestrahlung des Agars abbricht und sofort *Bact. coli* aufbringt, nimmt zunächst (Abb. 1) etwa linear mit der Dosis zu und erreicht dann einen Grenzwert, der mit der Dosisleistung ansteigt. Läßt man zwischen dem Abbruch der Bestrahlung und dem Aufbringen des *Bact. coli* Zeit verstreichen (rechte Seite der Figur), so ergibt sich, daß die Schädigungsrate mit der Wartezeit rasch abnimmt, also die entstandenen IS (inaktivierenden Substanzen) im Agar eine entsprechend begrenzte Lebenszeit besitzen. Infolgedessen bildet sich im Agar mit fortlaufender Bestrahlung eine Gleichgewichtszustand aus, bei dem in der Zeiteinheit ebenso viele IS-Moleküle verschwinden, wie gebildet werden. Dabei ist vor allem charakteristisch, daß die Zeit, die zum Aufbau der IS-Gleichgewichtskonzentration benötigt wird, viel länger ist als die Abklingzeit, d. h. die Lebensdauer der IS im Agar.

Analoge Verhältnisse wie bei der Vorbestrahlung des Agars mit Röntgenstrahlen ergeben sich auch bei Anwendung von α-Strahlen und UV (Tab. 1).

Bei Anwendung von Röntgenstrahlen ist die Konzentration der IS stark vom Sauerstoffgehalt des Agars während der Bestrahlung abhängig, bei Anwendung von α-Strahlen ist das Ergebnis völlig unabhängig vom Sauerstoffgehalt (Abb. 2).

Die IS haben bei Anwendung von Röntgenstrahlen dieselben Abklingzeiten nach Abschalten der Bestrahlung wie bei Anwendung der α-Strahlen. Man wird deshalb zunächst vermuten, daß bei beiden Strahlenarten dieselben IS gebildet werden. Versuche mit anderen

Tabelle 1

Strahlenart	Dosisleistung [rad/min]	Aufbaudosis $D^1/_2$ [krad]	Aufbauzeit $t^1/_2$ [sec]	Grenzwert der Überlebenden [%]	Abklingzeit*) $T^1/_2$ [sec]
Röntgen	12000	10,2	50	52	12
	1200	9,4	470	75	6
α-Strahlen**)	7000	8,0	67	48	12
	700	5,5	458	73	6,5
UV	1800 erg/cm²·min	375 erg/cm²·min	12,5	43	11
	180 erg/cm²·min	270 erg/cm²·min	90	70	5

*) Für die Differenz Grenzschädigung minus Restschädigung
**) Dosis und Dosisleistung sind an der Agaroberfläche angegeben

Bakterien (*Staph. cocc. aureus*) haben jedoch gezeigt, daß dies keineswegs der Fall ist. Wie aus Abb. 3 hervorgeht, haben nach Anwendung von Röntgenstrahlen die IS bei *Staph. cocc. aureus* etwa die doppelte Inaktivierungswahrscheinlichkeit wie bei *Bact. coli.* Nach Vorbestrahlung des Agars mit α-Strahlen und UV erhält man jedoch im Falle des *Staph. cocc. aureus* überhaupt keine nachweisbare Schädigung. Es ist daher nur der Schluß möglich, daß durch α-Strahlen und UV andere IS gebildet werden als durch Röntgenstrahlen und die

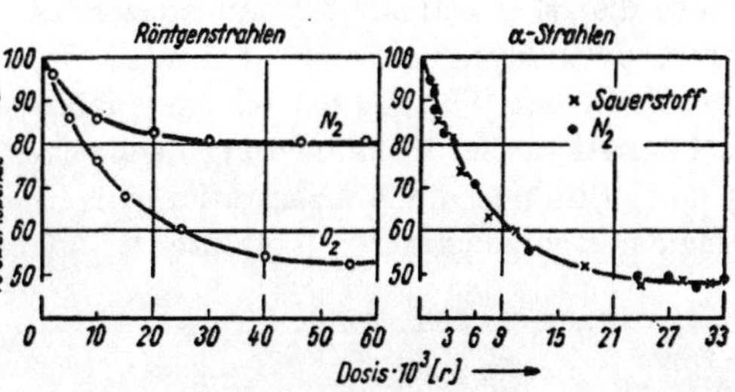

Abb. 2. Bildung der IS im Agar, in Abhängigkeit vom O²-Gehalt

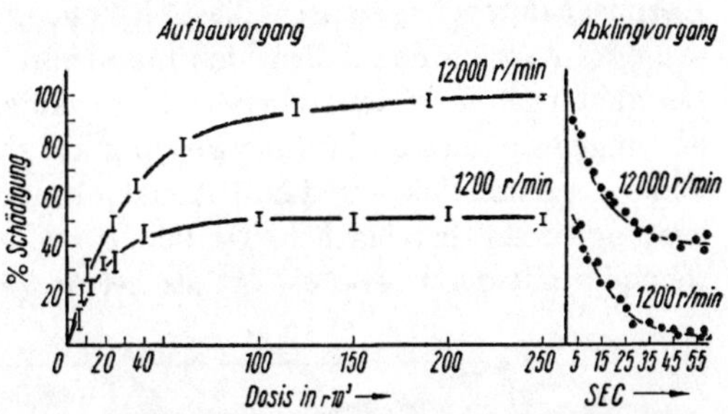

Abb. 3. *Staph. cocc. aureus* – Röntgenstrahlen, 30 kV

durch die α-Strahlen und UV erzeugten IS bei *Staph. cocc. aureus* weit weniger wirksam sind als bei *Bact. coli*; vielleicht werden sie in der Zellmembran von *Staph. cocc. aureus* unwirksam gemacht.

Wie schon erwähnt, ist für die Vorgänge, die zur Entstehung der IS führen, charakteristisch, daß wenigstens bei energiereichen Strahlen immer die Zeit, die zum Aufbau der Gleichgewichtskonzentratien an IS benötigt wird, viel länger ist als die Lebensdauer der IS im Agar. Man muß daraus schließen, daß zunächst ein nur wenig biologisch aktives Vorprodukt gebildet wird, welches eine relativ große Lebensdauer besitzt, und daraus erst durch weitere Reaktionen die biologisch hochwirksamen Substanzen entstehen.

Besonders interessant ist ein Vergleich der biologischen Wirksamkeit unserer IS mit der biologischen Wirksamkeit der organischen Peroxyde, wie sie besonders von LATARJET angewandt worden sind. Eines der wirksamsten der bisher angewandten organischen Peroxyde ist das Disuccinoyl-Peroxyd. Nach den Angaben in der Literatur kann man abschätzen, daß etwa 10^{11} dieser Moleküle zu einer *Coli*-Zelle diffundieren und mit ihr in Berührung kommen müssen, wenn ihre Inaktivierung erfolgen soll. Wir haben die Wirkungswahrscheinlichkeit unserer IS auf *Bact. coli*, so wie sie bei der Bestrahlung des Agars durch Röntgenstrahlen entsteht, abgeschätzt, und zwar haben wir für diesen Zweck die Bedingungen zugrunde gelegt, wie sie bei einer Dosisleistung von 12000 r/Minute im linearen Teil der Dosiswirkungskurve (Abb. 1, linke Seite) bestehen. Die Zahl der IS-Moleküle, die zu einer nach der Bestrahlung aufgebrachten *Coli*-Zelle gelangen, ist $W = 2\pi Agl^2$, wobei A die Konzentration der IS-Moleküle im Agar bei Beendigung der Bestrahlung, g der Radius der *Coli*-Zelle und l die Diffusionslänge im Agar ist. Die Diffusionslänge l im Agar haben wir durch besondere Versuche mit Agarschichten verschiedener Dicke (Abb. 4) zu etwa 100 μ bestimmt. Die Konzentration der IS-Molekühle ergibt sich aus der Reaktionsausbeute. Diese kann nach den obigen Bemerkungen über die Reaktionen höchstens 1/10 pro 100 eV betragen.[1]) Mit allen diesen Zahlenwerten errechnet man für die zur Inaktivierung der Zelle notwendige Zahl zu ihr hindiffundierender IS-Moleküle $< 10^8$. Also ist die Wirkungswahrscheinlichkeit der IS-Moleküle bei der Inaktivierung des *Bact. coli* mindestens 3 Zehnerpotenzen höher als die Wirkungswahrscheinlichkeit der Disuccinoyl-Peroxyd-Moleküle.

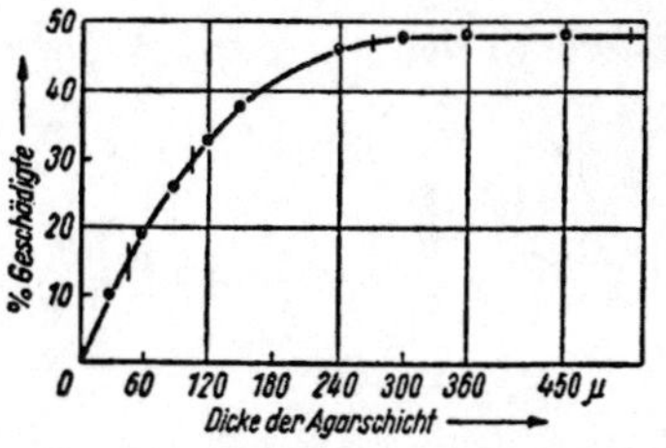

Abb. 4. Geschädigte *Coli*-Bakterien als Funktion der Agarschichtdicke bei Röntgenbestrahlung

Wird *Bact. coli* auf dem Agar bestrahlt, d. h. also eine direkte Bestrahlung des *Bact. coli* auf dem Agar vorgenommen, so erhält man eine Inaktivierung nicht nur als Folge der direkten Korpuskeldurchgänge durch die *Coli*-Zellen, sondern auch durch die IS-Moleküle, welche aus dem Agar zu den Zellen hindiffundieren. Da sich die IS erst allmählich mit steigender Bestrahlungszeit bilden, überwiegt bei kurzzeitiger Bestrahlung mit sehr hohen Dosisleistungen die direkte Strahlenwirkung auf die *Coli*-Zelle. Die Wirkung der im Agar gebildeten IS kann sich nur durch Abnahme der Halbwertsdosis bei Verminderung derDosisleistung bemerkbar machen. Da bei Anwendung der α-Strahlen die Konzentration der IS schneller mit der Dosis ansteigt als bei Anwendung der Röntgenstrahlen (vgl. Tab. 1), an-

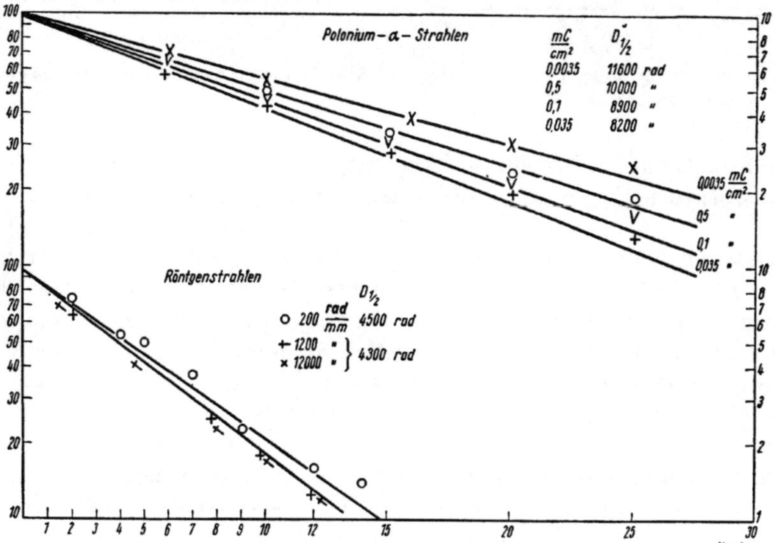

Abb. 5. *B. coli* — Direkte Bestrahlung mit Röntgenstrahlen und Polonium-α-Strahlen

[1]) Es wird dabei angenommen, daß die Ausbeute für die Bildung des Vorproduktes ≈ 1 pro 100 eV ist.

dererseits aber die Halbwertsdosen bei direkter Bestrahlung für α-Strahlen höher sind als für Röntgenstrahlen, sind bei den α-Strahlen solche Abhängigkeiten der Halbwertsdosen von der Dosisleistung in viel stärkerem Maße zu erwarten als bei der Anwendung von Röntgenstrahlen. In der Tat tritt bei direkter Bestrahlung des *Bact. coli* auf Agar mit α-Strahlen eine Abhängigkeit der Halbwertsdosis von der Dosisleistung auf, während sie sich bei Anwendung von Röntgenstrahlen nicht bemerkbar macht (Abb. 5). (Die geringfügige Abhängigkeit bei den Röntgenstrahlen ist dadurch bedingt, daß bei der kleinsten Dosisleistung während der Bestrahlung das Wachstum beginnt.)

Zusammenfassend stellen wir fest, daß bei Bestrahlung von Agar mit energiereichen Strahlen und auch mit UV inaktivierende Substanzen mit ganz beträchtlichen biologischen Wirkungswahrscheinlichkeiten erzeugt werden. Sie werden durch komplizierte Mehrtrefferreaktionen gebildet und erfordern eine hohe Dosisleistung. Ihre Entstehung ist also sehr stark von der Dosisleistung abhängig. Umgekehrt muß sich die Mitwirkung von solchen diffusiblen Substanzen, wie sie hier zur Untersuchung gelangt sind, bei letalen und ebenso mutagenen Strahlenreaktionen in der Existenz einer beträchtlichen Abhängigkeit von der Dosisleistung bemerkbar machen. Da solche Abhängigkeiten von der Dosisleistung bei der Erzeugung von Mutationen durch energiereiche Strahlen im allgemeinen nicht auftreten[1]), sind wir der Meinung, daß unsere Versuchsergebnisse mindestens nicht als unmittelbare Stütze der These, daß vor allem strahleninduzierte organische Peroxyde die Mutation erzeugen, dienen können.

---

[1]) Wir denken dabei insbesondere an die bekannten Versuche über rezessive Letalmutationen bei *Drosophila*.

Literatur

SOMMERMEYER, K., und H. MAGNUS, 1960: Vergleichende Untersuchungen über die Strahlenempfindlichkeit von *Bacterium coli* und von Hefe insbesondere bei Anwendung von energiereichen Strahlen. Z. Naturforsch. 15b, 770–778.

Zur Frage der Auslösung von Chromosomenaberrationen durch freie Radikale des Wassers bei *Vicia faba* L.

A. Michaelis und R. Rieger

Institut für Kulturpflanzenforschung der Deutschen Akademie der Wissenschaften, Gatersleben, Kreis Aschersleben

Die mit der Einstrahlung ionisierender Strahlenarten auf biologische Objekte verbundene Energieübertragung führt zu Anregungen und Ionisationen von Atomen und Molekülen. Die Folge sind u. a. chemische Veränderungen der sog. Target-Moleküle, wobei der Begriff „Target" spezifische Funktionseinheiten charakterisiert, die am physiologischen und biochemischen Geschehen der Zelle beteiligt sind und deren Zerstörung oder Veränderung an der endgültigen Strahlenwirkung unmittelbaren Anteil hat. Nach den Vorstellungen der klassischen Treffer- und Treffbereichstheorie wird vorausgesetzt, daß die für die einzuleitenden biologischen Reaktionen (wie Mutationen und Chromosomenbrüche) entscheidende Wirkung (Primärionisation) am Ort der Energieabsorption, im Treffbereich (Target), erfolgt. Ausgehend von den Befunden der Strahlenchemie des Wassers ist es andererseits möglich, daß durch Energiezufuhr an einem Orte x diffusible Reaktionsprodukte des Wassers gebildet werden, die an einem anderen Orte y (etwa dem Chromosom) die für den Eintritt von Mutationen bzw. Chromosomenbrüchen notwendigen Reaktionen einleiten. Im Gegensatz zur „direkten" läge hier eine „indirekte" Strahlenwirkung über diffusible Substanzen (Zersetzungsprodukte des Wassers) vor.

Für die Bedeutung indirekter Wirkung werden folgende Argumente angeführt:

1. Die Gasatmosphäre und Temperatur zur Zeit der Bestrahlung nehmen Einfluß auf die Strahlenempfindlichkeit.
2. Durch Strahlenschutzstoffe ist ein aktiver Strahlenschutz möglich.
3. Durch Einfrieren des Materials wird die Strahlenempfindlichkeit verringert.
4. Es gibt strahlenbiologische Nachwirkungen, die experimentell bewiesen, aber von der Treffbereichstheorie nicht in Betracht gezogen wurden.

Während man für einige Zeit geneigt war, indirekten Strahlenwirkungen die Hauptbedeutung bei der Induktion von Strahlenschäden zuzuschreiben, hat sich mittlerweile herausgestellt, daß es nicht möglich ist, indirekte Wirkungen in vivo allein daraus abzuleiten, daß Außenfaktoren die primären Strahlenwirkungen zu modifizieren in der Lage sind. Die Fragwürdigkeit einer ganzen Reihe der für indirekte Wirkung über diffusible Wasserzersetzungsprodukte sprechenden Argumente wurde u. a. von Butler (1956), Alper (1956), Alexander (1960) und Zimmer (1956, 1960) dargelegt, und heute deutet sich wieder eine stärkere Betonung der direkten Wirkung an. Sicherlich sind „direkte" und „indirekte" Wirkungen an den Strahlenschäden beteiligt, und es dürfte tatsächlich darum gehen, „den relativen Anteil der offenbar in vielen Fällen nebeneinander bestehenden Mechanismen"

(ZIMMER, 1960) festzustellen, wobei Ionisationen in der Hydrationshülle der Target-Moleküle in die direkte Wirkung eingeschlossen werden sollten, da diese einen Teil des Moleküls bildet und in ihr keine freie Diffusion erfolgt (EHRENBERG, 1959; ALEXANDER, 1960).

Bei der Einstrahlung ionisierender Strahlenarten auf Wasser werden nach ALLEN et al. (1952) eine Radikalreaktion ($H_2O \rightarrow H^{\cdot} + OH^-$) und eine Molekülreaktion ($2\,H_2O \rightarrow H_2 + H_2O_2$) unterschieden. Strahlenarten mit niedrigem linearem Energietransfer (LET) führen vorwiegend zur Bildung von chemisch aktiven H- und OH-Radikalen, die sich untereinander und mit anderen Atomen und Molekülen vereinigen. Nach den Reaktionen können chemisch aktive Sekundär- und Tertiärprodukte gebildet werden, von denen in unserem Zusammenhang nur das bei Sauerstoffgegenwart gebildete HO_2 interessiert. $H^{\cdot}$ und $OH^{\cdot}$ entstehen entweder durch Zerfall von ionisierten Wassermolekülen (H_2O^- oder H_2O^+) oder direkt durch Anregungen oder Elektronenstoß. Auf die Darstellung von Einzelheiten der Reaktionen muß hier verzichtet werden (vgl. GILES, 1952; SPARROW und RUBIN, 1952; BURTON und MAGEE, 1956; BACQ und ALEXANDER, 1958; SCHOLES, 1959). Als die möglichen schädigenden Komponenten des bestrahlten Wassers sind die hochreaktiven Radikale $H^{\cdot}$, $OH^{\cdot}$, $HO_2^{\cdot}$ und evtl. H_2O_2 anzusehen. Naturgemäß sind im Falle indirekter Strahlenwirkung auch andere Energieträger (etwa langlebige organische Radikale, indirekt oder direkt entstandene organische Peroxyde) und durch Radikale eingeleitete Reaktionsketten denkbar, an deren Ende erst das biologisch entscheidende Molekül steht.

Möglichkeiten zum experimentellen Test der Radikale OH und HO_2 auf ihre Fähigkeit, Mutationen und Chromosomenaberrationen auszulösen und damit Einblick in die Bedeutung indirekter Strahlenwirkungen über diffusible Radikale des Wassers zu gewinnen, ergeben sich aus der Tatsache, daß sich derartige Radikale auch ohne Einwirkung ionisierender Strahlenarten chemisch herstellen lassen. Zu diesem Zweck kann UV auf H_2O_2 zur Einwirkung kommen ($H_2O_2 \overset{UV}{\rightarrow} 2\,OH^{\cdot}$), es können gemischte Redoxsysteme vom Typ H_2O_2 und $FeSO_4$ ($H_2O_2 + Fe^{++} \rightarrow OH^{\cdot} + Fe^{+++} + OH^-$; $H_2O_2 + OH^{\cdot} \rightarrow HO_2^{\cdot} + H_2O$) verwendet werden, oder es kann von der Ozonzersetzung in Lösung ($HO_3^+ + OH^- \rightarrow 2\,H_2O_2$; $O_3 + HO_2 \rightarrow OH + 2\,O_2$) Gebrauch gemacht werden.

In den folgenden Versuchen fand vor allem das System $H_2O_2 + FeSO_4$ zur Induktion der Radikale $OH^{\cdot}$ und $HO_2^{\cdot}$ Verwendung. Das Vermögen der Radikale zur Auslösung von Chromosomenaberrationen wurde am Meristem der Hauptwurzel von *Vicia faba* überprüft. Ähnliche Experimente wurden von PHILLIPS (1956) an der Gerste und von DUBININ et al. (1959) an der Zwiebel durchgeführt. In beiden Fällen wird berichtet, daß die Radikale OH und HO_2 Chromosomenaberrationen zu induzieren vermögen. Diese ersten experimentellen Befunde zur Wirkung von Radikalen, die auch bei der Radiolyse des Wassers entstehen, erschienen uns so bedeutsam, daß wir den Versuch unternahmen, sie an einem der Standardobjekte der Zytologie – *Vicia faba* – zu reproduzieren, um Hinweise auf ihre Allgemeingültigkeit zu erlangen.

Ergebnisse

Es ist nicht erforderlich, die Ergebnisse aller mit *Vicia faba* durchgeführten Versuche anzuführen, da keine prinzipiellen Unterschiede auftraten. Zur Erläuterung der Situation genügen einige charakteristische Daten aus Versuchen mit den höchsten, nicht-letalen Konzentrationen von $FeSO_4$ und H_2O_2. Außer $FeSO_4$ wurde auch das von PHILLIPS benutzte $Fe_2(SO_4)_3$ verwendet. In diesem Falle stellt sich das Gleichgewicht $Fe^{++} \rightleftharpoons Fe^{+++}$ sehr schnell ein. Die Ergebnisse decken sich mit den bei $FeSO_4$-Verwendung erhaltenen, und im folgenden werden nur Versuche mit $FeSO_4$ aufgeführt.

In einem ersten Versuchskomplex wurden *Vicia*-Samen 24 Std. lang mit Eisensulfat und verschiedenen H_2O_2-Konzentrationen allein und in Kombination behandelt. Das Verhältnis trockene Samen : Lösung (pH 5) betrug 1 : 5. Nach Abschluß der Behandlung bei 24°C wurden die Samen in belüfteten Exsikkatoren bei 24°C zum Auskeimen gebracht. Die Fixierungen erfolgten in Abständen von 24 Std. nach einer zweistündigen Colchicinbehandlung (0,05%). Die weitere Verarbeitung des Materials erfolgte nach der Feulgen-Quetschmethode. Zur Auswertung auf Chromosomenaberrationen kamen Metaphasen des Hauptwurzelmeristems.

Tabelle 1

Der prozentuale Anteil von Metaphasen mit Chromosomenaberrationen im Wurzelspitzenmeristem in verschiedenen Fe^{++}- und H_2O_2-Lösungen gequollener *Vicia-faba*-Samen (250 ausgewertete Metaphasen pro Teilversuch: a) 48 Std. Erholungszeit, 24 °C; b) 24 Std. Erholungszeit)

Behandlungsart	% Metaphasen mit Aberrationen	*B''*	*T'*	*DD*	*D*
Trockene Samen					
a) 24 Std. in Aqua dest.	21,2	10	40	3	1
24 Std. in 1% H_2O_2-Lösung	12,0	13	14	0	2
24 Std. in 0,05% Fe^{++}-Lösung	22,4	14	37	6	0
24 Std. in Aqua dest.	19,2	11	31	1	6
4 Std. in 0,05% Fe^{++}; 20 Std. in 0,05% Fe^{++} + 0,03% H_2O_2	9,2	10	9	2	0
4 Std. in 0,05% Fe^{++}; 20 Std. in 0,05% Fe^{++} + 0,3% H_2O_2	5,6	5	5	1	2
Samen 48 Std. vorgequollen					
b) 24 Std. in Aqua dest.	26,4	10	45	5	6
4 Std. in 0,05% Fe^{++}; 20 Std. in 0,05% Fe^{++} + 0,03% H_2O_2	14,0	11	16	2	3
4 Std. in 0,05% Fe^{++}; 20 Std. in 0,05% Fe^{++} + 0,3% H_2O_2	11,2	7	15	2	5

Tab. 1 gibt einen Überblick über die Ergebnisse nach Behandlung trockener und 48 Std. vorgequollener Samen. Nach allen Behandlungsweisen traten Chromosomenaberrationen auf. Die Reunionen waren ausschließlich auf chromatidaler Basis erfolgt. In die Auswertung einbezogen wurden Isolocusbrüche (B''), Chromatidentranslokationen (T'), Duplikationsdeletionen (DD) und Deletionen (D). Der höchste prozentuale Anteil von Metaphasen mit Chromosomenaberrationen wurde in den Meristemen der Kontrollen und der mit $FeSO_4$ behandelten Samen gefunden. Kam H_2O_2 allein oder in Kombination mit $FeSO_4$ zur Einwirkung, wurde der Aberrationsprozentsatz im Vergleich mit den Kontrollen gedrückt, während bei einer Aberrationsauslösung durch die im Verlauf der kombinierten Einwirkung von $FeSO_4$ und H_2O_2 entstehenden Radikale eine Steigerung zu erwarten gewesen wäre, wenn die Radikale Aberrationen zu induzieren in der Lage wären. Die in der Tabelle angegebenen Werte stellen das Aberrationsmaximum dar und wurden für trockene Samen nach 48, für vorgequollene nach 24 Std. Erholungszeit gefunden. Daß H_2O_2 allein bzw. in Kombination mit KCN zur Verhinderung einer enzymatischen Zerlegung durch Katalase ohne radiomimetische Wirkung bleibt, wurde bereits von Loveless (1950) und Kihlman (1957) festgestellt und kann hier bestätigt werden. Im Falle kombinierter Samenbehandlung mit $FeSO_4$ und H_2O_2 war es bedeutungslos, ob beide Substanzen sofort kombiniert zur Einwirkung kamen, ob erst mit $FeSO_4$ und dann mit $FeSO_4 + H_2O_2$, oder erst mit $FeSO_4$ und dann mit H_2O_2 behandelt wurde. In keinem Falle wurden die Aberrationsprozentsätze der Kontrollen überschritten.

Die im Versuch aufgefundenen Chromosomenaberrationen sind das Ergebnis der Entstehung von radiomimetisch wirkendem Äthylalkohol. Die Alkoholbildung ist die Folge der mit 24stündiger Samenquellung unter Wasser bzw. in den Behandlungslösungen verbundenen partiellen Anaerobiose (RIEGER und MICHAELIS, 1958, 1959, 1960; MICHAELIS, RAMSHORN und RIEGER, 1959). Die im Vergleich mit unseren ersten Versuchen (RIEGER und MICHAELIS, 1958) höhere Aberrationsrate dürfte auf die Verwendung praktisch völlig entlüfteten destillierten Wassers zurückzuführen sein. Die Senkung der Aberrationsrate bei Einwirkung von H_2O_2 bzw. $FeSO_4 + H_2O_2$ ist wahrscheinlich die Folge der Sauerstofffreisetzung durch enzymatische Zerlegung von H_2O_2. Die Verteilung der Aberrationen über den Chromosomensatz war im Gegensatz zu den Befunden nach Einwirkung von Röntgenstrahlen (RIEGER und MICHAELIS, 1959) deutlich präferentiell und identisch mit der nach Alkoholeinwirkung gefundenen.

Der in Tab. 1 aufgeführte Versuch mit trockenen Samen wurde mit den gleichen Konzentrationen in Gaswaschtürmen unter ständiger Luftzufuhr während der 24stündigen Behandlungszeit wiederholt. Unter diesen modifizierten Bedingungen wurde der Prozentsatz der auf Grund von Stoffwechselumsteuerungen bei partieller Anaerobiose auftretenden Chromosomenaberrationen erheblich gesenkt, da die Gärungsvorgänge und damit die Alkoholproduktion durch die Belüftung stark zurückgedrängt wurden (Tab. 2). Auf Grund der Entwicklungsbeschleunigung der Samen bei ausreichender Sauerstoffzufuhr konnten die ersten Fixierungen bereits nach 24 Std. Erholungszeit durchgeführt werden. Eine Aberrationsauslösung durch freie Radikale war auch unter diesen Bedingungen nicht nachweisbar.

Tabelle 2

Behandlung	Zellzahl	% Metaphasen mit Aberrationen	B''	T'	DD	D
H_2O (Kontrolle)	500	2,8	8	4	2	0
0,05% $FeSO_4$ + 0,3% H_2O_2	500	1,0	4	1	0	0

Naturgemäß sind die zum Erreichen des Erfolgsortes notwendigen Diffusionswege der Radikale im Falle von Samenbehandlung relativ groß und könnten eine mögliche radiomimetische Wirkung verhindern. Aus diesem Grunde wurden in einem zweiten Versuchskomplex Hauptwurzeln mit einer durchschnittlichen Länge von 2 cm für unterschiedliche Zeiten mit H_2O_2, $FeSO_4$ und $FeSO_4 + H_2O_2$ behandelt. Wie Tab. 3 ausweist, führte auch hier

Tabelle 3

Der prozentuale Anteil von Metaphasen mit Chromosomenaberrationen im Wurzelspitzenmeristem nach Einwirkung von $FeSO_4$ und H_2O_2 auf die Hauptwurzeln von *Vicia faba* (250 ausgewertete Metaphasen pro Versuchsstufe)

Behandlungsart	Metaphasen mit Aberrationen [%]
24 Std. 0,03 % H_2O_2	0
24 Std. 0,005% Fe^{++}	0
1 Std. 0,005% Fe^{++}; 1 Std. 0,03% H_2O_2	0
6 Std. 0,005% Fe^{++}; 18 Std. 0,03% H_2O_2	0,6
4 Std. 0,005% Fe^{++}; 20 Std. 0,03% H_2O_2 + 0,005% Fe^{++}	0,8
4 Std. 1 mmol KCN;	0,7
1 Std. 0,005% Fe^{++} + 1 mmol KCN; 1 Std. 0,03% H_2O_2	1,3
2 Std. 0,005% Fe^{++} + 1 mmol KCN; 2 Std. 0,03% H_2O_2	0
4 Std. 0,005% Fe^{++} + 1 mmol KCN; 4 Std. 0,03% H_2O_2	1,3
6 Std. 0,005% Fe^{++} + 1 mmol KCN; 18 Std. 0,03% H_2O_2	0,6
1 Std. 0,005% Fe^{++}; 0,5 Std. 1 mmol KCN; 1 Std. 0,03% H_2O_2	0,6
6 Std. 0,005% Fe^{++}; 1 Std. 1 mmol KCN; 18 Std. 0,03% H_2O_2	letal

kombinierte bzw. aufeinander folgende Einwirkung von $FeSO_4$ und H_2O_2 zu keiner signifikanten Steigerung des Anteiles von Zellen mit Chromosomenaberrationen. Um eine enzymatische Zerlegung des H_2O_2 zu verhindern, wurde außerdem in einigen Versuchen KCN zugegeben bzw. zwischen $FeSO_4$ und H_2O eingeschoben. Auch unter diesen Umständen wurden keine Chromosomenaberrationen induziert.

Alle bisher aufgeführten Versuche waren mit der Schwierigkeit belastet, das $FeSO_4$ vollständig in Lösung zu bringen. Aus diesem Grunde wurde zusätzlich H_2O_2 (0,003 und 0,0003%), in dem sich *Vicia-faba*-Samen mit 3 bis 5 cm langen, dicht unter der Lösungsoberfläche befindlichen Hauptwurzeln befanden, mit dem Gesamtspektrum einer Quarzlampe 0,5, 2 und 6 Std. lang bestrahlt. Unter dem Einfluß des UV-Lichtes (68 erg/cm² × sec) werden OH-Radikale erzeugt ($H_2O_2 \rightarrow 2\,OH^{\cdot}$). Im zugehörigen Kontrollversuch wurden in Aqua dest. befindliche Wurzeln bestrahlt. Fixiert wurde 16 Std. nach Behandlungsende. Induzierte Chromosomenaberrationen wurden auch hier nicht gefunden.

Diskussion

Da etwa 80% der Zelle aus Wasser bestehen, liegt es auf Grund der Befunde zur Strahlenchemie des Wassers nahe, den unter Bestrahlungseinfluß entstehenden hochreaktiven Wasserzersetzungsprodukten $H^{\cdot}$, $OH^{\cdot}$, $O_2H^{\cdot}$ und H_2O_2 eine bedeutende Rolle bei der Entstehung genetischer Strahleneffekte zuzuschreiben. Ohne Zweifel bedarf es aber großer Vorsicht bei der Übertragung radiochemischer Erfahrungen auf die Verhältnisse im lebenden Organismus, da keine experimentellen Belege für das tatsächliche Wirkungsvermögen der Radikale vorliegen. So hängt die Bedeutung der durch ionisierende Strahlenarten innerhalb der Zelle induzierten freien Radikale im Falle indirekter Strahlenwirkungen sicherlich von ihrer Stabilität und ihrem Diffusionsvermögen ab. In der Zelle ist der Anteil diffusionshemmender Bestandteile so hoch, daß die Migration der Radikale stark eingeschränkt wird. Für Hydroxyl- und Perhydroxyradikale, die durch Radiolyse des Wassers entstehen, beträgt der mittlere Diffusionsweg, den sie im Plasma zurücklegen können, ohne mit einem anderen Molekül zu reagieren, nach HUTCHINSON (1960) nicht mehr als etwa 30 Å (einen Bruchteil des Chromatidendurchmessers), während die durchschnittliche Wasserhülle um ein beliebiges Makromolekül der Zelle bei etwa 20 bis 40 Å liegt. Voraussetzung einer Radikalreaktion mit einem Target-Molekül ist damit unter allen Umständen die Entstehung der Radikale in unmittelbarer Nähe des Wirkungsortes.

Eine meßbare Wirkung dieser Radikale im Rahmen indirekter Strahleneffekte ist nach ALPER (1956) deshalb im Grunde nur in wässerigen Systemen zu erwarten, die kein oder nur sehr wenig organisches Material enthalten, das als Radikalfalle wirkt und um die freien Radikale konkurriert. Zusätzlich besteht noch eine Konkurrenz zwischen den Reaktionen $H + O_2 \rightarrow HO_2$ bzw. H + organisches Material, wodurch die HO_2-Ausbeute in der Zelle kleiner wird als in Lösungen ohne H-Akzeptoren. Nach DAINTON (1959) ist HO_2 außerdem oberhalb eines pH von ca. 2 instabil und dissoziiert in H^+ und O_2^-.

Es erscheint auf Grund dieses Sachverhaltes a priori nicht sehr wahrscheinlich, daß freie Radikale als Produkte der Wasserzersetzung über größere Entfernungen hin eine bedeutende Rolle als Vermittler indirekter Strahleneffekte in der Zelle spielen.

Unter diesen Umständen und in Anbetracht der Schwierigkeit, die Bedeutung direkter und indirekter Strahleneffekte im Einzelfall bei biologischem Material abzuschätzen, erscheinen die Versuche von PHILLIPS (1956) und von DUBININ et al. (1959) an der Gerste bzw. Zwiebel um so interessanter. In beiden Fällen gelang es, nach Einwirkung gemischter Redoxsysteme ($Fe_2[SO_4]_3$ bzw. $FeSO_4$ und H_2O_2) Chromosomenaberrationen zu induzieren, deren Auslösung mit den entstehenden OH- und HO_2-Radikalen erklärt wird. Auch die von NATARAJAN und SWAMINATHAN (1958) nach Einwirkung röntgenbestrahlten Wassers auf Zwiebeln und Weizen beschriebenen Chromosomenaberrationen werden über strahleninduzierte freie Radikale des Wassers gedeutet. Diese Befunde dürfen als erste klare, experimentelle Beweise für eine Beteiligung durch Radiolyse des Wassers entstandener Radikale am Strahlenschaden der Zelle gewertet werden.

Wie die im experimentellen Teil aufgeführten Ergebnisse zeigen, dürfte es sich nicht um eine allgemeingültige Erscheinung handeln, da es bei *Vicia faba* nicht gelang, die bei Gerste und Zwiebel erhaltenen Befunde zu reproduzieren. Versuche mit negativem Ausgang sind naturgemäß mit Vorsicht zu deuten, da nicht auszuschließen ist, daß unter bestimmten Umständen auch positive Resultate zu erhalten sind. Trotzdem halten wir uns im Falle *Vicia faba* zu der Schlußfolgerung berechtigt, daß durch freie Radikale vom Typ OH und HO_2 bzw. ihre etwaigen Reaktionsprodukte bei Benutzung der angegebenen chemischen Systeme zum Radikalaufbau keine Chromosomenaberrationen ausgelöst werden können. Es läge auf Grund dieser Befunde nahe, die Möglichkeit zur Aberrationsauslösung (Aussagen zur möglichen Induktion von Genmutationen können nicht gemacht werden) durch strahleninduzierte Radikale vom gleichen Typ für *Vicia faba* zu verneinen. Da in diesem Falle aber günstigere Diffusionsverhältnisse vorliegen mögen, weil die Radikale dichter am Wirkungsort entstehen, dürfte Zurückhaltung angebracht sein.

Im Hinblick auf die radiomimetische Wirkung strahleninduzierter freier Radikale können unter diesen Umständen nur positive Versuchsergebnisse als beweiskräftig angesehen werden, da die zur Radikalerzeugung verwendeten chemischen Systeme möglicherweise gar nicht in der Lage sind, Strahleneffekte zu kopieren. Es darf aus diesem Grunde nicht geschlußfolgert werden, daß der negative Versuchsausgang beweist, daß strahleninduzierte OH- und HO_2-Radikale für genetische und cytologische Strahleneffekte bedeutungslos sind.

Ohne großes Bedenken vergleichbar erscheinen demgegenüber die an *Vicia*, *Hordeum* und *Allium* erhobenen Befunde nach Einwirkung chemisch erzeugter Radikale, die es verhindern, den Radikalen OH und HO_2 generell das Vermögen zur Auslösung von Chromosomenaberrationen zuzuschreiben. Den von PHILLIPS (1956) durchgeführten Gerstenversuch haben wir ergebnislos zu reproduzieren versucht. Wir erhielten zwar Letalitätseffekte, aber keine Chromosomenaberrationen.

Inwieweit die Sachlage noch dadurch kompliziert wird, daß das Versuchsergebnis durch das jeweilige zur Radikalerzeugung verwendete chemische System beeinflußt wird, muß dahingestellt bleiben. Hinweise in dieser Richtung könnten darin gesehen werden, daß FETNER (1958) die Induktion von Chromosomenaberrationen bei *Vicia faba* durch Ozonzersetzungsprodukte in Lösung (OH und HO_2) anhand von Anaphase-Auswertungen angibt. Dabei sollen nach Samenbehandlung alle Aberrationen auf chromosomaler Basis erfolgt sein, müßten also vor der Reduplikation der Chromosomen im Interphasekern eingetreten sein. Die unterschiedlichen Befunde zur Auslösung von Chromosomenaberrationen durch die Radikale OH und HO_2 am gleichen Objekt dürften weitere Experimente in diesem Zusammenhang wünschenswert erscheinen lassen.

Summary

Vicia faba seeds and main roots were exposed to HO and HO_2 radicals produced with Fenton's reagent ($Fe\ SO_4 + H_2O_2$). The same radicals are induced in the course of radiodecomposition of water with ionizing radiation and the purpose of the experiments was to find out whether these radicals (and H_2O_2) are able to induce chromosome aberrations when chemically produced. In contrast to experiments with barley and *Allium* done in the USA and USSR with the same system we failed to induce chromosome aberrations within a wide array of experimental conditions. Both the radicals and H_2O_2 proved ineffective in *Vicia faba*. The results are discussed in connection with their significance for indirect radiation effects via water decomposition products.

Literatur

Alexander, P., 1960: Protection of macromolecules in vitro against damage by ionizing radiations. In: A. Hollaender (Herausg.), Radiation protection and recovery. Pergamon Press, 3—44.

Alper, T., 1956: The modification of damage caused by primary ionization of biological targets. Rad. Res. **5**, 573—586.

Bacq, C. M., und P. Alexander, 1958: Grundlagen der Strahlenbiologie. Thieme, Stuttgart.

Burton, M., und I. L. Magee, 1956: Einige chemische Aspekte der Strahlenbiologie. Naturwissenschaften **43**, 433—442.

Butler, I. A. V., 1956: The action of ionizing radiations on biological material. Facts and theories. Rad. Res. **4**, 20—32.

Dainton, F. S., 1959: Chemical effects of radiation. Rad. Res. (Suppl. 1), 1—25.

Dubinin, N. P., B. N. Sidorov und N. N. Sokolov, 1958: Genetischer Effekt freier Radikale (russisch). Ber. Akad. Wiss. UdSSR **128**, 172—175.

Ehrenberg, L., 1959: Radiobiological mechanisms of genetic effects. Rad. Res. (Suppl. 1), 102—123.

Fetner, R. H., 1958: Chromosome breakage in *Vicia faba* by ozone. Nature (London) **181**, 504—505.

Giles, N. H., 1952: Recent evidence on the mechanism of chromosome aberration production by ionizing radiation. In: Nickson, I, I. Symposium on radiobiology. Chapman and Hall, London, 267—284.

Guzman-Barrow, E. S., 1954: The role of free radicals and oxygen in reactions produced by ionizing radiations. Rad. Res. **1**, 109—124.

Howard-Flanders, P., 1959: Primary physical and chemical processes in radiobiology. In: Radiation biology and cancer, University of Texas Press, Austin, 29.

Hutchinson, F., 1958: Magnitude of the indirect effect in cells. Rad. Res. **9**, 132 (Abstr.).

Kihlman, B., 1957: Experimentally induced chromosome aberrations in plants. I. The production of chromosome aberrations by cyanide and other heavy metal complexing agents. J. Biophys. Biochem. Cytol. **3**, 363—380.

Loveless, A., 1951: Qualitative Aspects of the chemistry and biology of radiomimetic substances. Nature (London) **167**, 338—342.

Michaelis, A., und R. Rieger, 1958: Cytologische und stoffwechselphysiologische Untersuchungen am aktiven Meristem der Wurzelspitze von *Vicia faba*. II. Chromosoma **9**, 514—536.

Michaelis, A., K. Ramshorn und R. Rieger, 1959: Äthylalkohol — radiomimetisches Agens bei *Vicia faba*. Naturwissenschaften **46**, 381.

Natarajan, A. T., and M. S. Swaminathan, 1958: Indirect effects of radiation and chromosome breakage. Indian J. Genet. **18**, 220—223.

Phillips, L. L., 1956: Effect of free radicals on chromosomes of barley. Science (Washington) **124**, 889—890.

Rieger, R., und A. Michaelis, 1958: Cytologische und stoffwechselphysiologische Untersuchungen am aktiven Meristem der Wurzelspitze von *Vicia faba*. I. Chromosoma **9**, 238—257.

Rieger, R., und A. Michaelis, 1959: Vergleichende Untersuchungen zur Verteilung durch verschiedene Mutagene induzierter Brüche über den Chromosomensatz von *Vicia faba* L. Chromosoma **10**, 163—178.

Scholes, G., 1959: Radiosensitivity of the model cell. In: T. G. Hennessy, Proceedings of a conference on radiobiol. at the intra-cellular level. Pergamon Press, London, 29—34.

Sparrow, A. M., and B. A. Rubin, 1952: Effects of radiations on biological systems. In: Survey of biol. Progress. Vol. II, 1—52.

Zimmer, K. G., 1956: The development of quantum biology during the last decade. Acta radiol. (Stockholm) **46**, 595—602.

Zimmer, K. G., 1960: Studien zur quantitativen Strahlenbiologie. Abhdlg. d. Math.-Naturwiss. Kl., Akad. d. Wiss. u. Literatur, Mainz, Jahrg. 1960, Nr. 3, 825.

Metabolic Factors Concerned in the Expression of X-radiation Induced Changes in DNA of Nucleated Cells

M. G. Ord

Department of Biochemistry, University of Oxford, England

Interest in this subject started with the observations of Ord and Stocken (1958) that ^{32}P incorporation into the DNA of rat thymus gland was depressed linearly with doses of X-radiation up to 200 r. Creasey and Stocken (1959) found that nuclear phosphorylation (Allfrey, Mirsky, and Osawa, 1955) was extremely sensitive to X-rays, complete inhibition being produced with 100 r. Irradiation of nuclei in vitro confirmed the great sensitivity of the process. In contrast to this, Myers (1960) showed that co-enzyme I synthesis in vivo by rat thymus nuclei was unaffected after 1500 r total body radiation. Very little is yet known about the mechanism of nuclear phosphorylation but the work of Allfrey and Mirsky (1959) indicated that conditions permitting the reaction were essential for protein and RNA synthesis in isolated nuclei.

The studies of Errera and his colleagues (1959) on protein synthesis by isolated nuclei have established that 300 r in vitro depresses adenine and amino-acid uptake. Sodium and potassium retention by nuclei is reduced by exposures in the same range, given either in vivo or in vitro (Creasey, 1960) and this loss is accompanied by a fall in bound nucleotide (Creasey; Ord and Stocken, unpublished). Another reaction which is probably associated with nuclear function and which is sensitive to 200 r in vivo is the formation of thymidylate kinase in regenerating liver in response to partial hepatectomy (Jaffe, Lajtha, Lascelles, Ord, and Stocken, 1959). At present the causation of these different changes is unknown but in view of the alterations in DNA-nucleoprotein (Cole and Ellis, 1954) and in DNA extractability (Hagen, 1960) post-radiation it is tempting to suggest that all the changes arise from one source – a disorganisation of the nuclear structure.

Because of the lack of knowledge about nuclear organisation particularly with respect to low molecular weight compounds, Dr. Stocken and I have recently commenced a study of the behaviour of acid-soluble phosphate compounds in normal thymus nuclei. The isolation procedure of Creasey and Stocken (1959) was followed exactly and the purity of the nuclear fraction was checked by phase contrast microscopy in albumin solution (Barer, Joseph, and Esnouf, 1956) and by electron microscopy. Both sucrose/$CaCl_2$ and a KCl-rich medium have been used, with so far no detectable differences in behaviour, but most of the results were obtained with the ionic medium.

The method of isolation of the nuclei is critical. The rate of removal of the thymus from the rat after death affects the amount of ATP bound by the nuclei (Ord and Stocken, 1961) which in turn affects their properties. In our early experiments the thymus was removed within 20 sec. after death by cervical fracture. In these nuclear preparations there was an acid-labile P/inorganic P ratio of about 5: 1, with up to 0.32 μmole acid-labile P/mg DNA P. Fractionation of the acid-soluble material on Dowex showed that 70% of the acid-labile P was ATP, and centrifugation of the standard nuclear preparation showed that the ATP was firmly bound to the nuclei whereas about 50% of the inorganic phosphate was in solution.

If the nuclei were incubated with ^{32}P in vitro a reversible uptake occured at 0° which was not temperature sensitive. Extraction of the nuclei with M NaCl indicated that the inorganic phosphate was associated with nucleoprotein; examination of the acid extract from such nucleoprotein showed the presence of arginine which was also taken up by the nuclei at 0°.

Preliminary experiments have shown that both ATP and arginine bound to the nucleus can be utilised for incorporation into its acid insoluble fraction. This utilisation of ATP depends critically on the method of removal of the thymus and is shown when nuclei are prepared after rapid removal of the gland, giving a high $P_{15}: P_i$ ratio. If the removal is delayed for 5–10 minutes the ratio falls to about 1. So far the formation of ATP – i. e. nuclear phosphorylation – has been demonstrated with nuclei having a P_{15}/P_i ratio of this order. This suggests that the level of ATP in the nuclei at the start of the incubation is a determining factor in their subsequent behaviour.

Unfortunately it is not yet known whether the nuclear reactions so far described occur throughout the life cycle of the cell or whether they are restricted for example to the presynthetic or synthetic periods of interphase. The possibility of variation in radiosensitivity of nuclei with respect to these properties in addition to the well-established variation in sensitivity of DNA synthesis cannot therefore be excluded.

References

Allfrey, V. G., A. E. Mirsky, and S. Osawa, 1955: Protein synthesis in isolated nuclei. Nature (London) **176**, 1042–1049.

Allfrey, V. G., and A. E. Mirsky, 1959: Contrasts in the metabolic stability of different nucleotides in the ribonucleic acids of isolated nuclei. Proc. nat. Acad. Sci. U. S. **45**, 1325–1344.

Barer, R., S. Joseph, and M. P. Esnouf, 1956: A method for distinguishing intact cells from free nuclei. Science (Washington) **123**, 24–25.

Cole, L. J., and M. E. Ellis, 1954: Studies on the chemical nature of the radiation protection factor in mouse species. Rad. Res. **1**, 347–357.

Creasey, W. A., 1960: Changes in the sodium and potassium contents of cell nuclei after irradiation. Biochim. biophysica Acta **38**, 181–182.

Creasey, W. A., and L. A. Stocken, 1959: The effect of ionizing radiation on nuclear phosphorylation in the radio-sensitive tissues of the rat. Biochem. J. **72**, 519–523.

Hagen, U., 1960: Labilization of deoxyribonucleic acid in the thymus nucleo-protein after whole body irradiation. Nature (London) **187**, 1123–1124.

Jaffe, J. J., L. G. Lajtha, J. Lascelles, M. G. Ord, and L. A. Stocken, 1959: The effects of X-radiation on the processes leading to DNA-synthesis in regenerating rat liver. Internat. J. Radiation Biol. **1**, 241–246.

Logan, R., M. Errera, and A. Ficq, 1959: The effect of X-rays and ultraviolet light on the uptake in vitro of [8-^{14}C]-adenine and [2-^{14}C]-phenylalanine by isolated nuclei. Biochim. biophysica Acta **32**, 147–155.

Myers, D. K., 1960: Synthesis of diphosphopyridine nucleotide in irradiated rat. Nature (London) **187**, 1124–1125.

Ord, M. G., and L. A. Stocken, 1958: Studies in synthesis of deoxyribonucleic acid. – The radio biochemical lesion in animal cells. Nature (London) **182**, 1787–1790.

Ord, M. G., and L. A. Stocken, 1961: Acid soluble phosphate metabolism in nuclei from rat thymus gland. Biochem. J. **88**, 1–8.

Dose Rate, Cyanide, and some other Factors Influencing Repair of Mutational Radiation Damage in *Drosophila*

F. H. Sobels

Department of Radiation Genetics at the State University of Leyden, Wassenaarseweg 62

In recent years, a great number of experimental data has been accumulated indicating, that the process of mutation is not irreversibly fixed at the time of irradiation, but that depending on the state of the irradiated cells, repair of pre-mutational damage is possible.

Until now the most complete evidence for the operation of repair mechanisms is derived from studies with *Paramecium* (Kimball et al., 1959a, b, 1960, 1961) and bacteria (Witkin, 1956, 1959; Doudney and Haas, 1958, 1959, 1960; Lieb, 1960).

In higher organisms, as the mouse and *Drosophila*, processes of natural repair have been postulated as an explanation for the observation that chronic radiation, given at a low dose rate produces fewer mutations in immature germ cells, as mouse spermatogonia and oocytes (Russell et al., 1958, 1959, 1960), *Drosophila* oogonia (Oster et al., 1959) and spermatogonia (Purdom and McSheehy, 1961) than exposures to the same dose at a high dose rate.

For *Drosophila* spermatids and spermatocytes, repair processes have been inferred to occur on the basis of observations after post-treatment with cyanide. In our earlier experiments it has been observed that, in these stages showing peak sensitivity to radiation, post-treatment with cyanide results in a significant increase of the mutation frequency, if X-radiation is given at a high dose rate of 2200 r/min. Post-treatment following radiation at lower dose rates of 600 r/min. or 150 r/min., however, does not notably change the frequency of mutation (Sobels, 1958 a, b). As to the genetic changes involved in this phenomenen, evidence was obtained that the increase in lethal frequency due to cyanide refers to gene mutations and possibly small deletions (Sobels, 1960). The observation that, at doses producing the same genetic effect, cyanide is equally effective in raising the mutation rate after X-irradiation in pure nitrogen, as in oxygen, made it unlikely that the post-treatment effects arise by an accumulation of radiation-produced peroxides. Alternatively, I have assumed that cyanide inhibits a system, responsible for repair of part of the pre-mutational radiation damage.

Here, I want to report on some new observations with respect to the peculiar dose rate effect which is observed only after post-treatment with cyanide. Secondly, the finding that cyanide can produce both enhancement and lowering of the mutation frequency induced by high-intensity radiation in spermatids will be discussed. In addition, I shall briefly review a number of recent findings, as observed after post-treatments with cyanide and nitrogen, after dose fractionation, and after pre-treatments with chloramphenicol and ribonuclease that bear to my opinion, relation to modification of premutational damage in different stages of *Drosophila* spermatogenesis (see also Sobels, 1961 a; Sobels and Tates, 1961; Tates and Sobels, 1961).

Material and Methods

In a number of the earlier experiments wild type, Oregon-K males were treated and tested for the incidence of sex- linked lethals by means of the Basc, or Muller-5 method. The mutational response in successive stages of spermatogenesis was studied by remating the treated males every three days to fresh females, or alternatively, brood changes were made on the lst, 4th, 6th, and 8th day after treatment. More recently we have used males carrying a ring-shaped X-chromosome. This procedure greatly increases the resolving power for detecting lethals originating as point mutations, or at least as one-hit effects. The treated one-day old X^{C_2} yB males were remated every two days to three, and in later experiments to six fresh females of the genetic constitution y sc^{s1} In 49 sc^{8}; bw; st p^{P}; Dr. I. I. OSTER kindly provided us with the stocks for this mating procedure.

X-irradiation was carried out by means of an ENRAF-machine at 100 kV and 4 mA. Low dose rates ranged from 216 to 565 r/min. High dose rates were 37 r/sec. without additional filtration in the earlier experiments, and 45 r/sec. with 1 mm Al-filtration, or 55 r/sec. without filtration in the more recent experiments. Dosimetry was carried out during exposure of the flies to high-intensity radiation; slight variations in dose rate were observed from experiment to experiment.

Hydrocyanic acid was applied at a constant rate of flow of 100 ml/min. during eight minutes at a concentration equivalent to 35 mg KCN. Recovery from poisoning by cyanide, given in air, took about 2 hours. The nitrogen was purified by passing it through washing flasks with sintered bottoms, containing "O_2-multirapid" (Union Apparate Baugesellschaft m. b. H., Karlsruhe).

If flies were both pre-treated with HCN and exposed to radiation in either O_2 or N_2, the following procedure was applied. First, the flies were kept for 15 minutes in a stream of O_2 or N_2 at a rate of flow of 1–2 liter per minute. In addition, to ensure better penetration of the gases into the flies, the pressure was increased by 0.1 atm four times per minute. Subsequent adjustment of the gas stream, so as to pass it at a rate of flow of 100 ml/min. through the cyanide apparatus took about 6 minutes. Then a mixture of either O_2 or purified N_2 with HCN was passed over the flies for 7–8 minutes. In N_2 a lower concentration of KCN was used than in O_2. Before exposure to radiation, O_2 or N_2 was passed over the flies once again for 3 minutes at a rate of flow of 1 liter per minute. During the whole procedure the flies were kept in a small air-tight perspex chamber. Controll flies, not receiving HCN, were kept in the respective gases during comparable periods of time.

For the fractionation experiments, flies were exposed to a total dose of 1350 r at a dose rate of 55 r/sec., given either at once, or in five fractions with intervals of two hours between the separate fractions. In order to avoid faulty dosimetry the flies were irradiated in a very small volume, so that an ionization chamber could be mounted immediately on top of the radiation vessel. To prevent anoxia, a stream of air was passed over the flies before and during the irradiation. In one experiment each fraction was followed by 25 minutes post-treatment with nitrogen.

Chloramphenicol solutions of 1 mg/ml, 0,02 mg/ml and 0,06 mg/ml and ribonuclease solutions of 0,05 mg/ml were administered by injection of approximately 0.25 μl per fly into the abdominal cavity. The exact concentrations of the injected chemicals were determined by weighing batches of 10 flies before and after the injections.

Results

1. The effect of dose rate in the presence of cyanide

Indications for an effect of dose rate in the presence of metabolic inhibitors were first obtained by CLARK (1956). CLARK observed that pre-treatment with sodium azide was more effective in raising the mutation rate in mature sperm after exposure to radiation at a dose rate of 2000 r/min., than after exposure at a dose rate of 100 r/min. Recently, CLARK (1961) reported that this effect is more pronounced in spermatids than in mature spermatozoa.

Our observation that post-treatment with cyanide only results in an increase of the mutation frequency after radiation given at a dose rate of 2200 r/min. (Table 2, line1), but not after radiation at a dose rate of 590 r/min. (Table 2, line 2), may refer to the same phenomenon, as that described by CLARK. It is surprising, however, that following post-treatment, differences in effect are found for dose rates differing not more than from 590 r/min. to 2200 r/min.

The ineffectiveness of cyanide following low-intensity radiation may have various reasons. First, there is a difference in exposure time to give the same dose of radiation at the two different intensities. It is not unlikely that, as in other biological material (DOUDNEY and HAAS, 1958; KIMBALL et al., 1957; WITKIN, 1956), the completion of the mutation process, or its susceptibility to cyanide is restricted in time. Suppose that this critical time interval were of the same order of magnitude as the more protracted exposure time, required to give radiation at the low dose rate, the repair process may have become refractory already to the effect of cyanide. If this interpretation should be correct, one would expect that the application of cyanide at a longer time interval after radiation at high dose rate would be likewise ineffective.

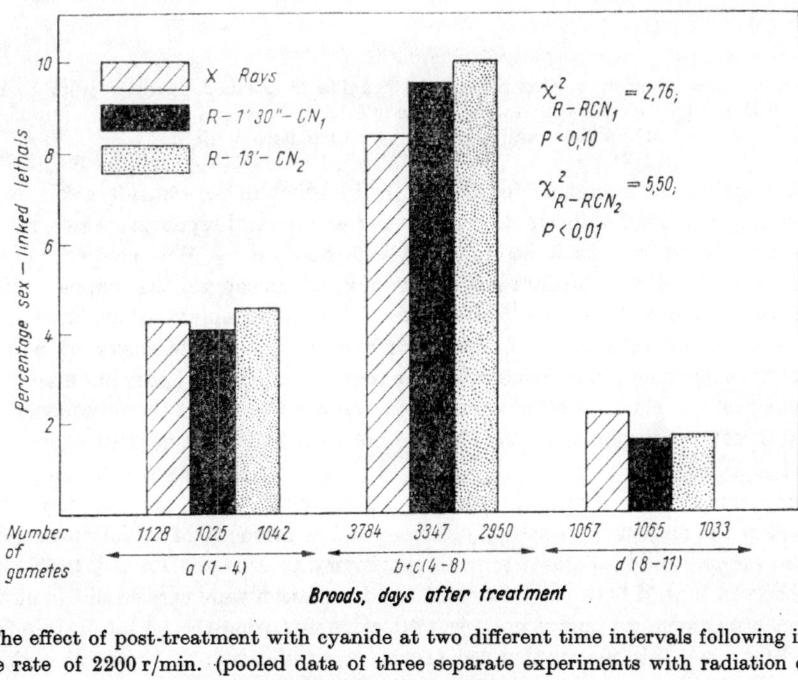

Fig. 1. The effect of post-treatment with cyanide at two different time intervals following irradiation at a dose rate of 2200 r/min. (pooled data of three separate experiments with radiation doses ranging from 1280—1730 r)

In the experiments to test this idea flies were post-treated at two different time intervals after exposure to high intensity radiation. In order to exaggerate this possible effect the difference in time between the two post-treatments of about twelve minutes was, in fact, greater than that between exposures at low and high dose rates. The data of three experiments of this kind showed no significant heterogeneity for stages with peak sensitivity. In consequence they were pooled; they are shown in Fig. 1.

It is seen that cyanide is definitely not less effective in enhancing the amount of mutation, if applied ten to fifteen minutes after exposure to high intensity radiation. In consequence the interpretation of the intensity effect, as outlined above, becomes very improbable. The observation that in these particular experiments cyanide, if following shortly after exposure to radiation, has only low effectiveness shall be discussed later.

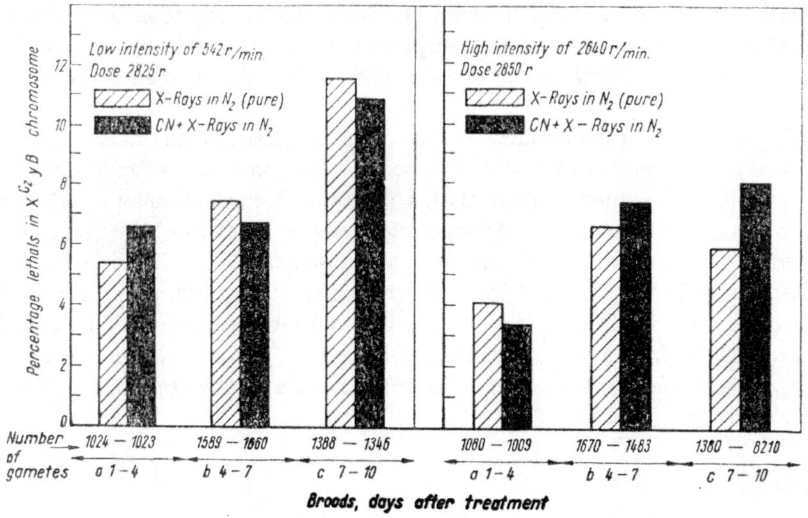

Fig. 2. The difference in effect of treatment with cyanide in purified nitrogen prior to radiation exposure in purified nitrogen on the mutation frequency in stages with peak sensitivity (broods b and c) at low (left) and high intensity of radiation (right)

As an alternative interpretation for the intensity effect under consideration, one could visualize some specific kind of interaction between dose rate and repair processes, resulting in a greater susceptibility to cyanide at high than at low dose rates. When experiments to test this hypothesis were started, the data presented above which suggest that exposure time does not play a part in the post-treatment effect, were still less complete. Since however, poisoning by cyanide lasts considerably longer than exposure at low intensity of radiation, it was thought that by using pre-treatment with cyanide, a time factor, possibly involved in the post-treatment effect, would be eliminated. Dose rate dependence in the presence of cyanide was studied therefore by comparing the effect of pre-treatment with cyanide at low and high intensity of radiation.

Earlier experiments had shown that pre-treatment with cyanide results in a significant enhancement of the mutation frequency produced at low intensity of radiation and that this effect is most pronounced in stages with peak sensitivity to X-rays (Sobels, 1955). Furthermore we observed that, if both pre-treatment and irradiation were carried out in either oxygen or purified nitrogen, cyanide is no longer effective in raising the mutation rate (Sobels, 1958, 1961 b). In fact this observation was taken as an indication that the enhancing effect of pre-treatment with cyanide, at low intensity of radiation in air, is caused by an accumulation of oxygen due to the inhibiting effect of cyanide on respiration. In consequence, it seemed that the question whether there is a specific susceptibility to cyanide at high intensity of radiation could be studied more effectively by pre-treating and irradiating the flies in either purified nitrogen or oxygen, that is under conditions where cyanide is ineffective, if preceding exposures at a low dose rate. The procedure used for such experiments is outlined under the section describing material and methods.

The data of a set of experiments in which X^{C_2} yB males were pre-treated and irradiated at the two contrasting dose rates in nitrogen, are shown in Fig. 2. It is seen that in stages with peak sensitivity, as sampled in broods b and c, cyanide had no appreciable effect, if preceding a low intensity of radiation (Fig. 2, left). After exposure to radiation, given at the high intensity (Fig. 2, right), however, there is a clear tendency that pre-treatment raises the mutation

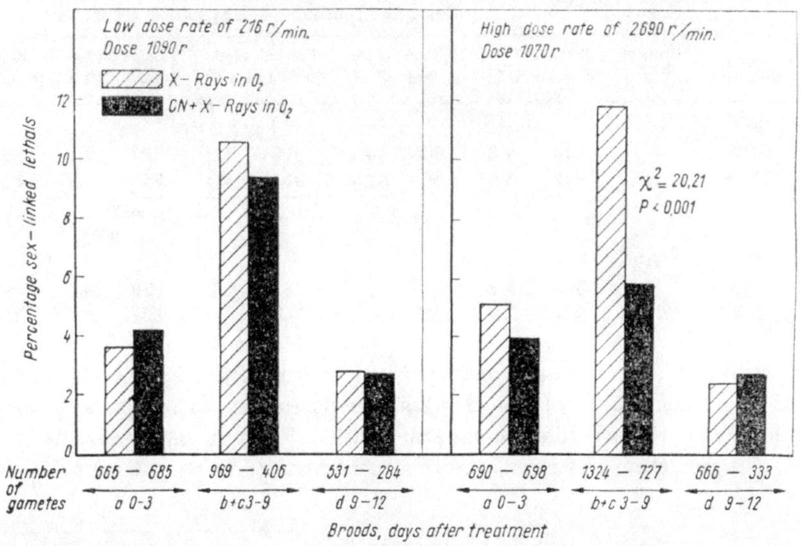

Fig. 3. The difference in effect of treatment with cyanide in oxygen prior to radiation in oxygen on the mutation frequency in stages with peak sensitivity (broods b and c) at low (left) and high intensity of radiation (right)

frequency ($\chi^2 = 3.62$; $0.05 < P < 0.10$). With respect to the effect of cyanide, the data of the experiments at the two contrasting dose rates show significant heterogeneity ($\chi^2 = 4.30$; $P < 0{,}05$). All other experimental conditions being exactly the same, this result indeed suggests that at the two dose rates under investigation, cyanide affected the mutation frequency in a different manner.

A similar experiment was carried out in oxygen, the data of which are presented in Fig. 3. Fig. 3, left, shows that, if radiation is given at a low dose rate in oxygen, pre-treatment does not notably influence the mutation frequency. This result corresponds with those of two earlier experiments showing no effect of cyanide pre-treatment in an oxygen atmosphere at low intensity of radiation (Sobels, 1958, 1961 b). The data presented in Fig. 3, right, show that, if all experimental conditions are the same, except for dose rate, being higher by about a factor 12, pre-treatment may result in a highly significant decrease of the mutation frequency produced in spermatids and spermatocytes. The finding that cyanide can produce both enhancement and lowering of the mutation frequency induced by high-intensity radiation raises several questions. They will be discussed later in connection with similar observations after post-treatment with cyanide. In this context, I want to point out only that it is not likely that the reduction of the mutation frequency due to cyanide preceding high-intensity radiation in oxygen, is brought about by anoxia because the flies used for the experiment with low-intensity radiation were treated in exactly the same manner and their mutation frequencies hardly show an effect of the pre-treatment. Such an interpretation is also improbable because, as will be shown below, a similar reduction due to cyanide has been observed after pre-treatment and high-intensity radiation in nitrogen (Table 1, upper part).

Table 1

The effect of pre-treatment with HCN, in either N_2 or O_2, on the frequencies of sex-linked lethals in a X^{C_2} yB chromosome produced by X-irradiation in either N_2 or O_2, respectively at a dose rate of 2880 r/min (100 kV, 4 mA, 1 mm Al filter)

Experiment	Gas: Dose of X-rays [r]	Gas: Dose of HCN [mg KCN]	Brood a (0–2) [n chr.]	Brood a (0–2) [% l.]	Brood b (2–4) [n chr.]	Brood b (2–4) [% l.]	Brood c (4–6) [n chr.]	Brood c (4–6) [% l.]	Brood d (6–8) [n chr.]	Brood d (6–8) [% l.]	Brood e (8–10) [n chr.]	Brood e (8–10) [% l.]
	N_2	N_2										
Radiation	2208	—	874	4.3	864	4.3	873	7.2	823	7.3	858	1.9
CN + Rad.	2215	8.5	862	3.7	861	4.8	855	4.3	845	5.7	857	2.5
							Broods c + d		$\chi^2 = 7.94$ P < 0.01			
	O_2	O_2										
Radiation	746	—	869	4.3	495	3.8	768	5.9	539	6.1	898	2.7
CN + Rad.	746	17.5	869	2.9	853	4.6	873	7.0	502	6.2	876	2.9

The pre-treatment experiments in nitrogen and oxygen with high-intensity of radiation were repeated by using two- day brood changes. The doses were chosen, so as to produce the same genetic effect in the two different gas atmospheres. From the data, presented in Table 1, upper part, it is seen that in nitrogen, pre-treatment resulted in a significant decrease of the

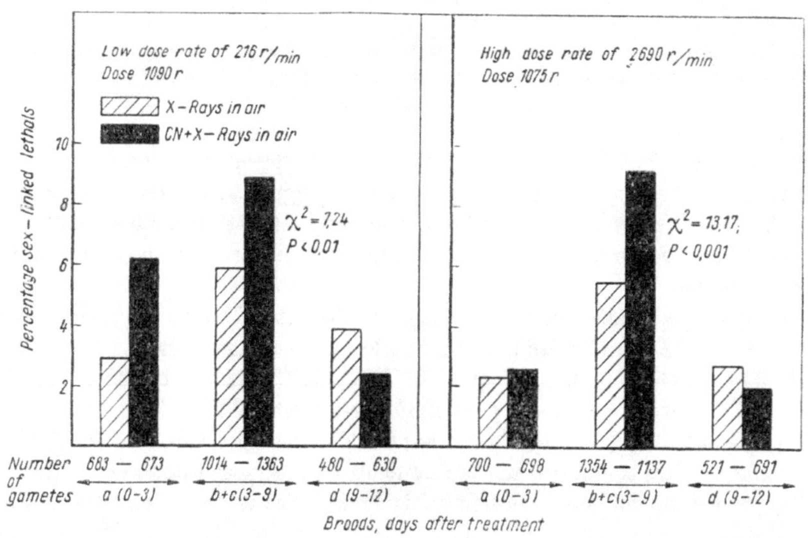

Fig. 4. The effect of pre-treatment with cyanide in air on the frequency of sex-linked lethals induced by exposure to radiation in air at low (left) and high (right) dose rates. In contrast to the results in N_2 and O_2, no significant difference in stages with peak sensitivity is observed at the two contrasting dose rates

mutation frequency in stages with peak sensitivity to radiation, as sampled in broods c and d. This result is in contrast to that of the preceding experiment in nitrogen, shown in Fig. 2, right, but it is in line with the cyanide effect in oxygen, as obtained in the experiment, shown in Fig. 3, right.

In view of the pronounced effect of cyanide in oxygen, as shown in the experiment of Fig. 3, right, it is surprising that in the replica (Table 1, lower part) pre-treatment failed to modify the mutation frequency. The material used for the two experiments differed only in the kind of males which were Oregon-K, with a normal rod-shaped X-chromosome in the first, and X^{C_2} yB with a closed X-chromosome in the second experiment. It seems unlikely that this difference, or that consisting in more frequent changes of broods, can be held responsible for the difference in effect of the pre-treatment. It will be pointed out later that it is possible to explain the lack of effectiveness of cyanide in this experiment and in the one shown in Fig. 1 (R–1′30″–CN), by assuming that cyanide can influence the mutation-frequency by modifying two contrasting processes involved in repair of premutational damage.

Since Clark (1956, 1961) reported an intensity effect in mature sperm, if prior to irradiation in air, males or inseminated females were injected with sodium azide, we were interested whether cyanide, if preceding high intensity radiation in air, would likewise more effectively increase the mutation rate, than if given prior to low-intensity radiation.

Fig. 4 shows the pooled data of two such sets of experiments. It is seen that in stages with peak sensitivity pre-treatment resulted in a significant enhancement of the mutation frequencies at both low and high intensity of radiation. With respect to the effect of cyanide, the difference between these two groups is not significant ($\chi^2 = 0.11$; $0.70 < P < 0.80$). Moreover, in stages sampled in the first brood, presumably corresponding to mature sperm, cyanide exerted a considerably greater effect, if preceding radiation given at low than that given at high intensity. The results for both sperm and the more sensitive stages are different from those obtained by Clark after pre-treatment with azide. It may be noted that in our earlier experiments the enhancing effect of cyanide has usually been more pronounced in spermatids and spermatocytes than in mature sperm (Sobels, 1955, 1961). It is possible that, in the low dose rate experiment under consideration, the increase of the mutation frequency in brood a has been brought about by a differential shift in the brood pattern, so that in the pre-treated group of this experiment more radiosensitive cells were sampled than in the corresponding group of the high-dose rate experiments. This interpretation would be in line with Clark's (1961) suggestion that high-intensity radiation by means of physiological stimulus, may increase the rate of sperm maturation.

In summarizing the outcome of the experiments described above, the following picture is obtained:

(1) If pre-treatment and radiation are carried out in air no different effect is observed at the high intensity of radiation. Since, however, pre-treatment in air results already in such a pronounced modification of the mutation frequency at low intensity of radiation, it is possible that in air the method is not sufficiently sensitive to show a different effect at the high dose rate.

(2) In purified nitrogen no effect of pre-treatment is observed after exposure to low intensity radiation. However, following exposure to high-intensity radiation a tendency to increase the mutation frequency was observed in one experiment, the data being significantly different from those obtained after low-intensity radiation. In a second experiment a significant decrease was observed.

(3) In oxygen pre-treatment did not significantly modify the mutation frequency in three independent experiments with low intensity of radiation. After exposure to high intensity of radiation a significant decrease of the mutation frequency was observed in one experiment and, for unknown reasons, cyanide had no effect in another experiment.

Considering the experiments in nitrogen and oxygen only, altogether a modification of the mutation frequency was observed in three out of four experiments, if pre-treatment was given prior to high-intensity radiation, but no effect was observed if it preceded low-intensity of radiation. The results of these experiments could be called at least suggestive for the existence of a dose rate dependence in the presence of cyanide.

A dose rate dependence would be expected if more than one hit contributes to the genetic event or if restitution of the damage occurs within a limited time interval. In this context, I want to point out that, at least part of the experiments (those shown in Fig. 2 and Table 1) were carried out with a ring-shaped X-chromosome, so that the lethals observed, to all probability arose as one-hit events. However, even in the case of the oxygen experiment (Fig. 3) for which a rod-shaped X-chromosome has been used, it seems unlikely that a specific effect of cyanide on position-effect lethals with two or multiple break origin, would have been responsible for the decrease observed after the pre-treatment. This is so, because in experiments specifically designed to study the effect of cyanide on the frequency of radiation-produced translocations a significant "increase" of the translocation frequency has consistently been observed at the two contrasting dose rates after both pre- and post-treatment with cyanide (Sobels, 1958a, 1960).

In consequence, the present observation suggesting a dose rate dependence in the presence of cyanide for a genetic effect which, to all probability, is brought about by a single hit to the chromosome, calls for a further interpretation. It would seem that, apart from the direct radiation damage to the chromosome, there is a second system which is only susceptible to high intensity of radiation in the presence of metabolic inhibitors. It is of interest that Clark (1961), on the basis of his experimental evidence about such an intensity effect, arrives at a similar conclusion. In view of our other observations on the existence of repair processes in stages characterized by peak sensitivity to the mutagenic action of X-irradiation, I would venture to speculate that in these cells high intensity of radiation influences the system associated with repair of part of the pre-mutational damage.

2. The contrasting effects of post-treatment with cyanide

From the data discussed in the preceding section it is seen that treatment with cyanide, if preceding high-intensity radiation, may result in either an increase or a decrease of the mutation frequency in stages with peak sensitivity. Since, lately similar effects were also observed after post-treatment with cyanide, I want to discuss briefly the relevant findings and the possible causes underlying these contrasting observations. The data of experiments on post-treatment with cyanide are presented in Table 2. They will be discussed below:

(1) In the first seven experiments on the effect of post-treatment with cyanide, an enhancement of the mutation rate has consistently been observed. In three separate experiments the difference in mutation frequency between flies which had received radiation only and those that were post-treated, was significant. In the other four experiments the difference was not significant, but in the direction of an increase. Since, with respect to the effect of cyanide, the heterogeneity of χ^2 for the seven separate experiments was not significant, the data were pooled. It is seen that then the enhancing effect, due to post-treatment, is highly significant.

(2) For sake of completeness, the pooled data of five experiments on post-treatment following low-intensity of radiation are also incorporated in the table. As has been pointed out before, post-treatment, under these conditions of radiation, does not result in an enhancement of the mutation frequency.

Table 2. A survey of data on the effect of post-treatment with cyanide on the frequency of mutation induced by high-intensity radiation (with the exception of 2) in stages with peak sensitivity

No.	Number of expe-riments	Brood chan-ges in days [nr. ♀♀/♂]	Average dose in r	Dose rate [r/min]	Gas	X-rays [n chr.]	X-rays [% leth.]	X-rays + HCN [n chr.]	X-rays + HCN [% leth.]	Effect	χ^2	P	Reference
1	7	3–2 — 3	1620	2200	Air	8583	8.8	7982	11.2	incr.	26.38	<0.001	Sobels (1960)
2*)	5	3–2 — 3	1390	590 or 154	Air	5595	8.3	6035	8.7	No	0.47	0.50	Sobels (1960)
3	2	3–3 — 3	1200	2240	O_2	2225	8.8	2238	11.9	incr.	11.74	<0.001	Sobels (1960)
4	2	3–2 — 3	3900	2240	N_2	2790	8.8	2802	11.8	incr.	14.26	<0.001	Sobels (1960)
5	3	3 — 3	1470	2200	Air	3784	8.4	3347	9.5	unsignific. incr.	2.76	<0.10	This paper, Fig. 1
6	2 Rad. 3 RCN	2 — 3	1540	3660	Air	1446	4.9	4290	8.8	incr.	22.59	<0.001	Sobels and Tates (1961)
7	1 Rad. 2 RCN	2 — 3	1525	3300	Air	1132	11.7	2861	8.4	decr.	11.07	<0.001	Sobels and Tates (1961)
8	1	2 — 6	415	2880	Air	1777	2.9	989	3.0	No	—	—	Sobels (1961a)
9	1	2 — 6	788	2620	Air	1025	6.4	1174	3.6	decr.	13.01	<0.001	Sobels (1961a)
10	1	2 — 6	825	2880	Air	1360	5.7	1707	4.9	unsignific. decr.	2.00	<0.20	Sobels (1961a)
11	1	2 — 6	825	2880	Air	1633	5.1	1741	4.3	,, ,,			Sobels (1961a)
12	1 b	2 — 6	1628	2620	Air	1857	6.5	864	3.1	decr.	10.90	<0.001	Sobels (1961a)
	1 c		1628	2620	Air	314	14.6	130	4.6	decr.	8.95	<0.01	Sobels (1961a)

*) Here the lack of effectiveness is shown, if HCN follows low-intensity radiation

(3 and 4) At doses that produce the same genetic effect, a highly significant increase due to post-treatment was likewise observed after exposure of a ring-shaped X^{C_2} yB chromosome to high-intensity radiation in either oxygen or nitrogen. Two separate experiments were carried out in either gas. All four showed a significant increase in consequence of the post-treatment.

(5) In the experiment mentioned in the preceding section (Fig. 1), which was designed to test whether the duration of the interval between irradiation and post-treatment plays a part in the enhancing effect, only a minor increase was observed in flies that were post-treated immediately following the completion of radiation.

(6) In a recent experiment we studied whether etherization, or changing the interval between radiation and post-treatment from 2 seconds to 2 minutes, would change the post-treatment effect. Changes of the interval, produced no noticeable effect. In flies that were etherized, cyanide produced a slightly greater enhancement in brood d (spermatocytes and perhaps early spermatids) than in unetherized flies. All three post-treated groups consistently showed an enhancement of the mutation rate in radiosensitive cells, as compared with that in the two groups which received radiation only. It may be noted, however, that in the latter groups the mutation frequency in stages with peak sensitivity is exceptionally low.

(7) On the other hand, a highly significant decrease was observed in an experiment that was carried out under exactly the same conditions, as the one described above (6). There is a remarkable correspondence between the mutation frequencies in the post-treated groups of the two experiments, but the mutation frequencies observed in broods c and d of flies that were exposed to radiation only, are considerably different. Since in the experiments described under 6 and 7 the mutation frequencies in broods a and e, corresponding to mature sperm and spermatogonia respectively, were exactly the same, it is intriguing what caused such pronounced differences in response in the more sensitive, intermediate broods.

(8) In view of the possibility that an irregular release of germ cells in the successive broods, resulting in an admixture of stages with different sensitivities had been responsible for this kind of discrepancy, six females per male per brood were used in a continuation of these experiments. It is seen that after exposure to 415 r post-treatment remained without effect.

(9 and 12) In an experiment in which the effect of post-treatment was studied after exposure to doses of about 800 and 1600 r a highly significant decrease was observed.

(10) A replica experiment in which likewise one-day, old males were exposed to 800 r failed to show a pronounced decrease.

(11) The same observation was made if 2-day, old males were used for treatment. However, with respect to the effect of cyanide, heterogeneity between the three experiments carried out with doses of 800 r (9, 10 and 11) is not significant ($X^2 = 5.21$, $P < 0.10$ for $N = 2$). After pooling the data a significant decrease due to post-treatment is observed ($X^2 = 7.72$; $P < 0.01$).

By grouping the experiments according to the effect of cyanide the following picture is obtained:

Increase significant	9
Increase, but not significant	7
Decrease, significant	3
Decrease, but not significant	2
No effect (400 r)	1

In this context, I want to mention that in the earlier experiments post-treatment consistently resulted in an increase of the mutation frequency, whereas in the more recent experiments, with the exception of the data mentioned under 6, a lowering of the radiation-induced mutation frequencies has been observed. The same is true for the experiments in which cyanide was given prior to high-intensity radiation (see Fig. 3 and Table 1). The main difference between the more recent and the earlier experiments consists in more frequent brood changes, with a greater number of females per male per brood. We have considered the possibility that a more efficient sampling of stages with a specific susceptibility to the effects of cyanide might have resulted in a proportionally greater selective elimination of cells with mutations. This assumption is not corroborated, however, by the observation that post-treatment fails to increase the mutation frequency after exposure to a low dose of 400 r. Also, one would, on this assumption, expect an increase of dominant lethals after exposure to higher doses of radiation. A preliminary experiment in which the effect of post-treatment on egg-hatchability was determined after exposure to 1620 r, failed to show, however, such an effect (SOBELS, 1961 a).

In an attempt to explain the contrasting effects of cyanide I want to mention first the recent discovery of SÄVHAGEN (1961 a) that after exposure to a dose of 1100 r in air, the frequency of chromosome loss is considerably greater in spermatids and spermatocytes of 0–1, day old males than in those of 3–4, day old males. After exposure to the same dose in oxygen, SÄVHAGEN observed, on the other hand, roughly the same mutation frequencies in progeny from the two kinds of males. The mutation frequencies obtained in oxygen were intermediate between those of young and older males after irradiation in air, so that in comparison to radiosensitivity in air, radiation in oxygen resulted in a decrease of the mutation frequency in 0–1, day old males and in an increase in 3–4, day old males.

In the light of these observations one wonders, whether perhaps by some unknown condition, the response of the males to radiation has varied either to the pattern of young, or to that of old males. And, if this were so, whether variation of the mutation frequencies in correlation with the age of flies that received radiation only, may have been responsible for the contrasting effects of post, or pre-treatment. For example, the discrepancies in mutational response, as observed in experiments 6 and 7, of flies which were exposed to radiation only, with the ensuing contrasts in the effect of post-treatment, resemble the relationship described by SÄVHAGEN after radiation in air and oxygen for males of different ages.

As to possible differences in the genetic constitution of the males that might have caused such variation, I want to mention that the majority of the earlier experiments has been carried out with 18–30 hour old males of a wild type, Oregon-K strain. For some experiments sc^8. Y/y In 49 B; bw^D males of comparable age were used. In all the more recent experiments, always X^{C_2} yB males were treated, with exception of the experiment shown in Fig. 3 that was carried out with Oregon-K males, so as to make it more comparable to the earlier experiments. The fact that also in this experiment, cyanide, given prior to high-intensity radiation, resulted in a highly significant decrease of the mutation frequency, does not support the assumption that in the other recent experiments only differences in genetic constitution of the males have been responsible for the lowering effect of cyanide. The observation that post-treatment of X^{C_2} yB males in either oxygen or nitrogen (headed under 3 and 4) resulted in a significant increase, likewise argues against such a hypothesis.

With respect to a possible effect of differences in age, I want to point out that for some of the earlier experiments, in which a highly significant increase was observed, males were used that were definitely not older than 20 hours. In this connection, I may refer also to the pre-treatment experiment, using high-intensity radiation, that was carried out in oxygen (Fig. 3, right). Here pre-treatment with cyanide resulted in a highly significant decrease of the mutation frequency. However, in comparison to the mutation frequencies, obtained in males from the same sample after exposure to the same dose in air (Fig. 4, right, hatched columns), an increase of the mutation frequency is observed in flies that received radiation in oxygen only (see Fig. 3, right, hatched columns). This paired comparison of the mutation frequencies in air and oxygen, in males that were similar in all respects, suggests that these males responded to radiation in air, as 3–4, day old males in SÄVHAGEN's experiments. In consequence, these observations do not favour the assumption that a lowering effect of cyanide would be specific for young males (in SÄVHAGEN's sense) and an enhancing effect, for old males.

For a further test, the effect of post-treatment has been studied in 24-hour, old and 48-hour, old X^{C_2} yB males. To eliminate possible disturbing effects of cell selection, a low dose of radiation, of 800 r was used. The results of these experiments (Table 2, headings 10 and 11), did not show, however, an appreciable difference in either the response to radiation alone, or to that of post-treatment with cyanide.

In the absence of significant evidence supporting this idea, I find it more satisfactory therefore to assume that, in analogy to observations of KIMBALL (KIMBALL et al., 1961, and personal communication) on the effect of metabolic inhibitors in *Paramecium*, cyanide may affect two contrasting processes, one being associated with the rate of loss, the other with the rate of fixation of pre-mutational damage. Inhibition of the first process, that is inhibition of metabolic repair, will tend to increase the mutation rate. On the other hand, slowing down the rate of fixation will result in a decrease of the mutation frequency. If depending on hitherto unknown experimental conditions, the effect of cyanide is more pronounced on either one, or the other process, both enhancement and lowering of the mutation frequency may result.

3. Other observations on the modification of repair processes

In connection with the results discussed above, a number of other recent observations which suggest a modification of repair processes in different stages of *Drosophila* spermatogenesis, will be discussed briefly (SOBELS, 1961 a; SOBELS and TATES, 1961; TATES and SOBELS, 1961). The main findings of these experiments are summarized in Fig. 5. It will be seen that most data are available for stages showing peak sensitivity to X-irradiation, presumably corresponding to spermatids and spermatocytes (cf. SÄVHAGEN, 1961b).

Experiments of my collaborator Mr. A. D. TATES showed that, in this stage, fractionation of the dose in five exposures separated by 2-hour intervals, of irradiation given at high intensity, significantly reduces the mutation frequency in comparison to that produced by unfractionated radiation. These observations in fact provide the most direct evidence for repair of pre-mutational damage in spermatids.

TATES also observed that post-treatment with purified nitrogen results in a significant enhancement of the mutation frequency. The effect is seen in spermatids, spermatocytes, and spermatogonia. Anoxia following high-intensity radiation, thus interferes in a similar way with recovery processes, as has been observed in the earlier experiments with cyanide. These effects are most readily explained on the assumption that both cyanide and nitrogen raise the mutation-frequency by inhibiting metabolic repair. In addition, at least in the experiments under discussion, the enhancing effect of nitrogen post-treatment extends to late spermatogonia.

Furthermore, TATES found that, if in fractionation experiments, each fraction is followed by nitrogen, a significant reduction of the mutation frequency results in comparison to that produced by post-treatment following unfractionated exposure. This finding supports the idea that fractionation favours repair.

In view of the findings that inhibition of protein synthesis by means of chloramphenicol, influences repair from pre-mutational radiation damage in microorganisms (WITKIN, 1956, 1958; DOUDNEY and HAAS, 1958; ALPER and GILLIES, 1960; LIEB, 1960), we studied the effect of pre-treatment with chloramphenicol in *Drosophila* males. Also, it has been suggested

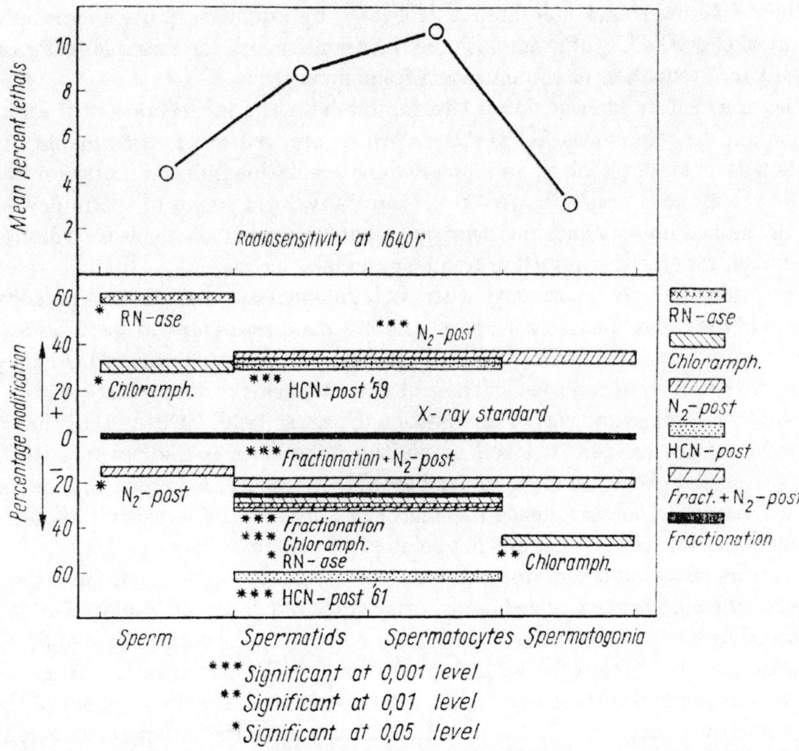

Fig. 5. A summary of the data on the modification of pre-mutational damage in *Drosophila* spermatogenesis. The subdivision in stages at the bottom is schematic. The top-graph shows weighted mean lethal frequencies after exposure to 1640 r, as a measure for differential radiosensitivity in the stages under consideration. The lower diagram shows the percentual modification of the mutation rates, after dose fractionation and treatment with the different chemicals, in comparison to the mutagenic action of radiation alone

that changes in the composition of proteins during the transition from spermatids to mature sperm (Felix, Fisher, and Krekels, 1956; Alfert, 1959) might be responsible for the differential radiosensitivity of these two stages (Oster, 1959). Since dose – fractionation, and various chemical co-treatments are only effective in modifying the radiation-induced mutation frequencies in spermatids and not in sperm, we were interested in a clarification of the causes underlying differential radiosensitivity in these two stages. The results of a number of such experiments showed that pre-treatment with chloramphenicol decreases the radiation-induced mutation frequency in spermatids, spermatocytes and spermatogonia, whereas a slight, but significant increase was observed in mature sperm.

The lowering effect of chloramphenicol in immature stages of sperm development could be explained, like that observed in the recent cyanide experiments, by assuming a delay of mutation fixation. When discussing mutation fixation, I want to emphasize that a reduction of the mutation rate, as observed after both pre-treatment with chloramphenicol and post-treatment with cyanide is expressed already in post-meiotic spermatids, that is in cells which no longer divide. All relevant studies on microorganisms consistently show, however, that the terminal event, involved in the mutation fixation of pre-mutation to mutation is associated with chromosome replication. At present we do not know the equivalent counter part of such a process in the development of post-meiotic germ cells of *Drosophila*. The observation that chloramphenicol has an effect in meiotic stages and spermatogonia is not so surprising because, in these dividing stages, chloramphenicol may, by inhibiting protein synthesis, either delay cell division or DNA synthesis; both modifications, which for example in *Paramecium*, tend to result in a reduction of the mutation frequency.

Ribonuclease which is likewise known to inhibit both the incorporation of amino acids into proteins and protein synthesis, was also used in pre-treatment experiments. As in the experiments with chloramphenicol, an increase of the radiation induced mutation frequency, though more pronounced, was observed in the most advanced stages of sperm development. Moreover, an indication was obtained that pre-treatment with ribonuclease, like that with chloramphenicol, reduces the mutation frequency in spermatids.

The observation that pre-treatments with both ribonuclease and chloramphenicol result in an increase of the radiosensitivity in sperm and in a decrease of that in spermatids, support the idea that amino acid incorporation into proteins and protein synthesis may play a part in the origin of the different radiosensitivities of these two stages of spermatogenesis. In connection with these findings, an earlier observation (SOBELS, 1958, 1960, 1961b) that nitrogen is more effective than oxygen in levelling off the differences in radiosensitivity between sperm and spermatids, becomes more understandable because it is known that protein synthesis requires oxygen. In consequence this factor in different radiosensitivity is presumably hampered under anoxic conditions, but not so at exposures in oxygen.

The contrasting effects of chloramphenicol and ribonuclease in spermatids and sperm show resemblence to the modification of radiosensitivity in *E. coli* B and its radioresistant mutant B/r, as observed by ALPER and GILLIES (1960) after changes in the composition of the medium, oxygen tension and post-treatment with chloramphenicol. With respect to changes in radiosensitivity by oxygen and nitrogen and after pre-treatment with chloramphenicol the radiosensitive spermatids behave like *E. coli* B and the more radioresistant sperm as B/r. It would be of interest to know whether a common biochemical pattern underlies this similarity in response to radiation and factors modifying radiation damage in these widely different cells.

Summary

1. The intensity effect which is observed after post-treatment with cyanide has been subjected to further analysis. The effects of post-treatment at different time intervals after radiation suggest that differences in exposure time required for administering radiation at high and low intensities, are not responsible for the ineffectiveness of cyanide following low-intensity radiation. The outcome of pre-treatment experiments under conditions (N_2 or O_2), where cyanide is ineffective, if preceding low-intensity radiation, suggests that there is a specific dose rate dependence in the presence of cyanide. It is assumed that in the presence of cyanide high-intensity radiation, apart from direct damage to the chromosomes, affects a second susceptible system which may be associated with repair processes.

2. Data showing that post-treatment with cyanide may result in either enhancement or lowering of the radiation induced mutation frequencies in stages with peak sensitivity are reviewed. A possible interpretation for these apparent discrepancies in the effects of cyanide is given in terms of a model involving two contrasting processes, one being associated with the rate of loss of pre-mutational damage, the other with the rate required for its fixation to mutation.

3. A number of other observations on modification of radiation induced frequencies after dose fractionation, post-treatment with nitrogen and pre-treatment with both chlorampheni-

col and ribonuclease are reported. They suggest that modification of pre-mutational damage is possible in spermatids, meiotic stages and spermatogonia. The findings after pre-treatment with chloramphenicol and ribonuclease in sperm and spermatids suggest that protein synthesis is involved in the different radiosensitivities of these two stages.

Acknowledgement

These studies were supported by the Health Research Council T. N. O. and by the Institute for Radiopathology and Radiation Protection.

Many thanks are due to Miss C. Bennema, Miss M. A. H. Lommerse, Miss F. J. de Ruiter, Miss A. Goedhart, Mr. J. I. M. van Hooft and Mr. J. J. Lambooy, for their help in completing the labourconsuming mutation experiments, and to Miss M. Verdegaal for her assistance in the preparation of the manuscript.

References

Alfert, M., 1959: Cytochemische Untersuchungen an basischen Kernproteinen während der Gametenbildung, Befruchtung und Entwicklung. In: Chemie der Genetik. Springer-Verlag, Berlin, pp. 73–84.

Alper, T., and N. E. Gillies, 1960: The influence of metabolic events immediately after radiation on the difference in sensitivity between *E. coli* B and *E. coli* B/r. In: A. A. Buzzati-Traverso (Ed.), Immediate and low level effects of ionizing radiations. Taylor & Francis Ltd., London, pp. 305–310.

Clark, A. M., 1956: Genetic effects of X-rays in relation to dose rate in *Drosophila*. Nature (London) **177**, 787.

Clark, A. M., 1961: Modification of genetic response to X-irradiation in *Drosophila*. Proc. of the Third Australasian Conference on Radiation Biology, Sydney, August 1960, Butterworths, London, pp. 216–228.

Doudney, C. O., and F. L. Haas, 1958: Modification of ultraviolet-induced mutation frequency and survival in bacteria by post-irradiation treatment. Proc. nat. Acad. Sci. U. S. **44**, 390–401.

Doudney, C. O., and F. L. Haas, 1959: Mutation induction and micro-molecular synthesis in bacteria. Proc. nat. Acad. Sci. U. S. **45**, 709–722.

Doudney, C. O., and F. L. Haas, 1960: Some biochemical aspects of the post-irradiation modification of ultraviolet-induced mutation frequency in bacteria. Genetics **45**, 1481–1502.

Felix, K., H. Fisher, and A. Krekels, 1956: Protamines and nucleo-protamines. In: J. E. V. Buttler (Ed.), Progress in biophysics and biophysical chemistry. Pergamon Press, London, **6**, 1–23.

Kimball, R. F., N. Gaither, and S. M. Wilson, 1957: Post-irradiation modification of mutagenesis in *Paramecium* by streptomycin. Genetics **42**, 661–669.

Kimball, R. F., N. Gaither, and S. M. Wilson, 1959a: Reduction of mutation by post-irradiation treatment after ultraviolet and various kinds of ionizing radiations. Rad. Res. **10**, 490–497.

Kimball, R. F., N. Gaither, and S. M. Wilson, 1959b: Recovery in stationary-phase *Paramecia* from radiation effects leading to mutation. Proc. nat. Acad. Sci. U. S. **45**, 833–839.

Kimball, R. F., and S. W. Perdue, 1960: The relation of the refractory period for mutation induction and nuclear synthesis in *Paramecium aurelia*, syngen' 4. Genetics **45**, 996.

Kimball, R. F., N. Gaither, and S. W. Perdue, 1961: Metabolic repair of pre-mutational damage in *Paramecium*. Internat. J. Rad. Biol. **3**, 138–148.

Lieb, M., 1960: Deoxyribonucleic acid synthesis and ultraviolet-induced mutation. Biochim. biophysica Acta **37**, 155–157.

Oster, I. I., 1959: The spectrum of sensitivity of *Drosophila* germ cell stages to X-irradiation. Rad. Biol., Proc. of the Second Australasian Conference on Radiation Biology, edited J. H. Martin. Butterworths Scientific Publications, London, pp. 253–267.

Oster, I. I., S. Zimmering, and H. J. Muller, 1959: Evidence of the lower mutagenicity of chronic than intense radiation in *Drosophila* gonia. Science (Washington) **130**, 1923.

Purdom, C. E., and T. W. McSheehy, 1961: Radiation intensity and the induction of mutation. Internat. J. Rad. Biol. **3**, 579–586.

Russell, W. L., L. B. Russell, and M. B. Cupp, 1959: Dependence of mutation frequency on radiation dose rate in female mice. Proc. nat. Acad. Sci. U. S. **45**, 18–23.

Russell, W. L., L. B. Russell, and E. M. Kelly, 1958: Radiation dose rate and mutation frequency. Science (Washington) **128**, 1546–1550.

Russell, W. L., L. B. Russell, and E. M. Kelly, 1960: Dependence of mutation rate on radiation intensity. In: A. A. Buzzati-Traverso (Ed.), Symposium on the immediate and low level effects of ionizing radiation. Taylor & Francis Ltd., London, pp. 311–320.

Sävhagen, R., 1961a: The effect of oxygen concentration on the frequency of induced XO males and non-disjunction females after irradiation of *Drosophila* males. Hereditas **47**, 163–189.

Sävhagen, R., 1961b: The relationship between type of aberration and sensitivity pattern in irradiated *Drosophila melanogaster* males. Hereditas **47**, 190–196.

Sobels, F. H., 1955: The effect of pre-treatment with cyanide and azide on the rate of X-ray induced mutation in *Drosophila*. Z. indukt. Abstamm.- Vererb.lehre **86**, 339–404.

Sobels, F. H., 1958a: Presumptive indication of radiation-produced peroxide as shown by its genetic effects in *Drosophila*. In: R. Latarjet (Ed.), Les peroxydes en radiobiologie. Masson et Cie, Paris, January 1957, pp. 74–85.

Sobels, F. H., 1958b: The enhancing effect of post-treatment with cyanide on the mutagenic action of X-rays in *Drosophila*. Rad. Res. **9**, 186.

Sobels, F. H., 1958c: The effect of pre-treatment with cyanide on radiosensitivity in nitrogen and oxygen. Drosophila Inform. Serv. **32**, 159–161.

Sobels, F. H., 1960: Chemical steps involved in the production of mutations and chromosome aberrations by X-irradiation in *Drosophila*. 1. The effect of post-treatment with cyanide in relation to dose rate and oxygen tension. Internat. J. Rad. Biol. **2**, 68–90.

Sobels, F. H., 1961a: Modification of pre-mutational radiation damage in *Drosophila*. Proc. Symposium "Radiation effects and milieu", Montreux, June 1961, Strahlentherapie, supplement (in the press).

Sobels, F. H., 1961b: Chemische Beeinflussung des röntgeninduzierten Mutationsprozesses bei *Drosophila*. Naturwissenschaften **48**, 146–155.

Sobels, F. H., and A. D. Tates, 1961: Recovery from pre-mutational damage of X-irradiation in *Drosophila* spermatogenesis. Proc. Symposium on repair from damage, April 1961, Gatlinburgh, Tenn., in: J. Cellul Comp. Physiol., supplement to vol. **58**, no. 3, 189–196.

Tates, A. D., and F. H. Sobels, 1961: Modification of the genetic radiation damage in *Drosophila* by post-treatment with nitrogen and fractionation of the dose. Internat. J. Rad. Biol. **1**, 553–554.

Witkin, E. M., 1956: Time, temperature and protein synthesis: A study of ultraviolet-induced mutation in bacteria. Cold Spring Harbor Symposia on Quant. Biol. **21**, 123–140.

Witkin, E. M., 1959: Post-irradiation metabolism and the timing of ultra-violet induced mutations in bacteria. Proc. Tenth Internat. Congr. Genetics, Montreal, **2**, 280–299.

The Effect of Cupferron, Carbon monoxide, and 2.4-Dinitrophenol on the Production of Chromosomal Aberrations by X-rays in *Vicia faba*

B. A. Kihlman

Institute of Physiological Botany, University of Uppsala, Sweden

Introduction

In previous experiments, the frequencies of chromosomal aberrations produced by X-rays at low oxygen tensions in the root-tips of *Vicia faba* were found to be much higher in the presence than in the absence of respiratory inhibitors such as cupferron, cyanide, azide, and hydrogen sulfide (Kihlman, 1958 a, b). The inhibitors in question were active only in the presence of oxygen. The effect of the inhibitors appeared essentially to be that they caused the oxygen effect (Thoday and Read, 1947) to appear at much lower concentrations of oxygen than it does in untreated roots. In the latter, the frequencies of aberrations were practically unaffected by oxygen at concentrations below 5% in the gas phase, and the maximum effect was not obtained until the oxygen concentration exceeded 35%. In the presence of cupferron, which at pH 5.8 strongly inhibits respiration in a concentration of 400 μM, an enhancement of the effect was obtained when the gas phase contained as little as 0.1% oxygen, and at 2% oxygen the effect exceeded that produced at 100% oxygen by the same dose of X-rays in the absence of cupferron (Kihlman, 1959). It was suggested that the chemicals in question enhance the X-ray sensitivity at low oxygen tensions because they prevent the formation of an oxygen gradient in the root as a result of their inhibitory action on respiration.

Conclusive evidence in favour of the oxygen gradient hypothesis was still lacking, however. There also remained unexplained the fact that the frequencies of aberrations produced by a given dose of X-rays in pure oxygen is considerably higher in the presence than in the absence of cupferron (Kihlman, 1958b, 1959). If cupferron influenced X-ray sensitivity only by eliminating an oxygen gradient, the inhibitor should be ineffective at oxygen concentrations above 21%, since at these high oxygen concentrations, the gradient should have no influence on the X-ray sensitivity.

Finally, results have recently been published which appear to be inconsistent with the oxygen gradient hypothesis as an explanation for the cupferron effect. The results referred to are those obtained by Beatty and Beatty (1957, 1960) in their studies on the effect of carbon monoxide on the frequencies of "two hit" aberrations produced by X-rays in the microspores of *Tradescantia paludosa*.

Beatty and Beatty found that carbon monoxide was ineffective in the absence of oxygen but had an enhancing effect on X-ray sensitivity when oxygen was present. The curves obtained when the effects in the presence and absence of carbon monoxide were plotted against oxygen concentration are, in fact, very similar to those obtained in experiments in which bean roottips were exposed to X-rays at various concentrations of oxygen in the presence and absence of cupferron.

Beatty and Beatty, however, have not interpreted their results on the basis of the oxygen gradient hypothesis. Experiments by Wolff and Luippold (1955) have indicated that inhibitors of respiration and/or oxidative phosphorylation produce a delay in the rejoining of chromosome breaks, and these authors, therefore, conclude that respiration energy is needed for rejoining. If rejoining is delayed, the number of "two hit" aberrations produced by X-rays at low dose rates increases, since more breaks are open at the same time and able to interact. Beatty and Beatty believe that carbon monoxide is effective in their X-ray experiments because it produces a delay of the rejoining of breaks by inhibiting respiration. Their hypothesis is supported by their finding that carbon monoxide is as effective when given immediately after, as when given before and during, irradiation (Beatty and Beatty, 1957). Subsequently they found that post-treatments do not become ineffective until the period between irradiation and carbon monoxide treatment is 10 min or more.

Because of the facts discussed above, further studies on the effects of cupferron appeared necessary. Some comparative experiments with carbon monoxide as a respiratory inhibitor seemed particularly desirable, as well as the use of an uncoupling agent such as dinitrophenol to distinguish between effects due to an inhibition of respiration and effects due to an inhibition of phosphorylation. By using such concentrations of dinitrophenol which inhibit phosphorylation only, it should be possible to produce a delay of rejoining without influencing the oxygen gradient.

Material and methods

The experimental material consisted of lateral roots of the broad bean, *Vicia faba*. During treatment and irradiation, the seedlings were in *Lucite* vessels (Kihlman, 1959), with their roots immersed in M/150 or M/75 phosphate buffer solutions, the pH of which were 5.8 in the cupferron and dinitrophenol experiments, and 6.8 in the carbon monoxide experiments. Unless otherwise stated, the temperature during treatment was 19–21°C and the duration of the treatments 30 min. When the treatments were finished, the roots were washed thoroughly and the seedlings transferred to tubes containing tap water. During the recovery period, the seedlings were kept in the dark at 19°C. The roots were fixed 18 hr after the treatment in an alcohol-acetic acid mixture (3:1). Before fixation, the roots were treated for 4 hr with 0.05% colchicine. The slides were prepared and analysed as described previously (Kihlman, 1959).

X-rays were generated at 180 kV, 5 mA in a Müller therapy tube, type 71780, and radiation was filtered through 1 mm Al. The dose was measured by a Philips ionization chamber, type 37480.

The gases and gas mixtures used were oxygen-free nitrogen and 1% oxygen + 99% carbon monoxide from British Oxygen Co. Ltd., and 1% oxygen + 99% nitrogen, 2% oxygen + 98% nitrogen, and 100% oxygen from the Swedish Gasaccumulator Company.

The chemicals used were 2.4-dinitrophenol (DNP) from Fluka A. G., and cupferron or N-nitrosophenylhydroxylamine-ammonium salt from E. Merck A. G., Darmstadt.

Results and discussion

1. Cupferron and carbon monoxide

As appears from Table 1, the frequencies of chromosomal aberrations produced by X-rays in the dark were much higher when the gas phase consisted of 1% oxygen (O_2) and 99% carbon monoxide (CO) than when the gas phase was 1% O_2 + 99% nitrogen (N_2). Other experiments (KIHLMAN, 1961) showed that CO enhanced X-ray sensitivity only in the presence of O_2. The data in Table 1 further show that although the increase produced by CO was considerable,

Table 1

The effect of carbon monoxide*) and cupferron in the dark and in the light on the frequencies of chromosomal aberrations produced by X-rays in the presence of 1 % oxygen (14μ M in solution)

Treatment	X-ray dosage [r]	Dose rate [r/min]	Temperature during treatment [°C]	Number of metaphases analysed	Abnormal metaphases [%]	Aberrations per 100 cells I Isolocus breaks	II Exchanges	I + II
99% N_2 + 1% O_2, dark	92	13	14	200	13.5	8.5	3.0	11.5
99% CO + 1% O_2, dark	92	13	14	200	30.5	22.0	12.5	34.5
99% CO + 1% O_2, light	92	13	13	200	12.0	10.5	2.0	12.5
99% N_2 + 1% O_2, dark	70	14	19	100	8.0	9.0	0	9.0
99% N_2 + 1% O_2, dark 400 μM cupferron	70	14	19	100	36.0	37.0	7.0	44.0
99% N_2 + 1% O_2, light, 400 μM cupferron	70	14	19	100	41.0	24.0	19.0	43.0

*) Data from KIHLMAN, 1961

it was nevertheless smaller than that produced by 400 μM cupferron at the same oxygen concentration. In contrast to the effect of cupferron, that of CO was light reversible.

The light source used in the experiments was a Philips "Attralux" lamp, type 13378E/44, 24 V, 150 W. To prevent heating of the solution surrounding the roots, the treatment vessel was placed in a glass walled water bath. The lamp and the X ray tube were on opposite sides of the water bath.

The fact that the carbon monoxide effect was reversed by light, indicates that cytochrome oxidase is involved.

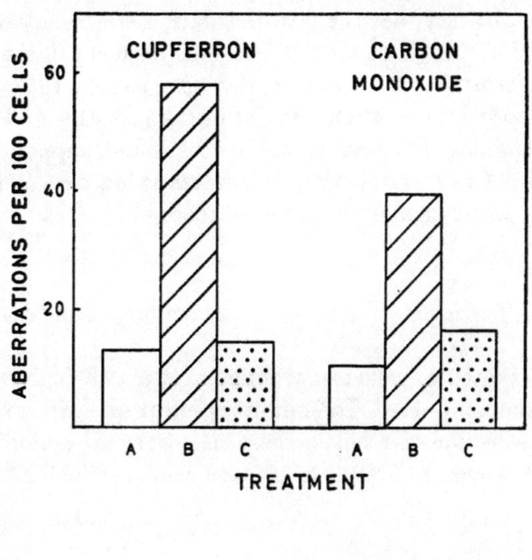

Fig. 1. The frequency of isolocus breaks + exchanges produced at 20 °C by 81 r of X-rays in the presence of 1 % oxygen and after the following treatments:
A = no inhibitor present before, during or after irradiation;
B = inhibitor before and during irradiation;
C = inhibitor immediately after irradiation. Inhibitor: 400 μM cupferron or 1000 μM carbon monoxide. (Diagram based on data from KIHLMAN, 1961)

In Fig. 1, the frequencies of aberrations produced by 81 r of X-rays under various experimental conditions are shown. In the "A" experiments, no inhibitor was present before, during, or after the irradiation. In the "B" experiments, cupferron or carbon monoxide were present during, as well as for $\frac{1}{2}$ hr before, irradiation. In the "C" experiments, irradiation was immediately followed by a $\frac{1}{2}$-hr treatment with cupferron or carbon monoxide. In all experiments, 1% O_2 was bubbled through the solution surrounding the roots for a $\frac{1}{2}$-hr period before and after irradiation as well as during the irradiation.

When cupferron was given as a post-treatment a concentrated O_2-free solution of cupferron was added with a pipette within 20 sec after the end of the irradiation. In the "B" experiment, the cupferron-containing solution in the vessel was replaced by a buffer solution within 40 sec after the end of the irradiation. In the experiments where CO was used as the inhibitor, the switch-over from a flow of gas consisting of 1% O_2 in N_2 to a flow of gas consisting of 1% O_2 in CO and vice versa was accomplished within 20 sec.

It is evident from Fig. 1 that the cupferron and carbon monoxide treatments were highly effective when given before and during irradiation. Post-treatments with the inhibitors in question, on the other hand, did not produce any significant increase in the frequencies of aberrations. The possibility of a post-treatment effect cannot be excluded on the basis of this evidence alone, however. It is possible that it takes some time for the inhibitors to become effective, although they are applied within a few seconds after the irradiation. This possibility, however, does not seem to be a very likely explanation for the lack of a post-treatment effect. It has been found that the frequencies of aberrations produced by 108 r of X-rays were almost doubled when the change from a gas-mixture consisting of 1% O_2 in N_2 to a gas-micture consisting of 1% O_2 in CO was made 70 sec before irradiation. To obtain the full CO effect, the change to the CO-containing gas had to be made between 2 and 5 min before irradiation (KIHLMAN, 1961). Similar experiments with cupferron as the inhibitor showed that the X-ray effect was doubled when the roots were pre-treated with the inhibitor for 1 min, but that pretreatments of more than 15 min were required to obtain the full cupferron effect. The results indicate that the inhibitor treatments become effective rapidly enough for a possible post-treatment effect to appear in the experiments presented in Fig. 1.

The results of the experiments described above support the conclusion previously drawn (KIHLMAN, 1959) that the enhancement of the frequencies of X-ray induced chromosomal aberrations, which is obtained with cupferron at low oxygen tensions, is due to the inhibitory action of cupferron on respiration.

The fact that the post-irradiation treatments with cupferron and CO were negative, indicated that, in contrast to the enhancement of the X-ray sensitivity produced in *Tradescantia* microspores by CO (BEATTY and BEATTY, 1957), the enhancement of the X-ray effect produced by cupferron and CO in the root-tips of *Vicia faba* cannot be due to an inhibition of rejoining. The results, on the other hand, support the idea that respiratory inhibitors enhance the X-ray sensitivity of bean roots at low oxygen tensions because they prevent the formation of an oxygen gradient in the roots.

2. Cupferron and 2.4-dinitrophenol

Of the so-called uncoupling agents, 2.4-dinitrophenol (DNP) is the best known and most commonly used. Its inhibitory effect on oxidative phosphorylation becomes evident at concentrations far below those necessary to inhibit respiration (LOOMIS and LIPMANN, 1948). However, if DNP is to be used as a specific inhibitor of oxidative phosphorylation, its effect on the respiratory activity of the experimental material has to be known. In Fig. 2 the effects of various concentrations of DNP on oxygen consumption of excised bean roots are expressed in per cent of the control.

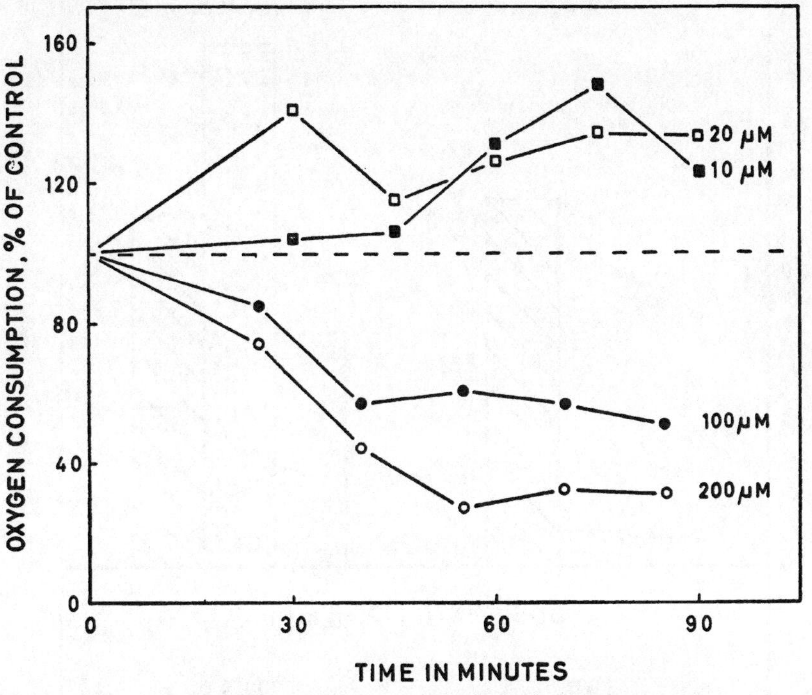

Fig. 2. The effect of various concentrations of 2.4-dinitrophenol at pH 5.8 on oxygen consumption of excised bean root tips

The oxygen uptake was determined at 22 °C in a Warburg apparatus. Twenty lateral roots-tips, 5 mm long, of *Vicia faba* were used in each vessel. The phosphate buffer used had a pH of 5.8, a concentration of 0.03 M and contained 0.5% glucose. The values in the figure give the oxygen consumption in glucose-buffer + inhibitor expressed as % of the consumption in the glucose-buffer controls. A correction was made for differences in the initial rates (before the addition of the inhibitor).

At concentrations of 20 or 10 μM, DNP stimulated respiration, whereas an inhibitory effect was produced by 100 and 200 μM of DNP. The inhibition increased with time, being 15% after a 25 min treatment with 100 μM DNP and about 40% after a 55 min treatment with the same concentration. The corresponding values for 200 μM DNP were 25 and 73%, respectively. The experiments show that if an effect on respiration is to be avoided, DNP should not be used in concentrations higher than 100 μM and the period of treatment should not exceed 30 min.

If the enhancing effect of cupferron on the frequencies of aberrations produced by X-rays at low oxygen tensions was due to its effect on oxidative phosphorylation, DNP should be as effective as cupferron. Furthermore, since inhibitors of oxidative phosphorylation are supposed to enhance the frequencies of two-hit aberrations by preventing breaks from rejoining, these inhibitors should become ineffective at high dose rates, when the majority of breaks are open simultaneously anyway. However, it is evident from Fig. 3 that a pronounced enhancement of the frequency of chromatid interchanges produced by 70 r of X-rays in the presence of 2% oxygen was obtained with cupferron both at low and high dose rates. It is also evident that a corresponding enhancement was not obtained with DNP. (The inhibitors were

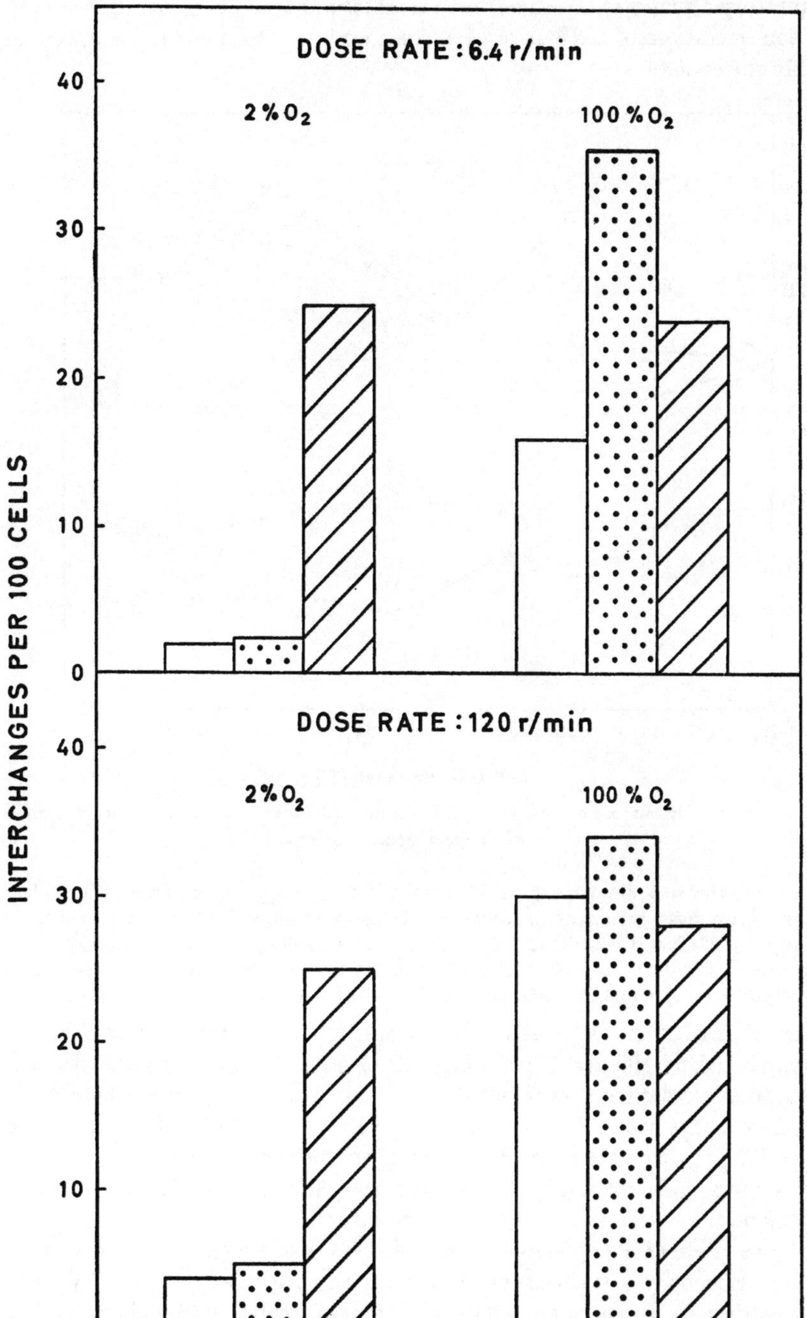

Fig. 3. The effect of $^1/_2$-hour treatments with 400 μM cupferron and 100 μM 2.4-dinitrophenol on the frequency of chromatid interchanges produced by 70 r of X-rays. After each treatment 200 cells were analysed. C = control; DNP = 2.4-dinitrophenol; CF = cupferron

present during irradiation as well as before irradiation for a period long enough to give a total inhibitor-treatment lasting 30 min.) These results strongly suggest that the effect of cupferron on X-ray sensitivity at low oxygen tensions is not due to its effect on oxidative phosphorylation.

Table 2

The effect of pre-treatments with 2.4-dinitrophenol (DNP) on the frequencies of chromosomal aberrations produced by 2-hour treatments with 10^{-2} M 8-ethoxycaffeine (EOC) at 12 °C and by 150 r of X-rays at 20 °C. Gas phase: air in the EOC-experiments; 2% oxygen in the X-ray experiments. After each treatment, 100 metaphases were analysed

Pre-treatment	Mutagen	Abnormal metaphases [%]	Aberrations per 100 cells I Isolocus breaks	II Exchanges	I + II
Buffer, pH 5.8, 1 hr	EOC	26	25	4	29
Buffer, pH 5.8, 1 hr	EOC	37	33	5	38
50 μM DNP, pH 5.8, 1 hr	EOC	1	1	0	1
50 μM DNP, pH 5.8, 1 hr	EOC	3	2	1	3
Buffer, pH 5.8, 1 hr	X-rays	34	26	10	36
Buffer, pH 5.8, 1 hr	X-rays	27	21	10	31
50 μM DNP, pH 5.8, 1 hr	X-rays	33	25	17	42
50 μM DNP, pH 5.8, 1 hr	X-rays	30	17	20	37

A similar conclusion can be drawn from the results reported in Table 2. The radiomimetic effect of 8-ethoxycaffeine (EOC), which appears to be dependent on oxidative phosphorylation (KIHLMAN, 1955), was almost completely inhibited by a 1-hour pre-treatment with 50 μM DNP. This indicates that the DNP pre-treatment effectively inhibited the formation of energy-rich phosphate. If the enhancement of X-ray sensitivity by cupferron was due to an inhibition of oxidative phosphorylation, such a pre-treatment would be expected to produce a considerable increase of the X-ray effect at 2% oxygen. As appears in Table 2, this was not the case. There was, however, an increase in the number of exchanges, which indicates that the inhibition of phosphorylation is not entirely without influence on the X-ray effect. The nature of this influence appears more clearly in Fig. 3.

When the gas phase consists of pure oxygen, an enhancement of the frequency of X-ray induced exchanges can hardly be attributed to the prevention of the formation of an oxygen gradient. And yet cupferron enhanced the frequency of interchanges produced at low dose rates (Fig. 3). When the dose rate was increased, the enhacement disappeared. The disappearance was due not to a decrease of the number of exchanges produced in the presence of cupferron, but to an increase of the number of exchanges produced in the absence of the inhibitor. This is exactly what would be expected if cupferron eliminated the dose rate (intensity) effect by preventing the repair of chromosome damage. That this, indeed may be the explanation is further indicated by the fact that in this case DNP not only acted in the same way, but that it was even more effective (Fig. 3).

In Table 3 the results of other similar X-ray experiments are summarized. The results are essentially the same as those shown in Fig. 3. At low dose rates the frequency of exchanges is markedly enhanced both by cupferron and by DNP; at high dose rates the enhancement disappears in the case of cupferron and is much less marked in the case of DNP. The fact that the enhancement does not disappear completely in the case of DNP indicates that in the absence of an effective inhibitor, some repair occurs during the period of irradiation even at these comparatively high dose rates.

Table 3

The effect of cupferron (CF) and 2.4-dinitrophenol (DNP) on the frequencies of chromosomal aberrations produced by X-rays at "low" and "high" dose rates. Gas phase: 100% oxygen

Chemical Name	Chemical Concentration [μ]	X-ray dosage [r]	Duration of irradiation [min]	Number of metaphases analysed	Abnormal metaphases [%]	Aberrations per 100 cells I Isolocus breaks	Aberrations per 100 cells II Exchanges	Aberrations per 100 cells I + II
—	—	81	11	200	47.5	37.0	22.5	59.5
CF	400	81	11	200	58.0	35.5	44.0	79.5
—	—	65	$9^{1}/_{2}$	200	40.0	35.0	9.5	44.5
DNP	100	65	$9^{1}/_{2}$	200	55.0	40.5	33.0	73.5
—	—	81	$^{3}/_{4}$	200	57.5	53.5	30.0	83.5
CF	400	81	$^{3}/_{4}$	200	63.0	58.0	28.0	86.0
—	—	65	$^{1}/_{2}$	200	47.5	42.0	21.5	63.5
DNP	100	65	$^{1}/_{2}$	200	59.5	41.5	33.5	75.0
CF	400	—	—	100	1.0	1.0	0	1.0
DNP	100	—	—	100	0	0	0	0

It may be mentioned that a dose rate effect within such a short period as 10 min has previously been reported by Catcheside et al. (1946) for chromatid exchanges in *Tradescantia* microspores, by Wolff and Luippold (1956) for chromosome exchanges in *Vicia faba*, and by Cohn (1958) for chromosome exchanges in *Allium cepa*. The studies of Wolff (1954) and Savage et al. (1960) have shown that X-ray irradiation produces in *Vicia* another type of damage, which is capable of reacting under exchange formation for a period of a least 1 hr.

Another fact evident in Table 3 is that the isolocus breaks do not seem to be affected by the inhibitor treatments. But since only two-hit aberrations are expected to be influenced, this is not so surprising.

Thus the facts obtained appear to be in agreement with the hypothesis that cupferron and DNP enhance the frequency of X-ray-induced exchanges at high oxygen tensions and low dose rates, because they inhibit the formation of the energy-rich phophate needed for repair of chromosome damage. However, another possibility, which cannot yet be excluded is that the enhancement is a reflection of an effect of the inhibitors on the rate of cell division. If cupferron and DNP inhibited mitosis markedly, a proportion of the cells scored in the treated roots would have been irradiated at a stage closer to metaphase than the corresponding cells in the control roots. Since at a constant dose a higher frequency of exchanges is produced in cells irradiated in late interphase than in cells irradiated in early interphase (Revell, 1961), inhibition of mitosis would result in an apparent enhancement of the production of exchanges. However, in previous experiments with cupferron, treatments of the kind used here did not produce any marked inhibitory effect on mitosis (Kihlman, 1959). Another fact which does not seem to favour this interpretation is the disappearance of the inhibition at higher dose rates. More experimental evidence is needed, however, before any definite conclusions can be drawn.

Summary and Conclusions

The effects of cupferron, carbon monoxide and 2.4-dinitrophenol (DNP) on the frequencies of chromatid aberrations produced by X-rays in the root-tips of *Vicia faba* were studied.

At an oxygen concentration of 1% in the gas phase, cupferron and carbon monoxide produced a strong increase in the frequencies of aberrations when they were present before and during irradiation. No enhancement was obtained when the inhibitor treatments were applied immediately after the X-ray irradiation. These results agree with the idea that respiratory inhibitors enhance the X-ray sensitivity of bean roots at low oxygen tensions because they prevent the formation of an oxygen gradient in the roots.

In contrast to the cupferron effect, the enhancement produced by carbon monoxide was sensitive to light and occurred only in the dark.

At a dose rate of 6.4 r/min. and at an oxygen tension of 100% in the gas phase, both cupferron and DNP enhanced considerably the frequency of chromatid interchanges produced by a given dose of X-rays. When the dose rate was increased to 120 r/min., the effect obtained in the absence of the inhibitors increased, whereas that obtained in their presence remained constant. As a consequence, the enhancement produced by the inhibitors at low dose rates disappeared completely (cupferron) or was markedly reduced (DNP) at high dose rates.

In contrast to cupferron, DNP did not produce a strong enhancement of the X-ray effect at low oxygen tensions (2% in the gas phase). The enhancement produced by cupferron in the presence of 2% oxygen, was obtained both at low and high dose rates and was equally strong for all types of aberrations.

It is concluded that the very strong enhancement of the X-ray effect which cupferron produced at low oxygen tensions is due to its inhibitory effect on respiration. The ineffectiveness of DNP in these experiments indicates that phosphorylation is not involved. The results obtained are in agreement with the idea that cupferron enhances the effect of X-rays by eliminating an oxygen gradient, which is supposed to exist in the roots at low oxygen tensions.

In contrast to the strong enhancement of the X-ray effect produced by cupferron at low oxygen tensions, the considerably less marked enhancement obtained at high oxygen tensions and low dose rates could well be due to the inhibitory action of cupferron on oxidative phosphorylation. The finding that DNP acts in the same way as cupferron in these experiments is in agreement with this interpretation. The effect of cupferron and DNP on X-ray induced chromosome damage at high oxygen tension appears to be mainly that they cause the dose rate effect to disappear. This is in agreement with the idea that oxidative phosphorylation is needed for the repair of chromosome damage.

Acknowledgement

This investigation was supported by a grant from the Swedish Natural Science Research Council. I am indebted to Dr. J. D. Eisen for valuable discussions and critical reading of the manuscript. Thanks are also due to Mrs. Eva Olander for skilful technical assistance.

References

Beatty, A. V., and J. W. Beatty, 1957: Chromosome breakage and rejoining in *Tradescantia* microspores. Amer. J. Bot. **44**, 778–783.

Beatty, A. V., and J. W. Beatty, 1960: Postirradiative effects on chromosomal aberrations in *Tradescantia* microspores. Genetics **45**, 331–344.

Catcheside, D. G., D. E. Lea, and J. M. Thoday, 1946: The production of chromosome structural changes in *Tradescantia* microspores in relation to dosage, intensity and temperature. J. Genet. **47**, 137–149.

Cohn, N. S., 1958: An analysis of the rejoining of X-ray-induced broken ends of chromosomes in the root tips of *Allium cepa*. Genetics **43**, 362–373.

Kihlman, B. A., 1955: Oxygen and the production of chromosome aberrations by chemicals and X-rays. Hereditas **41**, 384–404.

Kihlman, B. A., 1958a: The effect of oxygen, nitric oxide, and respiratory inhibitors on the production of chromosome aberrations by X-rays. Exper. Cell Res. **14**, 639–642.

Kihlman, B. A., 1958b: Respiration and radiosensitivity of broad bean roots. Nature (London) **182**, 730–731.

Kihlman, B. A., 1959: The effect of respiratory inhibitors and chelating agents on the frequencies of chromosomal aberrations produced by X-rays in *Vicia*. J. Biophys. Biochem. Cytol. **5**, 479–490.

Kihlman, B. A., 1961: Cytological effects of phenylnitrosamines. III. The effect on X-ray sensitivity at low oxygen tensions. Radiation Botany **1**, 51–60.

Loomis, W. F., and F. Lipmann, 1948: Reversible inhibition of the coupling between phosphorylation and oxidation. J. Biol. Chem. **173**, 807–808.

Revell, S. H., 1961: An attempt at continuous metaphase estimation of chromatid and chromosome aberration frequencies in broad bean root-meristem cells in the period 2 to 23 hours after 50 r of X-rays. In: "Effects of Ionizing Radiations on Seeds", p. 229–242. Int. Atom. Energy Agency. Vienna 1961.

Savage, J. R. K., G. J. Neary, and H. J. Evans, 1960: The rejoining time of chromatid breaks induced by gamma radiation in *Vicia faba* root tips at 3 °C. J. Biophys. Biochem. Cytol. **7**, 79–85.

Thoday, J. M., and J. Read, 1947: Effect of oxygen on the frequency of chromosome aberrations produced by X-rays. Nature (London) **160**, 608.

Wolff, S., 1954: Delay of chromosome rejoining in *Vicia faba* induced by irradiation. Nature (London) **173**, 501–502.

Wolff, S., and H. E. Luippold, 1955: Metabolism of chromosome-break rejoining. Science (Washington) **122**, 231–232.

Wolff, S., and H. E. Luippold, 1956: The production of two chemically different types of chromosomal breaks by ionizing radiations. Proc. nat. Acad. Sci. U. S. **42**, 510–514.

Analysis of Radiation-Induced Genetic Damage in Seeds[1]

R. A. Nilan, C. F. Konzak, E. Froese-Gertzen, and N. S. Rao

Department of Agronomy, Washington State University, Pullman, Washington

The induction of beneficial mutations in the breeding of agricultural crops is of major interest in plant science centers throughout the world. These mutations have been induced by a variety of physical and chemical mutagens. Of these, radiation perhaps is the most well known; and in this field, significant advances already have been achieved through investigations of the process which induces genetic changes in seeds. However, in order to obtain a greater abundance and variety of mutations, work with these physical and chemical mutagens must be continued.

At Pullman, Washington, the induced-mutation process in seeds is being investigated by two distinct but related approaches. One is through studies of the influence of environmental factors, chiefly oxygen, on radiation-induced biological damage. The second approach is by comparative studies of the biological effects of certain alkylating chemical mutagens.

Results of the pre- and post-radiation environment studies now can be correlated with the effects of environmental factors on the production and decay of free-radicals. These radicals have been measured by electron spin resonance techniques. The combined biological and physical-chemical analysis of radiation damage in seeds has augmented our understanding of the initial events that lead to genetic damage following radiation absorption. This understanding of the mechanisms of radiation action has led to a high degree of control of radiation-induced damage in seeds. So far, this control is possible only in seeds.

Studies have shown that the chemical mutagens, diethyl sulfate and ethyl methane sulfonate, are not typical of radiomimetic agents. Unlike most radiomimetic agents, these chemicals induce a low level of cytogenetic effects that are similar to those obtained in radiation experiments conducted under special environmental conditions. Studies of the effects of these chemicals alone and in combination with X-rays already have provided new information on the nature and control of the induced-mutation process.

---

1) Scientific Paper No. 2159 Washington Agricultural Experiment Stations, Pullman, Washington. Research supported by Projects 1068 and 4068 and by U.S. Atomic Energy Commission Contract AT (45-1)-353, U.S. Public Health Service Grant A-2184, and funds provided for medical and biological research by the State of Washington Initiative Measure 171.

Review of Literature

Oxygen effect

As some important concepts have evolved from experiments on the oxygen effect in seeds, a brief review of recent developments in seed radiobiology is in order. Several years ago, it was found that as seeds decreased in moisture from 15% to 4% their radiosensitivity increased (EHRENBERG and NYBOM, 1954; CALDECOTT, 1954, 1955). Though difficult to interpret at the time, this phenomenon now appears to be determined by the availability and mobility of oxygen in relation to the life-time of induced radicals in seeds at the different moisture levels.

The finding of biological aftereffects in irradiated seeds (GUSTAFSSON, 1938; EHRENBERG and LUNDQVIST, 1957; CALDECOTT, JOHNSON, NORTH, and KONZAK, 1957; ADAMS and NILAN, 1958; JACKSON, 1958; SIRE and NILAN, 1958; CURTIS, DELIHAS, CALDECOTT, and KONZAK, 1958; CALDECOTT, 1959) indicated that under certain experimental conditions, processes causing radiation damage were maintained in the seed for a considerable time after the radiation treatment. This phenomenon is now believed to represent a prolongation of the early radiation-induced changes in the cell. Furthermore, it permits for an extended period detailed investigations of the influence of environmental factors such as oxygen on these early events. In a highly metabolic or so-called "wet" system these events and environmental interactions take place so quickly that there is little chance to identify or experiment with them.

Biological aftereffects of radiation were detected in dry (4–10% moisture), but not in moist (13–16% moisture) seeds (CURTIS, DELIHAS, CALDECOTT, and KONZAK, 1958). It was further shown that seeds stored at dry ice temperatures (–78 °C) maintained such aftereffects for extended periods (KONZAK, CURTIS, DELIHAS, and NILAN, 1960; CALDECOTT, 1961). In addition, oxygen increased these after-effects in dry seeds, but no change resulted from oxygen post-treatment of moist seeds. Thus, there was established a definite but undetermined relationship between oxygen and moisture in the development of aftereffects in irradiated seeds.

A physical-chemical basis for biological aftereffects was introduced when free radicals with relatively long half-lives were detected in seeds after irradiation (ZIMMER, EHRENBERG, and EHRENBERG, 1957; EHRENBERG and EHRENBERG, 1958; KIRBY-SMITH and RANDOLPH, 1960; CONGER and RANDOLPH, 1959; KLINGMÜLLER, LANE, SAXENA, and INGRAM, 1959; ZIMMER, 1959; MÜLLER and ZIMMER, 1961; A. EHRENBERG, 1961). Both the production and decay of the radicals appeared to be influenced by the environment of the seed during and after irradiation. A close interplay between the free radicals and biological aftereffects in seeds was demonstrated by the Swedish group (SPARRMAN, EHRENBERG, and EHRENBERG, 1959). They were able to conclude from experiments with nitric oxide (NO), a radical scavenger, that a large part of the biological aftereffects in seeds was due to radiation-induced free radicals. In fact, the data from the electron spin resonance studies remarkably paralleled those obtained from the studies of biological aftereffects of irradiated seeds given different oxygen, moisture, and temperature treatments. CONGER (1961) has presented data more recently that further support the opinion that there is a correlation between the induction and decay of longlived radicals and the biological aftereffects in seeds. Thus, radiation-induced free radicals are now considered an important component of one of the pathways of radiation damage in the seeds. Free radicals also have been shown to be important in the initiation of damage in dry spores of bacteria (POWERS, WEBB, and EHRET, 1960; POWERS, EHRET, and SMALLER, 1961).

The radical concept, supported by recent information on radical interactions provided by chemists (BASS and BROIDA, 1960; BLOIS, BROWN, LEMMON, LINDBLOM, and WEISSBLUTH,

1961), has provided a useful and so far fruitful basis for developing a hypothesis concerning the actions of oxygen and moisture in the development of radiation damage in seeds.

This hypothesis (L. EHRENBERG, 1959, 1960; KONZAK, CURTIS, DELIHAS, and NILAN, 1960; NILAN, KONZAK, LEGAULT, and HARLE, 1961; A. EHRENBERG, 1961) maintains that the oxygen-radical interaction in the irradiated dry seed produces reactive organic peroxy radicals which initiate radiation damage in the cell. The level of water in the seed influences this prime interaction by governing the mobility of the radicals and oxygen molecules in either radical oxygen or radical-radical interactions.

Chemical mutagens

Some of the biological effects, especially the genetic effects, of diethyl sulfate in barley seeds have been described previously (HEINER, KONZAK, NILAN, and LEGAULT, 1960; KONZAK, NILAN, LEGAULT, and HEINER, 1961; KONZAK, NILAN, HARLE, and HEINER, 1961). The action of ethyl methane sulfonate appears to be similar to that of diethyl sulfate (Table 1).

It is of particular interest that compared with X-rays, diethyl sulfate and ethyl methane sulfonate induce relatively higher frequencies of mutations but far fewer visible chromosome aberrations. The aberrations induced by the chemicals are mostly of the intra- rather than of the inter-chromosomal type (NILAN and KONZAK, 1961). Iso-chromosomes appear to be among the induced changes.

Table 1

A comparison of the biological effects of ethyl methane sulfonate and gamma rays on barley

Treatment			Seedling height [% of control]	Plant survival [%]	Spike fertility [%]	Mutated Plants [%]	Mutated Spikes [%]	Chromosome aberrations Bridges [per cell]	Chromosome aberrations Fragments [per cell]
Ethyl methane sulfonate 0.05 M	1 hr	30 °C	77.3	54.6	68.2	30.6	9.2	0.01	0.01
Ethyl methane sulfonate 0.05 M	2 hr	30 °C	59.4	43.0	36.9	82.5	30.6	0.00	0.05
H_2O	1 hr	30 °C	100.0	82.4	79.5	1.5	0.4	0.01	0.15
Gamma rays	60 Kr		76.8	75.8	—	32.5	12.6	0.52	3.4
Gamma rays	80 Kr		63.3	52.6	—	26.6	12.6	0.42	6.0

The high degree of specificity of these chemicals for mutation induction may be accounted for in part by the results of BAUTZ and FREESE (1960). In phage T_4 ethyl ethane sulfonate induces both transitions and transversions; mutations which are caused by single base pair changes in DNA. Such specific changes in the base pairs presumably do not cause chromosome aberrations.

Partitioning the types of genetic damage by chemicals provides a breakthrough toward understanding the induced mutation process in seeds. This understanding of the action of the chemicals will supply information and data that should lead to greater progress in the partitioning of radiation-induced damage. Attempts are being made to partition genetic damage in irradiated seeds by controlling the effects of certain environmental factors and by supplementary or post-treatments. However, only partial success in radiation experiments has been obtained with such treatments as heat shock (KONZAK, CURTIS, DELIHAS, and NILAN, 1960) and low temperature (NILAN, 1954) and oxygen pre- and post-treatments.

The effects of after-treatments and the analyses of the in vitro and in vivo reaction kinetics of diethyl sulfate and ethyl methane sulfonate are now under study in our laboratory. We hope to learn more about the action of these agents and to improve the precision with which treatments can be repeated. Preliminary results from attempts to obtain a measure of dosimetry with these chemicals have been described recently in other papers from our laboratory (Konzak, Nilan, Harle, and Heiner, 1961; Heiner, Konzak, Nilan, and Bartels, 1961).

A review of the literature reveals that very few systematic studies of the effects of treatments combining X-rays and chemical mutagens have been conducted. Oster (1958) has reviewed the pertinent literature on the subject and has presented data on the effects of X-rays in combination with urethane or mustard gas in *D. melanogaster* spermatozoa. The combination treatments had an additive effect on the formation of sex-linked lethal mutations and chromosome breaks but a synergistic effect on the formation of translocations. The Moutschen-Dahmens (1961) have studied the combined effect of X-rays and myleran on the chromosomes of *Vicia faba* and found that the chemical considerably reduced the chromosome aberrations induced by the X-rays. A synergistic effect of the two mutagens only occurred for the frequency of chromosome gaps.

Materials and methods

Barley seeds (*Hordeum vulgare* var. Himalaya) have been used in all of the experiments described herein. Barley is most useful for analyzing genetic effects of mutagen treatments since it has relatively large chromosomes and numerous well-defined heritable seedling characteristics. Furthermore, this crop plant is easily cultured and large populations can be grown both in the field and in the greenhouse.

At least ten different biological effects can be measured following mutagen treatment of the seed. The techniques for measuring these effects have been described in several papers (cf. Nilan and Konzak, 1961; Konzak, Nilan, Harle, and Heiner, 1961).

The effects measured include the frequencies of chromosome bridges and fragments (rods, isodiametrics and dots) in shoot-tips of the mutagen treated seed, of chromosome interchanges in the M_1 plant and the frequencies and proportions of chlorophyll-deficient seedling mutations among M_2 progeny. These and several other measures including rate of growth of M_1 seedlings and survival of M_1 plants furnish a broad basis for interpreting the action of the physical and chemical mutagens in the system. By measuring both chromosomal aberrations and genetic mutations induced by the same treatment, a more nearly complete understanding of the nature and control of the total genetic damage can be obtained.

In many mutagen treatments, especially those with radiation, several criteria of damage are affected to much the same degree. However, those treatments that cause a differential response to the criteria are most valuable since they provide much information from which the nature of the mechanisms that lead to mutagen damage in the seeds can be surmised.

The seeds used in the experiments are genetically uniform because they arose from a single plant and are regularly repurified. The variety Himalaya is hulless, hence complications due to the effect of hulls in mutagen treatments and in subsequent growth of shoots and seedlings are avoided. At Pullman, non-weathered disease-free seeds are usually produced and they are hand-threshed and sieved in order to secure non-damaged and uniformly-sized seeds that produce even germination and growth.

Oxygen effect

For the analysis of the oxygen effect in irradiated cells, the seed has several advantages. When resting and dry, it is like an organic crystalline substance, in which the influence of the oxygen can be studied before, during, or after irradiation without the complication of metabolism. Furthermore, the influence of oxygen can be analyzed under a wide and well-controlled range of moisture and temperature conditions. The seed survives high pressures (2000 lbs.) or complete removal of oxygen for extended periods. These severe environments cause little or no biological damage to the seeds. Upon controlled rehydration, the life processes of the seed are reactivated, whereupon the biological effects of treatments applied in the inactive state can be measured. A comprehensive and more detailed discussion of the advantages of the seed for this phase of radiobiological research has been published elsewhere (Nilan, Konzak, Legault, and Harle, 1961; Konzak, Nilan, Harle, and Heiner, 1961).

Techniques for controlling the levels of oxygen and moisture content of the seed before, during, and after irradiation, for post-irradiation hydration in the presence or absence of oxygen, and for measuring the moisture content of the seed have been previously described (Nilan, Konzak, Legault, and Harle, 1961; Konzak, Nilan, Harle, and Heiner, 1961; Nilan, Konzak, Harle, and Heiner, 1962). It should be emphasized here that to obtain and maintain seeds relatively free of oxygen has been most difficult. The best technique at present is to dry seeds over calcium oxide in a partial vacuum, evacuate them in glass tubes on a high vacuum system and irradiate them in sealed tubes. Transfer of the seeds between tubes and desiccators is performed under a controlled gaseous atmosphere in a specially designed chamber. The seeds are then rehydrated in water freed from oxygen by boiling and by bubbling with nitrogen gas that has first passed through a Hoskins furnace to remove the traces of oxygen.

Irradiation

All of the seeds in the radiation experiments (except Table 1) described herein, were irradiated using a Mitchell (Seifert) Isovolt 300 kV Constant Potential X-Ray Machine. The machine operated at 300 kV and 10 mA and delivered anywhere from 175 to 1000 r per minute depending upon the diameter of the irradiated sample.

Combined chemical and X-ray treatment

Seeds with 13% moisture were frozen (−78°C) and exposed to 10, 20, 30, or 40 kr of X-rays and then hydrated in 0.0076 M diethyl sulfate for two hours. The concentration was maintained by replacing the treatment solution after 30 minutes and again after 60 minutes of the two-hour treatment. In addition, the irradiated seeds were also rehydrated in the solution obtained after hydrolysis of diethyl sulfate for four days at room temperature. At this time no diethyl sulfate per se remained (cf. Konzak, Nilan, Harle, and Heiner, 1961).

Results and discussion

The oxygen effect

Storage of the oxygen response

Under certain environmental conditions the seed will "store" its reactive sites and its response to oxygen indefinitely. Seeds held at dry-ice temperatures (−78°C) retain their ability to respond to post-oxygen hydration (23°C) for several days (Caldecott, 1961). A recent experiment in our laboratory has demonstrated that gamma-irradiated seeds maintained under vacuum, and in very dry conditions (below 2% moisture), still showed a marked response to oxygen during rehydration after one year of storage.

The relation of moisture to the oxygen effect

As indicated previously, post-irradiation oxygen treatment during rehydration is effective in enhancing radiation damage in seeds of 4% moisture but not in those of 13% moisture. It was of concern, then, to determine the range of moisture levels in which pre-irradiation storage in oxygen is effective. In this way, additional data would be obtained for a further understanding of the oxygen and radical mobility.

A recent experiment has demonstrated that the critical moisture content at which aerobic rehydration post-treatment may be effective is between 10 and 12%. For this experiment, seeds with 10, 12, 14, and 16% moisture were stored in oxygen or nitrogen, frozen, and irradiated at 15, 30, 45, and 75 kr, then rehydrated in nitrogen-bubbled water.

The pre-irradiation storage in oxygen had some effect in seeds of all moisture levels tested. The greatest effect was in the 10% moisture seeds, whereas the seeds with approximately 12, 14, and 16% moisture exhibited a definite but weaker oxygen effect (Fig. 1). Similarly, seeds with 10% moisture, stored first in nitrogen and exposed to 45 kr, showed a definite post-radiation oxygen effect. No post-oxygen treatment effect was observed in the 12, 14, and 16% moisture seeds.

Thus, it appears that the moisture content between 10 and 12% is the critical point at which oxygen pre- and post-irradiation effects are revealed. Presumably, this narrow moisture

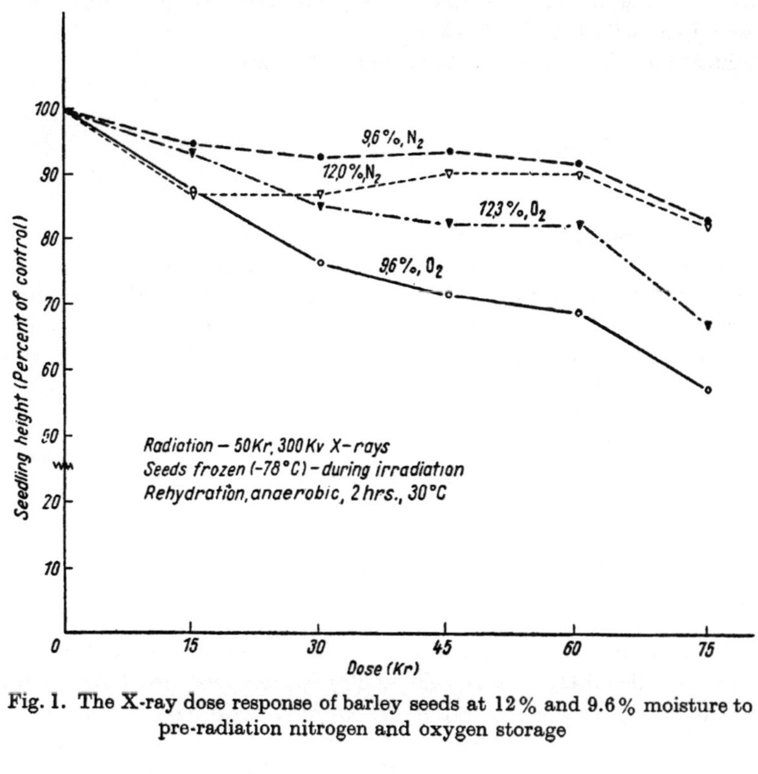

Fig. 1. The X-ray dose response of barley seeds at 12% and 9.6% moisture to pre-radiation nitrogen and oxygen storage

range permits the degree of molecular mobility that provides equal opportunity for either radical-oxygen (damaging) or radical-radical or radical-sulfydryl (non damaging) combinations.

The fact that an effect of pre-radiation storage in oxygen has been observed for moist seeds (14 and 16%) indicates that when oxygen molecules are present at the time of irradiation, they interact with the reactive sites in a high moisture environment. Evidently, moist seeds that are free of oxygen at the time of radiation do not show an oxygen post-radiation effect because the reactive sites are inactivated before the oxygen has a chance to be effective. Another possible explanation is that the moist seeds use up the oxygen in respiration, leaving none available for radical-oxygen interaction. This possibility currently is being tested in our laboratory by the use of respiration inhibitors.

In moist seeds, it has recently been demonstrated, however, that high-pressure oxygen as a post-radiation treatment is as effective as pre-radiation storage in an oxygen atmosphere. Seeds with 12% moisture were stored in oxygen or nitrogen, frozen (−78°C) and exposed to 50 kr. of X-rays. The seeds were then stored in oxygen at 2000 pounds pressure or in air for four hours at room temperature and rehydrated anaerobically at 30°C. The 2000 pounds oxygen post-treatment significantly increased the damage in the nitrogen pre-stored seeds (Fig. 2). This use of oxygen pressure appears to provide an additional means for obtaining data on the oxygen-responsive radicals and their half life under conditions prevailing in the seeds after irradiation.

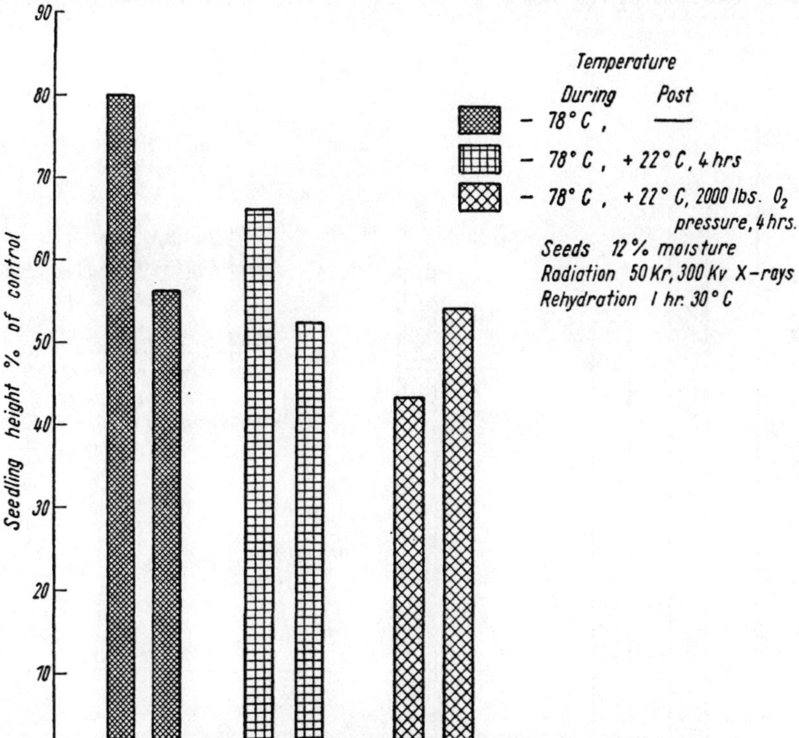

Fig. 2. The X-ray response of barley seeds prestored in oxygen and nitrogen and post-treated at 22 °C in air or under 2000 lbs. oxygen pressure for 4 hours

Other experiments have provided additional information about the interrelation of oxygen and moisture in controlling the fate of radiation-induced reactive sites. For instance, a definite oxygen dilution effect has been demonstrated in very dry seeds. The details of this experiment have been presented and discussed elsewhere (Nilan, Konzak, Legault, and Harle, 1961). The results showed that the growth of seedlings from irradiated seeds decreased with the increasing oxygen concentration in the hydration water.

Further studies have demonstrated the response of irradiated dry seeds (6% moisture) to post-irradiation oxygen treatment when the seeds have been soaked in water and then dried back to their original moisture prior to radiation treatments. A companion experiment also was conducted in which dry seeds were first irradiated, soaked in oxygen-free water and dried back to the original moisture content and then bubbled in oxygen and nitrogen. The first experiments described earlier (Nilan, Konzak, Legault, and Harle, 1961) showed that seeds soaked and then dried under oxygen-free conditions were more sensitive to oxygen than seeds of similar moisture content that were not soaked.

In a more recent study, dry seeds (4% moisture) were irradiated at −78°C and then rehydrated anaerobically and soaked in nitrogen-bubbled water for two hours. The seeds were then dried to 4% moisture under oxygen-free conditions, and rehydrated in oxygen and nitrogen. Thus, the seeds were dry at the time of the irradiation and as they went into the oxygen treatment. No oxygen effect occurred, however, in those seeds hydrated anaerobically before drying and then hydrated aerobically (Fig. 3). The anaerobic hydration apparently inactivated the reactive sites so the subsequent aerobic rehydration had no damaging effect.

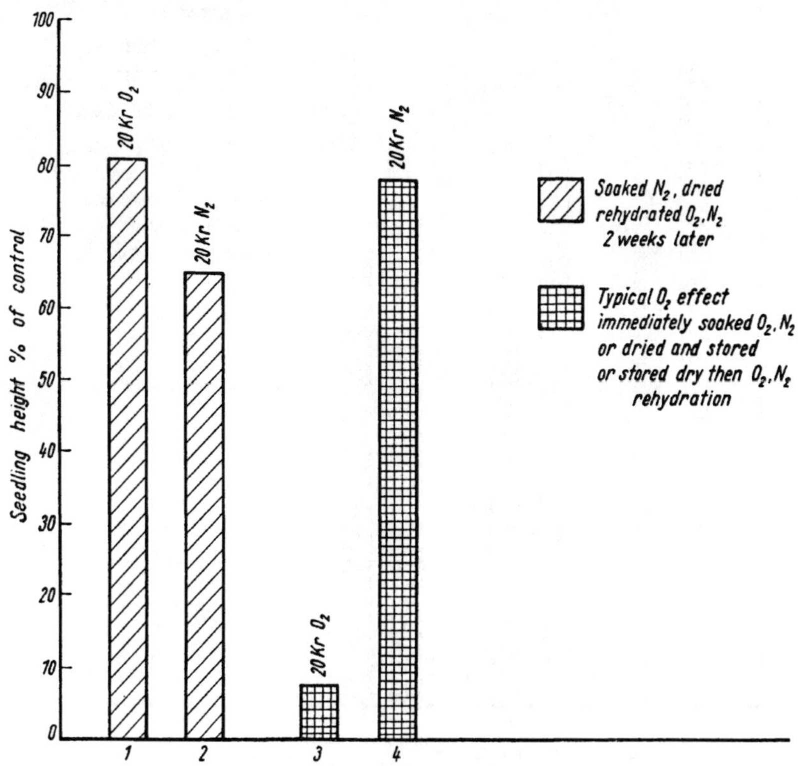

Fig. 3. The elimination of the oxygen sensitivity in dry barley seeds (4% moisture) by post-radiation anaerobic hydration prior to drying

The roles of oxygen and other factors in initial radiation events

From conclusions reached in previous papers (cf. Nilan, Konzak, Legault, and Harle, 1961; A. Ehrenberg, 1961; Conger, 1961) and from the results presented here, it is now possible to indicate more precisely the roles of several factors in the early events following photon absorption that lead to biological damage in the seed. These factors apparently govern the production and fate of the free radicals produced by the ionizing radiation.

According to present views, the radicals produced in the seed by radiation may react in the following ways:

1. Radical interaction with oxygen to form organic peroxy radicals. This is apparently the chief reaction in a pathway to biological damage in the seed.
2. Radical interaction with SH or NH_2 compounds, radical recombination, or radical breakdown. These reactions dispose of the radicals in a harmless fashion.

The present knowledge of these interactions permits certain interpretations of the role of oxygen and other factors in the production and fate of the radicals:

a) Kind and dose of radiation: Only with the sparsely ionizing radiations, as beta-, gamma-, and X-rays, can a large and distinct oxygen effect be demonstrated. Presumably, these radiations produce radicals far enough apart so that at least some are available for $R^{\cdot} + O_2$ interactions. The dose of these radiations determines the concentration of available radicals.
The densely ionizing neutrons, on the other hand, show little oxygen effect. This is probably because the radicals are produced so closely together that they recombine ($R^{\cdot} + R^{\cdot}$) before they interact with oxygen.

b) Atmosphere: governs the concentration of oxygen available for the $R^{\cdot} + O_2$ interaction.

c) Moisture: provides the medium for the radical interactions and controls solubility and mobility of oxygen molecules and mobility of radicals and cell substances. In dry seed, radical mobility and hence $R^{\cdot} + R^{\cdot}$ and $R^{\cdot}$ + XH interactions are inhibited and the potentiality for $R^{\cdot} + O_2$ interactions are increased. In moist seed, radical mobility and hence $R^{\cdot} + R^{\cdot}$ and $R^{\cdot}$ + XH interactions are greater and the potential $R^{\cdot} + O_2$ interactions are decreased.

d) Temperature: as developed in other papers (Konzak, Nilan, Harle, and Heiner, 1961; Nilan, Konzak, Harle, and Heiner, 1962). It appears that temperature governs the rate of reaction, the availability of water, and affects the mobility of oxygen molecules and radicals.

e) Time: The time between irradiation and hydration governs the extent of diffusion of reactive molecules and hence the number of $R^{\cdot} + O_2$, $R^{\cdot} + R^{\cdot}$ and $R^{\cdot}$ + XH interactions.

Combined X-ray and diethyl sulfate treatments

Height of M_1 seedlings and frequencies of chromosome aberrations in shoot tips grown from seeds X-rayed and then hydrated in a diethyl sulfate solution are presented in Tables 2 and 3. Both seedling height and chromosome aberration frequencies show a good radiation dose response. On the other hand, diethyl sulfate greatly reduced seedling height but induced very few chromosome aberrations.

Table 2. The combined effect of X-rays and diethyl sulfate on barley seeds (13% moisture)

Treatment	Chromosome aberrations [per cell][1]					Seedling height [cm]
	Bridges	Fragments Rods	Fragments Isodiam.	Dots	Total	
0 Kr	.00	.00	.00	.00	.00	13.1
Diethyl sulfate (D. S.)	.02	.04	.00	.00	.04	7.7
5 Kr	.02	.02	.02	.00	.04	13.1
5 Kr + D.S.	.04	.06	.02	.00	.08	7.0
10 Kr	.07	.00	.02	.02	.04	13.1
10 Kr + D. S.	.08	.22	.02	.00	.24	6.5
20 Kr	.18	.18	.15	.00	.33	11.9
20 Kr + D. S.	.12	.28	.36	.10	.74	6.1
30 Kr	.39	.26	.08	.06	.40	11.2
30 Kr + D. S.	.15	.56	.40	.23	1.19	5.8
40 Kr	.48	.33	.03	.03	.39	11.0
40 Kr + D. S.	.42	1.26	.32	.20	1.78	4.2

[1]) 400 cells scored per treatment except 40 kr = 366 cells and 40 kr + D. S. = 362 cells.

Combination treatments of X-rays and diethyl sulfate show a significant interaction (Tables 2 and 3). The seedling height is again greatly reduced compared to that shown by radiation or diethyl sulfate alone. Chromosome aberration frequencies show something of the cytogenetic nature of the interaction. It is interesting that dicentric chromosome bridges are not increased by the diethyl sulfate post-treatment but fragment frequencies are higher than would be expected from the additive effect of each mutagen. Furthermore, this increase can be attributed chiefly to an increase of rod fragments.

Table 3. The combined effects of X-rays and diethyl sulfate on barley seeds (13% moisture)

Treatment	Chromosome aberrations [per cell][1]					Seedling height (cm]
	Bridges	Fragments Rods	Fragments Isodiam.	Dots	Total	
Control	.00	.00	.00	.00	.00	11.9
Diethyl sulfate (D. S.) .0076 M	.01	.01	.00	.01	.02	7.0
Hydrolyzed products of diethyl sulfate (H. D. S.) .0076 M	.00	.00	.01	.00	.01	11.2
10 Kr	.03	.06	.04	.03	.13	10.5
10 Kr + D. S.	.06	.21	.05	.07	.33	6.6
10 Kr + H. D. S.	.04	.06	.06	.04	.16	11.3
20 Kr	.09	.23	.06	.10	.39	11.8
20 Kr + D. S.	.11	.40	.07	.12	.59	6.7
20 Kr + H. D. S.	.11	.29	.06	.16	.51	11.7
30 Kr	.24	.33	.22	.08	.63	10.8
30 Kr + D. S.	.28	.61	.23	.18	1.02	5.0
30 Kr + H. D. S.	.25	.37	.14	.12	.63	10.8

[1]) 400 cells scored per treatment.

It was thought at first that ethyl sulfuric acid formed as a byproduct of the hydrolysis of diethyl sulfate might induce the effect by chelation. However, the hydrolysis product has no effect on the radiation-induced aberration frequencies (Table 3). Thus, the increase in aberrations, especially fragments, from the chemical mutagen post-radiation treatment must be attributed to effects induced by diethyl sulfate itself.

At present there is no clear interpretation of the synergistic effect of radiation and diethyl sulfate on radiation-induced chromosome aberrations but several possibilities are being explored. The diethyl sulfate is probably altering chiefly the DNA bases, possibly guanine (BAUTZ and FREESE, 1960), whereas the radiation is attacking to a large extent the protein (cf. KIRBY-SMITH and RANDOLPH, 1960) of the chromosomes. It is hoped that the mutation analyses of these combined treatments may help to clarify the nature of the interactions observed and provide new knowledge of the mechanisms responsible for the induction of genetic damage in seeds.

References

ADAMS, J. D., and R. A. NILAN, 1958: After-effects of ionizing radiation in barley. II. Modification by storage of X-irradiated seeds in different concentrations of oxygen. Radiation Research 8, 111—122.

BASS, A. M., and H. P. BROIDA, 1960: Formation and Trapping of Free Radicals. Academic Press, New York, pp. 522.

BAUTZ, F., and E. FREESE, 1960: On the mutagenic effect of alkylating agents. Proc. nat. Acad. Sci. U. S. **46**, 1585—1594.

BLOIS, M. S. JR., H. W. BROWN, R. M. LEMMON, R. O. LINDBLOM, and M. WEISSBLUTH, 1961: Free Radicals in Biological Systems. Academic Press, New York, pp. 387.

CALDECOTT, R. S., 1954: Inverse relationship between the water content of seeds and their sensitivity to X-rays. Science (Washington) **120**, 809—810.

CALDECOTT, R. S., 1955: Effects of hydration on X-ray sensitivity in *Hordeum*. Radiation Research **3**, 316—330.

CALDECOTT, R. S., 1959: Post-irradiation modification of injury in barley — its basic and applied significance. Proc. 2nd Geneva Conf. on the Peaceful Uses of Atomic Energy **27**, 260—269.

CALDECOTT, R. S., 1961: Seedling height, oxygen availability, storage and temperature: Their relation to radiation-induced genetic and seedling injury in barley. "Effects of Ionizing Radiations on Seeds". IAEA, Vienna, STI/PUB/13, 3—24.

CALDECOTT, R. S., E. B. JOHNSON, D. T. NORTH, and C. F. KONZAK, 1957: Modification of radiation induced injury by post-treatment with oxygen. Proc. nat. Acad. Sci. U. S. **43**, 975—983.

CONGER, A. D., 1961: Biological aftereffect and long-lived free radicals in irradiated seeds. In: Recovery of Cells from Injury. Oak Ridge Nat. Lab., Biology Research Conf., Gatlinburg, Tenn. J. cellular comp. Physiol. **58** (Suppl. 1), 21—32.

CONGER, A. D., and M. L. RANDOLPH, 1959: Magnetic centers (free radicals) produced in cereal embryos by ionizing radiation. Radiation Research **11**, 54—66.

CURTIS, H. J., N. DELIHAS, R. S. CALDECOTT, and C. F. KONZAK, 1958: Modification of radiation damage in dormant seeds by storage. Radiation Research **8**, 526—534.

EHRENBERG, A., 1961: Research on free radicals in enzyme chemistry and in radiation biology. In: Free Radicals in Biological Systems. Academic Press, New York, pp. 337—350.

EHRENBERG, A., and L. EHRENBERG, 1958: The decay of X-ray induced free radicals in plant seeds and starch. Ark. Fysik **14**, 133—141.

EHRENBERG, L., 1959: Radiobiological mechanisms of genetic effects. A review of some current lines of research. Radiation Research (Suppl. 1), 102—123.

EHRENBERG, L., 1960: Induced mutation in plants: mechanisms and principles. Genetica Agraria **12**, 364—389.

EHRENBERG, L., and N. NYBOM, 1954: Ion density and biological effectiveness of radiations. Acta Agr. Scand. **4**, 396—418.

EHRENBERG, L., and U. LUNDQVIST, 1957: Post irradiation effects on X-ray-induced mutation in barley seeds. Hereditas **43**, 390—402.

GUSTAFSSON, Å., 1938: Studies on the genetic basis of chlorophyll formation and the mechanism of induced mutation. Hereditas **24**, 33—93.

HEINER, R. E., C. F. KONZAK, R. A. NILAN, and R. R. LEGAULT, 1960: Diverse ratios of mutations to chromosome aberrations in barley treated with diethyl sulfate and gamma rays. Proc. nat. Acad. Sci. U. S. **46**, 1215—1221.

HEINER, R. E., C. F. KONZAK, R. A. NILAN, and H. BARTELS, 1961: Effect of temperature on in vitro and in vivo reaction of diethyl sulfate. Nature (London) (in press).

JACKSON, W. D., 1958: The life-span of mutagens produced in cells by irradiation. In: Radiation Biology. Butterworths Scientific Publications, London, pp. 190—208.

KIRBY-SMITH, J. S., and M. L. RANDOLPH, 1960: Production and lifetimes of radiation-induced free radicals in some molecules of biological importance. Proc. of Symposium on "Immediate and Low Level Effects of Ionizing Radiations", Venice, June 1959, pp. 11—24.

KLINGMÜLLER, W., G. R. LANE, M. C. SAXENA, and D. J. E. INGRAM, 1959: Free radicals in X-rayed seeds of high and low water content, as measured by electron spin resonance. Nature (London) **184**, 464—465.

KONZAK, C. F., H. J. CURTIS, N. DELIHAS, and R. A. NILAN, 1960: Modification of radiation induced damage in barley seeds by thermal energy. Canad. J. Genet. Cytol. **2**, 129—141.

KONZAK, C. F., R. A. NILAN, J. R. HARLE, and R. E. HEINER, 1961: Control of the factors affecting the response of plants to mutagens. "Fundamental Aspects of Radiosensitivity". Brookhaven Biological Symposia **14**, 128—157.

KONZAK, C. F., R. A. NILAN, R. R. LEGAULT, and R. E. HEINER, 1961: Modification of induced genetic damage in seeds. "Effects of Ionizing Radiations on Seeds". IAEA, Vienna, STI/PUB/13, 155—169.

MOUTSCHEN-DAHMEN, J., and M. MOUTSCHEN-DAHMEN, 1961: Effets de traitements combinés du myleran avec d'autres poisons radiomimétiques et avec les rayons X. " Effects of Ionizing Radiations on Seeds". IAE, Vienna, STI/PUB/13, 333—344.

MÜLLER, A., and K. G. ZIMMER, 1961: Studies on radiation-produced free radicals in biological systems. In: Free Radicals in Biological Systems. Academic Press, New York, pp. 325—335.

NILAN, R. A., 1954: The relation of carbon-dioxide, oxygen, and low temperature to the injury and cytogenetic effects of X-rays in barley. Genetics **39**, 943—954.

NILAN, R. A., and C. F. KONZAK, 1961: Increasing the efficiency of mutation induction. Proc. of Symposium on "Mutation and Plant Breeding". Cornell University, Nov.-Dec. 1960, 437-460.

NILAN, R. A., C. F. KONZAK, J. R. HARLE, and R. E. HEINER, 1962: The interrelation of oxygen, moisture, and temperature in the production of radiation-induced genetic damage in plants. Proc. of Symposium on "Radiation Effects and Milieu", Montreux, Switzerland, May-June 1961 (in press).

NILAN, R. A., C. F. KONZAK, R. R. LEGAULT, and J. R. HARLE, 1961: The oxygen effect in barley seeds. "Effects of Ionizing Radiations on Seeds". IAEA, Vienna, STI/PUB/13, 139—154.

OSTER, I. I., 1958: Interactions between ionizing radiation and chemical mutagens. Z. indukt. Abstamm.- Vererb.lehre **89**, 1—6.

POWERS, E. L., R. B. WEBB, and C. F. EHRET, 1960: Storage, transfer, and utilization of energy from X-rays in dry bacterial spores. Radiation Research (Suppl. 2), 94—121.

POWERS, E. L., C. F. EHRET, and B. SMALLER, 1961: The role of free radicals in the lethal effects of X-rays in dry bacterial spores. In: Free Radicals in Biological Systems. Academic Press, New York, pp. 351—366.

SIRE, M. W., and R. A. NILAN, 1958: The relation of oxygen post-treatment and heterochromatin to X-ray-induced chromosome aberration frequencies in *Crepis capillaris*. Genetics **44**, 124—136.

SPARRMAN, B., L. EHRENBERG, and A. EHRENBERG, 1959: Scavenging of free radicals and radiation protection by nitric oxide in plant seeds. Acta chem. scand. **13**, 199—200.

ZIMMER, K. G., 1959: Evidence for free-radical production in living cells exposed to ionizing radiation. Radiation Research (Suppl. 1), 519—529.

ZIMMER, K. G., L. EHRENBERG, and A. EHRENBERG, 1957: Nachweis langlebiger magnetischer Zentren in bestrahlten biologischen Medien und deren Bedeutung für die Strahlenbiologie. Strahlentherapie **103**, 3—15.

Evidence for Two Resolvable Sites of Action of Radiation on Micro-Organisms

Tikvah Alper

Medical Research Council, Experimental Radiopathology Research Unit, Hammersmith Hospital, London, W. 12

Some of the evidence I shall present concerns comparisons of the effects of ionizing radiation and ultraviolet light at 2537 Å, and it is relevant that the UV is likely to exert its effects by specific absorption in the nucleic acids or their precursors. In contrast, ionizations will occur at random, the fraction occurring in the nucleic acid complex being the same as the fraction of the total cell contents represented by that complex. It is true also that UV at 2537 Å exerts its effects through excitations, which cannot be assumed to initiate the same sequence of chemical events as ionizations or "super-excitations". If the sequences of biochemical events following irradiation by UV and by X-rays are similar, this may be an indication of the primary disturbances having taken place in similar sites; differences may be an indication that different sites have been affected, though presumably differences may also arise if excitations and ionizations initiate different sequences of events, even when these occur at the same site.

In recent years much of our information about the mode of action of radiation has been derived from observations on how effects may be modified. Broadly speaking, there are two types of modification: one occurs at the physico-chemical level, presumably by influencing the reaction which takes place immediately after the original disturbance, while the other type of modification depends on influencing in some way the biochemical results of the initial chemical reaction. Examples of modification at the physico-chemical level occur much more frequently with ionizing than with UV radiation, notable examples being the oxygen effect and the possibility of chemical protection or sensitization. On the other hand, it is perhaps the effects of UV which are more readily influenced by agents acting through metabolic pathways. This sort of modification is perceived in a clear-cut way when effects of radiation are modified by methods of culture after irradiation, but in some cases the physiological state of the cell at the time of irradiation may determine a large part of its subsequent biochemical response. I intend in this paper to show how there may be interaction between physico-chemical and metabolic methods of modification, and how it may be deduced from these interactions that the observed lethal effect of ionizing radiations, at least on micro-organisms, is the summation of events which have taken place in at least two sites, or by at least two distinct modes of action.

Let us set up a model for this concept. For simplicity's sake I shall adopt the view that there are two sites, which I shall call *N* and *O*, such that the cell will be killed (i. e. rendered unable to reproduce) if either site is irreversibly damaged by an ionization. The familiar survival curve will in such a case represent the summation of two components, and, if the curve is of exponential form, its equation may be written:

$$f = e^{-(\lambda_N + \lambda_O)D} \tag{1}$$

If we can provide conditions in which the cell is completely rescued from the damage to site N, the equation for survival will become:

$$f = e^{-\lambda_O D} \tag{2}$$

Let us further suppose that these equations refer to survival when the cell has been irradiated in anoxic conditions, and that the presence of oxygen affects the response of the sites to a different extent. For irradiations carried out in the presence of oxygen, equation 1 would become:

$$f = e^{-(m_N \lambda_N + m_O \lambda_O)D} \tag{3}$$

m_N and m_O being respectively the oxygen enhancement ratios referring to sites N and O. If there were no cell killing by virtue of damage to site N, equation 2 would become:

$$f = e^{-m_O \lambda_O D} \tag{4}$$

Oxygen enhancement ratios are commonly observed by comparing the exponents of survival curves obtained for cells irradiated with and without oxygen, and the ratios which the observer would see, in this hypothetical case, would be respectively

$$\frac{m_N \lambda_N + m_O \lambda_O}{\lambda_N + \lambda_O}$$

and

$$\frac{m_O \lambda_O}{\lambda_O} = m_O;$$

these could be equal only if m_N and m_O were equal.

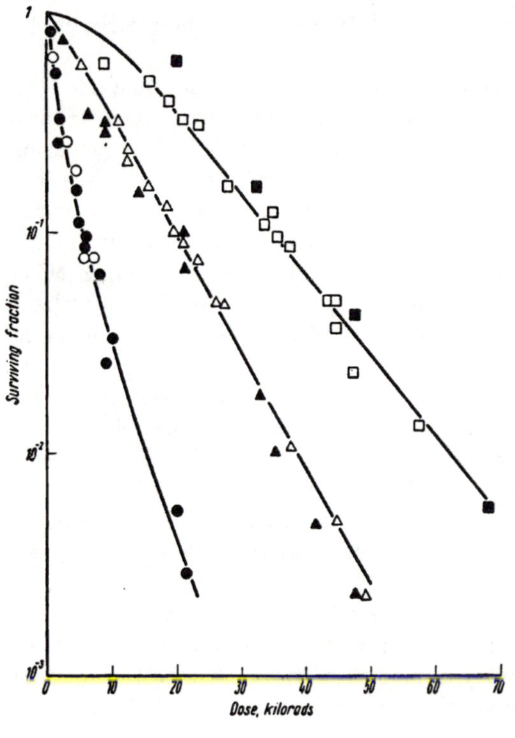

Now the strain *E. coli* B is one in which radiation effects are markedly dependent on postirradiation conditions of culture; if cells irradiated as one population are plated thereafter in different conditions, widely differing sensitivities are observed (e. g. Fig. 1). ALPER and GILLIES (1958) reported that treatments which changed the observed sensitivity concomitantly changed the oxygen enhancement ratio, as is demonstrated by Figs. 2 and 3. Some oxygen enhancement ratios corresponding to a wide range of sensitivities are shown in Table 1 (ALPER, 1961). It should be emphasized that the different values were derived by making colony counts on cells irradiated as one population but cultured in different conditions thereafter.

Fig. 1. *E. coli* B irradiated anaerobically, plated on "Difco Nutrient Agar". Preliminary incubation temperature: 19 °C, ○; 37 °C, △; 45 °C, □. Open and closed symbols referr especively to irradiations with 250 kV X-rays and 8 MeV electrons

If we suppose that the variation in culture conditions modifies the damage to only one of the postulated sites, say N, then we may assume that for any given conditions sensitivity is given by

$$\lambda = \lambda'_N + \lambda_O,$$

λ'_N depending on how much of the damage to N still remains unmodified. In aerobic conditions we shall observe sensitivity $M\lambda$ and oxygen enhancement ratio M.

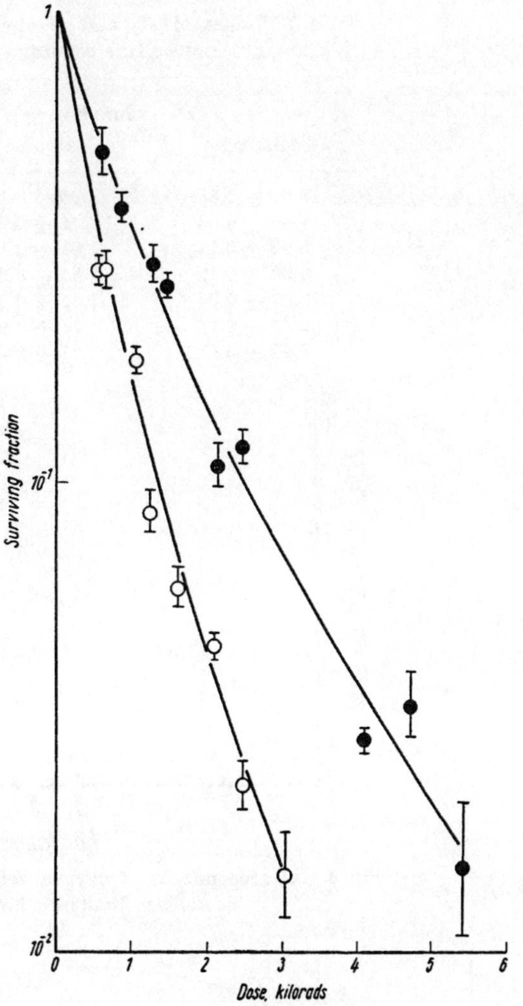

Fig. 2. *E. coli* B, plated on "Oxoid Blood Agar Base". Preliminary incubation temperature 19 °C. Open symbols – aerobic irradiations; closed symbols – anaerobic irradiations. 95% confidence limit indicated.

Now

$$M\lambda = m_N\,\lambda'_N + m_O\,\lambda_O$$

$$\therefore\ M = \frac{m_N\,\lambda'_N + m_O\,\lambda_O}{\lambda}$$

$$= \frac{m_N\,(\lambda - \lambda_O) + m_O\,\lambda_O}{\lambda}$$

$$= \lambda_0\,(m_O - m_N)\cdot 1/\lambda + m_N$$

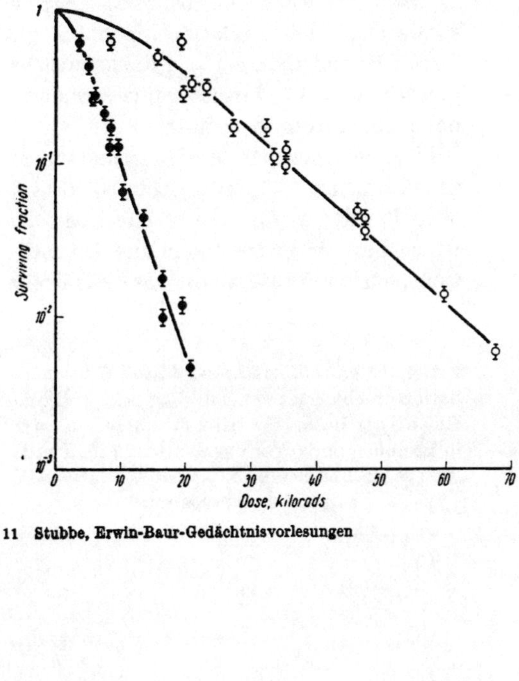

Fig. 3. *E. coli* B, plated on "Difco Nutrient Agar". Preliminary incubation temperature 45 °C. Open symbols – anaerobic irradiations; closed symbols – aerobic irradiations. 95% confidence limits indicated

Table 1. Values of D_O and of oxygen enhancement ratios for different culture conditions, *E. coli* B

D_O [kilorads] Anaerobic	Aerobic	Ratio m
12.00 ± 0.30	3.25 ± 0.10	3.67 ± 0.15
9.00 ± 0.34	2.71 ± 0.06	3.32 ± 0.13
8.72 ± 0.17	3.01 ± 0.13	2.90 ± 0.14
6.38 ± 0.13	2.34 ± 0.04	2.73 ± 0.08
2.47 ± 0.08	1.22 ± 0.05	2.02 ± 0.21
2.11 ± 0.11	1.18 ± 0.05	1.79 ± 0.11
0.98 ± 0.04	0.57 ± 0.03	1.63 ± 0.11

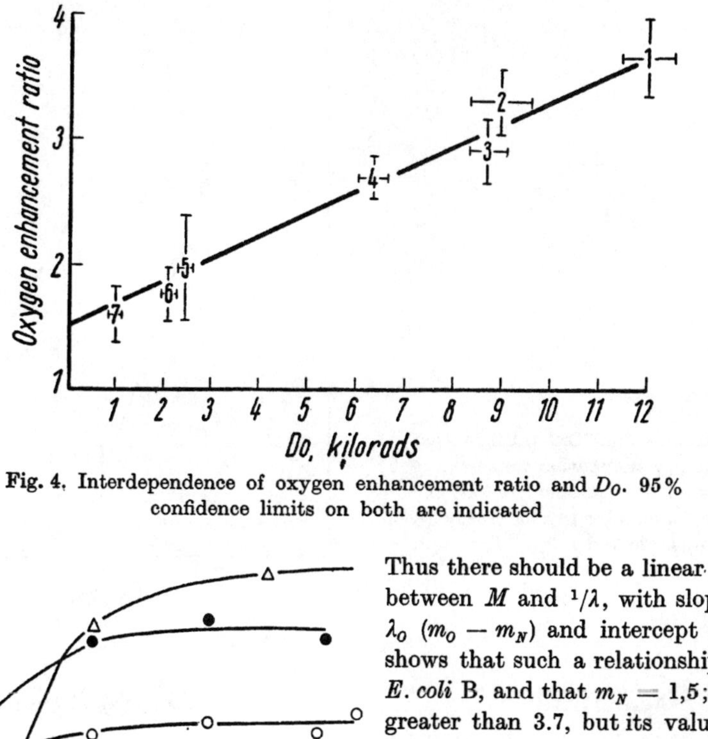

Fig. 4. Interdependence of oxygen enhancement ratio and D_O. 95% confidence limits on both are indicated

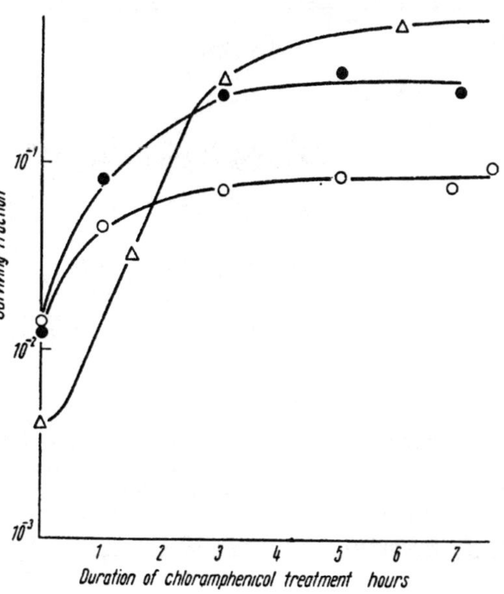

Thus there should be a linear relationship between M and $^1/\lambda$, with slope given by $\lambda_O\ (m_O - m_N)$ and intercept m_N. Fig. 4 shows that such a relationship holds, for *E. coli* B, and that $m_N = 1,5$; m_O must be greater than 3.7, but its value cannot be determined from these data.

If a given treatment modifies the damage to N, but not to O, as postulated above, then it will have a greater effect on cells which have been treated in anoxic conditions, since in those conditions a relatively

Fig. 5. Rescue of irradiated *E. coli* B by incubation on chloramphenicol, 5 μg/ml., + Oxoid Blood Agar Base. ○ = after 9.5 kilorads X-rays in presence of oxygen; ● = after 17.9 kilorads X-rays in absence of oxygen; △ = after UV at 2537 Å. 83 ergs/mm² smaller amount of the overall damage is due to events in O and the modifying treatment will have more scope for its action.

Let us assume that similarities between the action of ionizing and UV radiations are due to effects of both types of radiation on site N, damage to which is very little affected by the presence of oxygen. It could be expected that treatment which "rescues" the cells by allowing them to overcome the site N damage would rescue them also from UV, and this is found to be the case with *E. coli* strain B (Fig. 5).

Let us now examine the converse case, in which some form of treatment rescues the cell by virtue of allowing it to overcome damage to site o, which is greatly influenced by the presence of oxygen. We should then expect that

(1) the effect of the rescue would be much more marked if the irradiation had taken place in the presence of oxygen;

(2) the extent to which oxygen would be seen to have enhanced the effect of radiation would be less in the treated cells;

(3) the treatment would not be effective in rescuing cells from UV damage.

This state of affairs is precisely what occurs if the resistant mutant of *E. coli* B, B/r, is rescued by incubating irradiated cells for a while at 18°C (STAPLETON et al., 1955) The effectiveness of this method of restoration is much more marked if the cells have been irra-

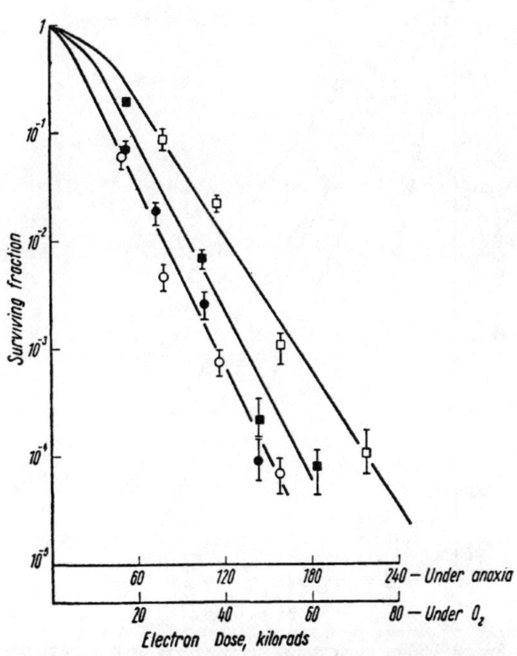

Fig. 6. Rescue of B/r, after 8 MeV electron irradiation, by preliminary incubation at 18 °C. Aerobic irradiations: ○, 37 °C; □, 18 °C. Anaerobic irradiations: ●, 37 °C; ■, 18 °C. Dose scale for aerobic irradiations three times that for anaerobic irradiations

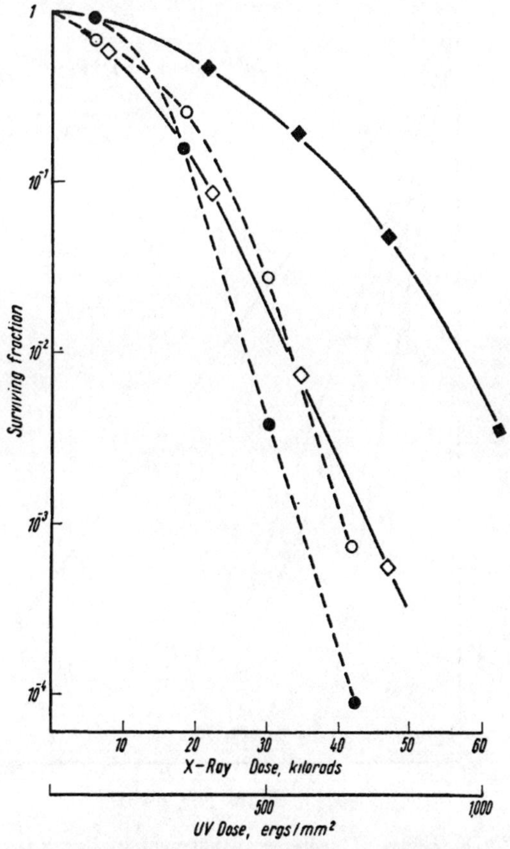

Fig. 7. Effect of 18 °C incubation on B/r exposed either to UV or to 8 MeV electrons. Preliminary incubation 37 °C: ◊, electrons; ○, UV. Preliminary incubation 18 °C: ♦, electrons; ●, UV

diated in the presence of oxygen, as is shown by Fig. 6; as a result, the oxygen enhancement ratio is reduced from 3, seen after normal post-irradiation treatment, to 2.4. After UV irradiation, the cells are not rescued at all by this treatment; indeed, colony counts are lower for the 18°C treated cells, because the extrapolation number for the survival curves pertaining to treated cells is reduced (Fig. 7). It seems plausible then to assume that, in terms of the model put forward, the 18° treatment rescues the cells from damage to site *O*, whereas most of the post-irradiation treatment of *E. coli* B seem to affect damage to site *N*. It is precisely in this respect that the strains differ. If conditions of culture are applied to strain B such that cells are rescued from damage to site *N*, then its sensitivity is hardly greater than that of the resistant mutant (ALPER and GILLIES, 1960).

Most post-irradiation treatments act in opposite senses in these two strains. Cysteine, however, which is a protective agent for both of them, acts in the same sense: it seems to protect (or perhaps restore) site *O*, so that its protective action is much more marked when the cells have been irradiated in the presence of oxygen, with the result that the oxygen enhancement ratio is much reduced if cysteine is present during irradiation. Furthermore, cells of *E. coli* B are more effectively protected by cysteine if post-irradiation treatment is such that relatively little damage to site *N* comes to light (Fig. 8). It is difficult to assess whether cysteine acts protectively at the physicochemical or at the metabolic level. It does not act strictly as a dose-reducing agent, as does oxygen (Fig. 9); and the work of KOHN and GUNTER (1959)

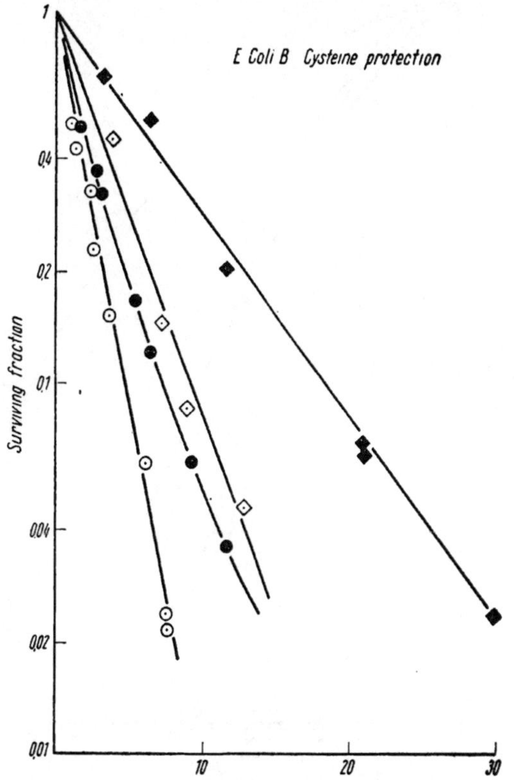

Fig. 8. Effect of cysteine on survival of *E. coli* B, cultured after irradiation at different preliminary incubation temperatures. Irradiated without cysteine: ○, 18 °C; ◇, 37 °C. Irradiated after incubation with cysteine: ●, 18 °C; ◆, 37 °C

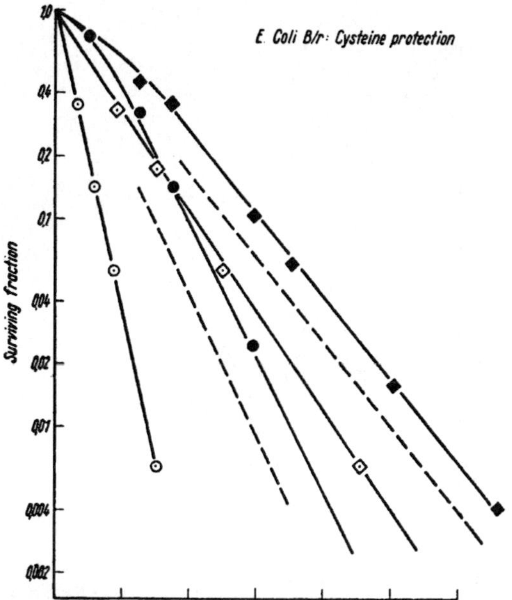

Fig. 9. Effect of cysteine on survival of *E. coli* B/r; irradiated in anoxic and in aerobic conditions. Irradiated without cysteine: ○, aerobic; ◇, anaerobic. Irradiated after incubation with cysteine: ●, aerobic; ◆, anaerobic. Dotted lines give slopes of cysteine curves extrapolated to origin

established that cysteine required a "reaction period" with cells before it would act protectively. However, the same authors also found that damage was reduced only if the reaction took place before the radiation.

It is risky to generalize about modes of action of radiation from experiments with the *E. coli* B family, since this is rather atypical in its marked response to post-irradiation culture conditions. However, evidence of a rather different nature, from studies with haploid yeast, also supports the inference that there are two sites of damage or two modes of action of ionizing radiation. Freshly harvested yeast cells form a population of mixed radiosensitivity, the budding or dividing cells requiring nearly eight times as many ionizations per lethal event as the much more sensitive single or resting cells. This holds if the radiation is delivered in aerobic conditions. In anoxic conditions, however, the corresponding ratio is only four to one; this is another way of expressing the fact that the oxygen enhancement ratio for the resistant cells is only half that for the sensitive cells (ALPER, 1959). In this case, the physiological condition which modifies radiosensitivity also changes the extent to which oxygen enhances it. It was found by ELKIND and BEAM that the relative effectiveness of α-rays and X-rays was greater than one for the resistent cells, but less than one for the sensitive cells. These observations together suggest that the most important target for damage in the resistant cells is akin to the site I have celled "*N*", which is (a) not much influenced by the presence of oxygen, (b) more effectively damaged by radiations of high than of low LET. In the sensitive cells, however, the damage to this target makes only a relatively small contribution, which is overshadowed by the damage to some other target. This second target is (a) greatly influenced by the presence of oxygen, (b) of such a nature that the effectiveness of α-particles is less than that of X-rays. Perhaps some molecule which is needed for the process of budding is the target in question; once the cell has budded, the future of further budding is not at risk because the cell can now synthesize this molecule anew. This view of the double-site nature of damage to haploid yeast cells is supported by UV survival curves on the same population (Fig. 10). It is seen that sensitivity to UV hardly differs for the two types of cell, and is perhaps acting therefore only, or mainly, on site *N*.

In general, it seems that site *N* for damage by ionizing radiation is roughly identifiable with a main site of damage by UV; and is perhaps therefore to the nucleic acid complex. This seems plausible for various reasons, not least of which is the fact that damage by X-rays to nucleic acid in vitro seems not to be oxygen-sensitive, and that UV damage is itself not oxygen-sensitive. The nature of *O* is obscure; but if *N* is nucleic acid, *O* is probably not nucleic acid. This leads us to the further inference that damage in the nucleic acid complex may be easily bypassed in many cells, for example by suitable post-irradiation treatment of *E. coli* B. Further, the resistant mutants of *E. Coli* B differ from the parent strain precisely in that this fraction of damage is to a great extent bypassed in quite ordinary culture conditions.

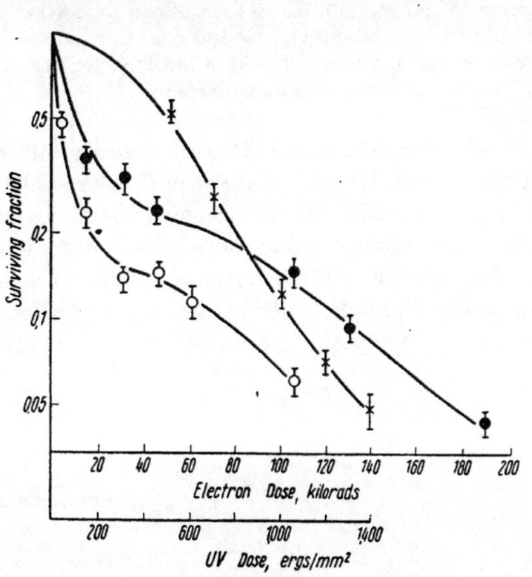

Fig. 10. Survival curves, freshly harvested haploid yeast. ○, X-rays, aerobic; ●, X-rays, anaerobic; ×, UV

The analysis made here depends on interaction between two modifying agents

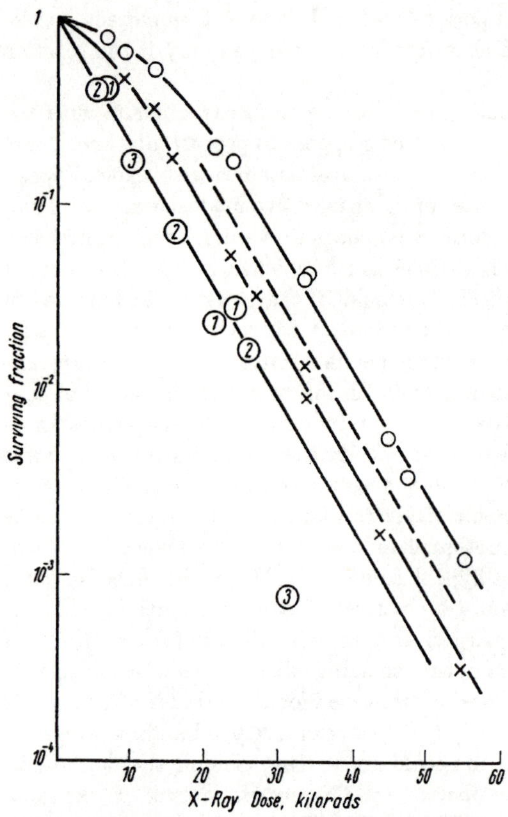

Fig. 11. Survival of B/r after irradiation, when plated: ○, on nutrient agar; ×, on salts-glucose agar; ① etc. on chloramphenicol agar for 2—3 hours after irradiation. Numbers indicate method of culture before irradiation: (1) stationary-phase cultures grown anaerobically; (2) log phase cultures; (3) stationary-phase cultures grown aerobically

which differ in nature, and on a comparison of the response of cells exposed to ionizing and ultraviolet radiation. Such data are not available from experiments on radiation mutagenesis. Post-irradiation conditions of culture modify mutation induction by UV in *E. coli* strains and in *Salmonella* (e. g. WITKIN, 1956); with *Paramecium*, KIMBALL and his co-workers (1959) have found that the genetic effects of both ionizing and UV radiation can be so modified. In general, inhibitory conditions are those in which fewest induced mutations come to light; more specifically, the inhibition of protein synthesis was found by WITKIN (1956) to reduce the number of mutations induced by UV in *Escherichia coli* strains and in *Salmonella*. There are some striking parallels between WITKIN's results for mutagenic effects and our own for lethal effects; but there are also striking differences, for example, in the action of chloramphenicol. This drug rescues *E. coli* B from lethal effects, after both X-rays and UV; and the cells are known to be rescued from mutagenic effects after UV. With the strain B/r, however, chloramphenicol may *increase* the lethal effect (ALPER and GILLIES, 1960) by reducing the extrapolation number to one (Fig. 11); whereas it diminishes mutagenic effects. Work of GILLIES (1961; 1962) with auxotrophic strains of *E. coli* B shows that when protein synthesis is specifically inhibited, the cells are rescued from the lethal effects of both UV and X-rays; and chloramphenicol rescues the cells from the effects of both types of radiation. It does at least seem likely that there is more than one site of radiation action in inducing mutations as well as for lethal effects, and it is of interest that WITKIN and THEIL (1960) have reported just such a conclusion from studies with UV on strain B/r: inhibition of protein synthesis affected reversions from auxotrophy to prototrophy, but not induction of streptomycin resistance or streptomycin independence.

Acknowledgement

I am grateful to Dr. N. E. GILLIES for permitting me to use some of his experimental data before publication.

References

ALPER, T., 1959: Variability in the oxygen effect observed with micro-organisms. Part I. Haploid yeast: single and budding cells. Internat. J. Rad. Biol. **1**, 414–419.

ALPER, T., 1961: Variability in the oxygen effect observed with micro-organisms. Part II. *Escherichia coli* B. Internat. J. Rad. Biol. **3**, 369–377.

ALPER, T., and GILLIES, N. E., 1958: Dependence of the observed oxygen effect on the post-irradiation treatment of micro-organisms. Nature (London) **181**, 961–963.

ALPER, T., and GILLIES, N. E., 1960: The relationship between growth and survival after irradiation of *Escherichia coli* Strain B and two resistant mutants. J. Gen. Microbiol. **22**, 113–128.

ELKIND, M. M., and C. A. BEAM, 1955: Variation of the biological effectiveness of X-rays and α-particles on haploid *Saccharomyces cerevisiae*. Radiation Research **3**, 88–104.

GILLIES, N. E., 1961: The use of auxotrophic mutants to study restoration in *Escherichia coli* B after ultraviolet irradiation. Internat. J. Rad. Biol. **3**, 379–387.

GILLIES, N. E., 1962: In preparation.

KIMBALL, R. F., N. GAITHER, and S. M. WILSON, 1959: Reduction of mutation by post-irradiation treatment after ultraviolet and various kinds of ionizing radiation. Radiation Research **10**, 490–497.

KOHN, H. I., and S. E. GUNTER, 1959: Factors influencing the radioprotective action of cysteine: effects in *E. coli* due to drug concentration, time and pH. Radiation Research **11**, 732–744.

STAPLETON, G. E., D. BILLEN, and A. HOLLAENDER, 1953: Recovery of X-irradiated bacteria at suboptimal incubation temperatures. J. Cellul. Comp. Physiol. **41**, 345–358.

WITKIN, E. M., 1956: Time, temperature and protein synthesis: A study of ultra-violet induced mutation in bacteria. Cold Spring Harbor Symposia on Quant. Biol. **22**, 123–140.

WITKIN, E. M., and E. C. THEIL, 1960: The effect of post-treatment with chloramphenicol on various ultraviolet-induced mutations in *Escherichia coli*. Proc. nat. Acad. Sci. U. S. **46**, 226–231.

Reversible Vorgänge im Genom des Phagen $\varkappa$ nach UV-Bestrahlung der extrazellulären Partikel

R. W. Kaplan

Institut für Mikrobiologie der Johann-Wolfgang-Goethe-Universität, Frankfurt (Main)

Bei der Auslösung von Mutationen in Zellen kann die Strahlung auf zwei verschiedenen Wegen wirken, 1. dem räumlich direkten, wenn die wirksame Strahlenenergie im Gen, also der DNS, selber absorbiert wird, 2. dem räumlich indirekten, wenn diese Absorption – der „Treffer" – außerhalb des Gens erfolgt und dann die Wirkung durch irgendwelche Energieträger an das mutierende Gen weitergegeben wird. Als Energieträger sind z. B. durch die Bestrahlung mutagen gewordene Molekeln, wie Peroxyde, Radikale u. ä., angenommen worden, andererseits auch geänderte metabolische Reaktionen, welche mutagene Stoffwechselprodukte erzeugen. Durch die zahlreichen Befunde über die Wirkung von Behandlungen nach der Bestrahlung, z. B. mit sichtbarem Licht, Wärme, Chemikalien, welche den Mutationsertrag vermindern und z. T. auch erhöhen, ist es klar, daß durch die Strahlentreffer in der Zelle – zumindest meist – nicht sofort fertige Mutationen entstehen, sondern zunächst prämutative Änderungen; diese sind teils reversibel, teils wird durch die Nachbehandlung die Chance ihrer Überführung zu fertigen Mutationen erhöht. Dies läßt zunächst die Hypothese der räumlich indirekten Wirkung wahrscheinlicher erscheinen. Jedoch kann die Tatsache prämutativer Zustände auch bei direkter Wirkung der Strahlen durch die folgende 3. Möglichkeit verstanden werden: Die Strahlenabsorption in der DNS erzeugt zunächst Läsionen in dieser, welche noch keine Mutationen sind. Diese Läsionen können jedoch später, z. B. bei der DNS-Reproduktion, mit einer gewissen Chance zu Mutationen – nach der verbreiteten Annahme Änderungen in der Basensequenz – führen, wenn sie bis dahin „überleben".

Eine Entscheidung zwischen der 2. und dieser 3. Hypothese „räumlich direkter, aber zeitlich indirekter" Wirkung könnte ermöglicht werden, wenn es gelänge, durch Bestrahlung extrazellulärer DNS Mutationen auszulösen und durch Nachbehandlungen die entstandenen prämutativen Läsionen zu revertieren oder auch, bei anderen Läsionsarten, die Perfektierung zu Mutationen zu fördern. Damit wären die von der 3. Hypothese geforderten prämutativen Stadien in der DNS als real erwiesen.

Leider zeigten sich die in Frage kommende extrazelluläre Erbsubstanz, nämlich die Transformations-DNS, die RNS des TMV sowie auch Phagen als meist resistent gegenüber Mutationsauslösung durch extrazelluläre Bestrahlung. Dies könnte wieder als Indiz für die Indirektheits-Hypothese genommen werden. Nur Krieg hatte bisher beim Phagen T4 Erfolg mit den sehr seltenen Mutationen r → +, allerdings bei sehr hoher Multiplizitätsreaktivierung.

Bei dem temperierten Phagen $\varkappa$ von *Serratia marcescens* beobachteten wir nun vor einiger Zeit ein Erscheinen von Klarlochmutanten (*c*- und *l*-Typen) mit Häufigkeiten bis 2% nach

UV-Bestrahlung extrazellulärer Partikel (Ellmauer und Kaplan, 1959). Die Plaques des Wildtyps bilden trübe Löcher mit violettem Hof im gelbroten Rasen des Indikatorstammes HY. Da Nitrit auch viele andere Mutationstypen bei diesem Phagen erzeugt, UV und Röntgenstrahlung jedoch fast ausschließlich Lochklärungs-Mutationen, wirken diese Strahlen stark elektiv auf die *c*- und *l*-Gene (Kaplan und Bose, 1961). Die Dosiseffektkurve dieser Mutationsauslösung durch UV zeigt 2 Treffer an, also 2 UV-Quanten pro Mutation; die UV-Inaktivierung erfordert nur 1 UV-Quant (Abb. 1). Winkler (unveröff.) erhielt bei manchen der *c*-Mutantenstämme durch Bestrahlung Rückmutationen zur Wildtyp-Trübheit des Loches. Diese *c*-Mutationen sind also keine stabilen Defizienzen. Er fand auch seltene Wirtsbereichsmutationen nach UV-Bestrahlung des freien Phagen und Wirtes. Da die Bestrahlung des Wirtes neben der des freien Phagen den *c*-Mutationsertrag erhöht (Winkler, unveröff.), besteht die Möglichkeit, daß diese *c*-Mutationen nicht (oder nicht nur) „echte" Mutationen darstellen. Auf Grund einer verbreiteten Hypothese über Erbmaterialaustausch zwischen Phagen und Wirt könnten sie auch auf „Einrekombinationen" von Wirtsgenomstücken in die *c*-Gene des Phagen zurückgehen. Diese Hypothese ist jedoch keineswegs gesichert; andererseits ist es nicht ausgeschlossen,

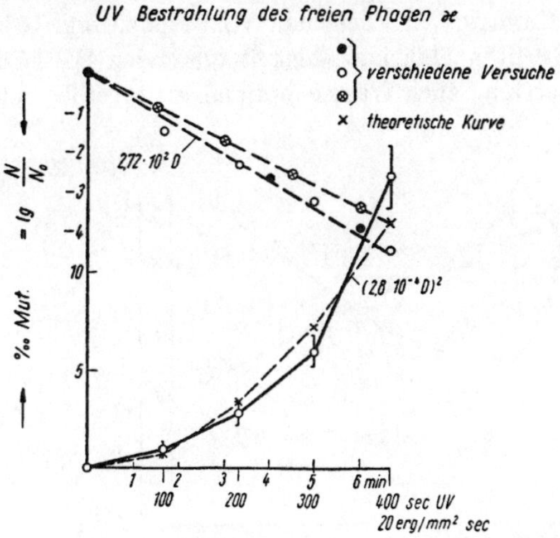

Abb. 1. UV-Bestrahlung des freien Phagen $\varkappa$. Dosisabhängigkeit von Inaktivierung (oben: $\lg N/N_0$) und *c*-Mutation (unten). Gestrichelt: 2-Trefferkurve für Mutation

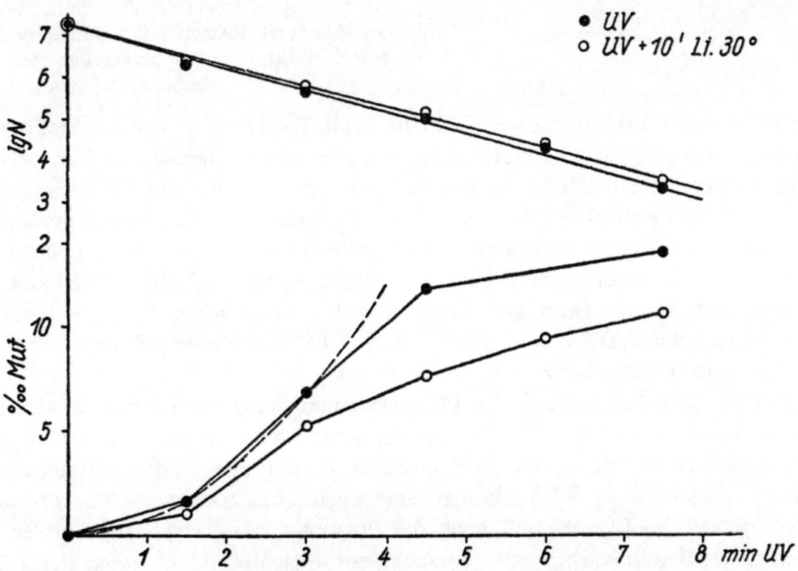

Abb. 2. UV-Dosiskurven von *c*-Mutation (unten) und Inaktivierung (oben) des Phagen $\varkappa$ mit (○) und ohne (●) Nachbelichtung (10 Min.)

daß auch im Falle der strahleninduzierten „echten" Mutationen von Zellen bei der Perfektierung der prämutativen Strahlenläsionen crossing-overähnliche Vorgänge (inhomologer Austausch) während der DNS-Replikation mitwirken. Bis auf weiteres wollen wir daher die *c*-Varianten als Mutationen bezeichnen.

Auf Grund der obigen Gedankengänge lag es nahe, eine Photoreversion der *c*-Mutationen durch Nachbelichtung des freien Phagen zu versuchen. Diese gelang tatsächlich, wobei im optimalen Falle etwa die Hälfte der *c*-Mutationen revertiert wurde. Die Reaktivierung der verlorenen Fähigkeit zur Plaquebildung (Inaktivierung) betrug dagegen nur etwa 1% (Kaplan, Winkler und Wolf-Ellmauer, 1960) (Abb. 2). Dies zeigt, daß die zum Verlust der Reproduktionsfähigkeit führenden UV-Läsionen von anderer Art sind als die prämutativen. Auch Wärme (optimal 40°C) erwies sich als revertierend.

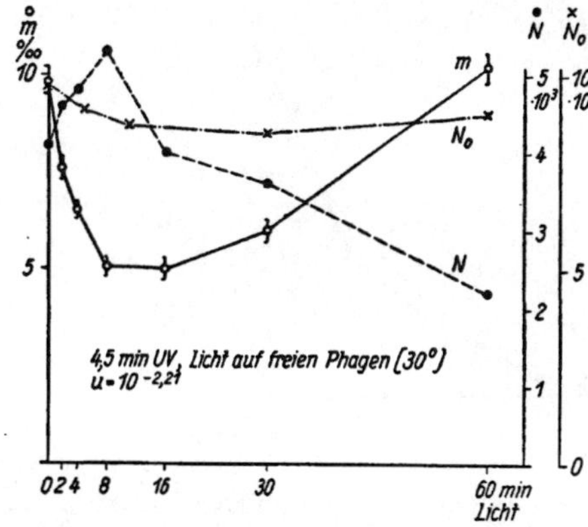

Abb. 3. *c*-Mutationsrate (m) und Plaquetiter (N_0 = ohne, N = mit UV) bei Bestrahlung des freien Phagen *x* mit einer UV-Dosis und variierter Nachbelichtung mit Blaulicht bei 30° C

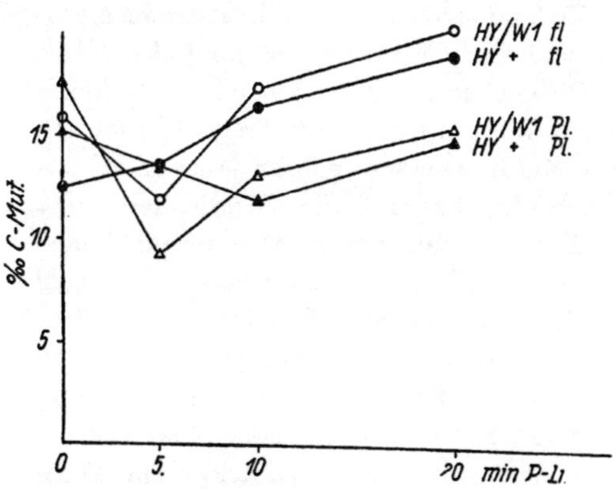

Abb. 4. *c*-Mutationsrate nach UV-Bestrahlung (4,5 min) und variierter Lichtdosis. Vier verschiedene Lysate: ○ auf weißer Wirtsmutante in flüssiger Kultur; ● auf rotem Wildtyp des Wirts in flüssiger Kultur; △ auf weißer Wirtsmutante durch Abwaschen von Platten mit konfluenter Lysis; ▲ auf rotem Wildtyp durch Abwaschen von Platten mit konfluenter Lysis

Die Lichtdosiskurve bei konstanter UV-Dosis (Abb. 3) zeigt einen etwa eintreffrigen Abfall der *c*-Mutationsrate und bei höheren Lichtdosen wieder einen Anstieg. Da Licht allein keine Mutationen auslöst, könnte dieser Anstieg auf der Überführung von UV-Läsionen in Prämutationen oder Mutationen beruhen, die ohne Licht nicht zu Mutationen geführt hätten. Diese unfertigen, „photoperfektierbaren" Prämutationen würden also einen dritten Typ von UV-Läsionen im Phagengenom darstellen. Es besteht jedoch auch die Möglichkeit, daß der Wiederanstieg des Mutationsertrages durch erneute Prämutation bereits (unvollständig) lichtrevertierter Phagenpartikel entsteht (Kaplan, 1960). Dies wird jedoch durch die folgenden Versuche unwahrscheinlicher.

Es zeigte sich, daß das Ausmaß der Photoreversion der *c*-Mutationen in verschiedenen Versuchen unterschiedlich war. Wenige ergaben sogar gar keine merkliche Photoreversion der Mutation; jedoch war die schwache Photoreaktivierung (der Plaquebildung) immer vorhanden. Die Ursachen dieser Schwankungen sind noch nicht geklärt. Es scheinen das Alter, die Herstellungsart des Lysates und auch die Plattungsbedingungen beteiligt zu sein. In Abb. 4 sind einige Ergebnisse eines der Versuche mit verschiedenen Lysaten dargestellt. Sie zeigt, daß bei dem einen gar keine Photoreversion, also ein Abfall der Lichtdosiskurve, bemerkbar ist; vielmehr steigt die Kurve sogleich bis zu einem Wert an, der unter dem auch von den übrigen Lysaten bei hohen Lichtdosen schließlich erreichten c-Mutationsgehalt liegt. Hier überwiegt also die „Photoperfektierung" der erwähnten „unfertigen" UV-Läsionen. Diese scheinen also tatsächlich ein vom UV erzeugter dritter Läsionstyp in der Phagen-DNS zu sein.

Außer diesem Typ gibt es noch eine vierte photoreversible UV-Läsionsart. Neben Tötung und c-Mutation verursacht UV-Bestrahlung auch eine starke Zunahme der kleinen Plaques, die oft gar kein Loch besitzen, sondern nur aus einem Farbfleck bestehen. Nachkulturen erweisen sie als nicht erblich; wahrscheinlich sind sie durch die bekannte Latenzzeitdehnung nach UV verursacht, möglicherweise liegt auch eine Verzögerung der DNS-Injektion z. B. infolge Schwanzproteinschadens, vor. Nach WINKLER (unveröff.) ist jedoch der Proteinschaden bei UV von 254 μm sehr gering. Die kleinen Plaques sind daher wahrscheinlich durch DNS-Läsionen verursacht. In Abb. 5 ist ein Versuch mit variierter UV-Dosis bei konstanter

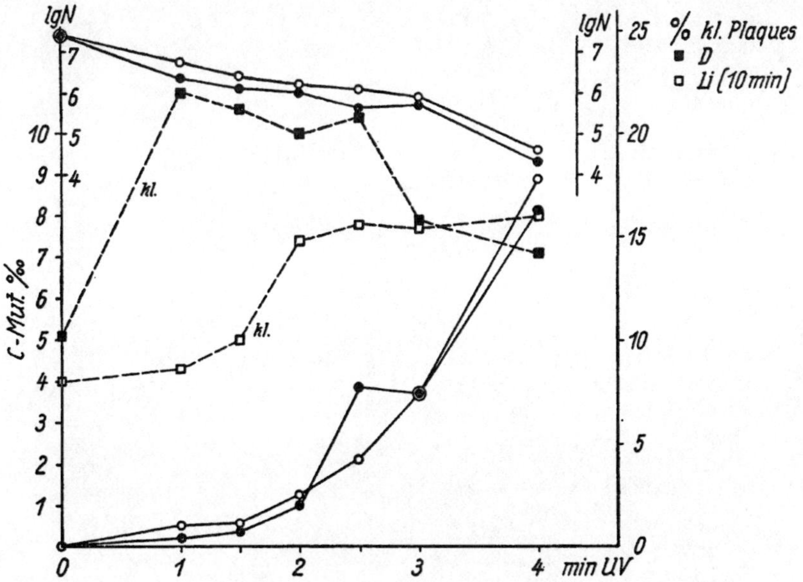

Abb. 5. UV-Dosiskurve von c-Mutation (unten), Inaktivierung (oben) sowie Kleinplaquebildung (Mitte) mit und ohne Nachbelichtung mit 10 min Licht. (Versuch ohne Mutationsreversion, aber mit Reaktivierung und Kleinplaque-Reversion)

Lichtdosis (10 min) dargestellt. Man erkennt deutlich die schwache, aber sichere Photoreaktivierung, das Fehlen der Photoreversion der Mutation in diesem Versuch sowie die starke Photoreversion der Kleinplaquebildung. Dieser Versuch mit fehlender Mutationsreversion wurde ausgewählt, um zu zeigen, daß die Mutationsreversion, die Reaktivierung und die Kleinplaque-Reversion voneinander unabhängige UV-Läsionen betreffen. Wenn es vorerst auch nur wahrscheinlich ist, daß auch die Latenzzeitdehnung auf Veränderungen im Genom beruht, so zeigen unsere Versuche doch eine Mannigfaltigkeit der strahleninduzierten chemischen Änderungsarten an der DNS an, welche biologische Effekte zur Folge haben. Wir dürfen also nicht nur von „der" biologischen UV-Wirkung sprechen, wie es oft geschieht. Bei der Aufklärung der chemischen Natur dieser verschiedenen Änderungen müssen die Photochemiker dem Genetiker helfen.

Literatur

Elmmauer, H., und R. W. Kaplan, 1959: Auslösung von Klarplaque-Mutationen durch UV-Bestrahlung des freien Phagen von *Serratia marcescens*. Naturwissenschaften **46**, 150.

Kaplan, R. W., 1960: Photoreversion of *c*-mutations induced by UV in the extracellular phage Kappa. Internat. Congr. Photobiol. Copenhagen 1960.

Kaplan, R. W., U. Winkler, and H. Wolf-Ellmauer, 1960: Induction and reversion of *c*-mutations by irradiations of the extracellular *ϰ*-Phage of *Serratia*. Nature (London) **186**, 330.

Kaplan, R. W., und S. K. Bose, 1961: Auslösung von Mutationen verschiedenen Typs durch Behandlung des extrazellulären Phagen *ϰ* von *Serratia* mit Nitrit. Z. allg. Mikrobiologie **1**, 274—285.

Krieg, D., 1958: Mutation induced in bacteriophage by ultraviolet light. Proc. 10th Int. Congr. Genetics II, 151—152.

Einfluß von Kälte oder Trockenheit während sowie von Behandlungen mit Biochemikalien nach der UV-Bestrahlung auf die Mutationsauslösung bei *Serratia*

R. W. Kaplan

Institut für Mikrobiologie der Johann-Wolfgang-Goethe-Universität, Frankfurt (Main)

Die Beeinflußbarkeit der strahleninduzierten Mutationsrate durch die Bedingungen während oder nach der Bestrahlung wird häufig als Beweis für räumlich indirekte Strahlenwirkung, z. B. über diffundierende Radikale oder mutagene Stoffwechseländerungen, angesehen. Jedoch kann ein während der Bestrahlung wirkender Faktor auch die Struktur und damit die Strahlensensibilität des direkt zu treffenden Gens selbst verändern, die Bedingungen nach der Bestrahlung können die Lebensdauer und Perfektierungsfähigkeit von prämutativen UV-Läsionen in der DNS beeinflussen. Im folgenden sollen einige Versuchsergebnisse von Doktoranden mitgeteilt werden, die für das Bild von den induzierten Mutationsvorgängen wichtig sein können. Als Objekt diente der Stamm CV von *Serratia marcescens*, bei dem drei verschiedene, häufige Farbmutationstypen ohne selektive Methode in demselben Zelltyp studiert werden können (Kaplan, 1959).

Daß diese Farbvarianten echte Genmutationen darstellen, darf aus allen bisherigen Beobachtungen geschlossen werden (Kaplan, 1959). Die Entstehung des *w*-Typs durch Mutation in wenigstens 5 verschiedenen Funktionsgenen (Cistren), die in eine Sequenz biochemischer Reaktionen der Pigmentbildung hintereinander eingreifen, wurde neuerdings von Winkler (1960) durch Versuche über gegenseitiges „Füttern" verschiedener *w*-Mutantengruppen mit diffusiblen Farbstoffvorstufen gezeigt. Die *s*-Kolonien gehen auf durch Mutation entstandene hochmutable Allele zurück, die während des Kolonienwachstums häufig zu verschiedenen stabileren Allelen (*w*-, *h*-, *r*-Typ) weiter mutieren, wodurch die Farbsektoren entstehen. Dies wird gestützt von Beobachtungen Winklers, wonach in jeder *s*-Kolonie *w*-Zellen von jeweils nur einem Funktionsgentyp vorhanden sind.

I. Wassergehalt während der Bestrahlung (K. Metzger)

In Fortsetzung von Studien von Kaplan und Kaplan (1956) an *Serratia* K wurden am Stamm CV nicht nur die UV-Effekte, sondern auch die Röntgenstrahlenwirkungen in Abhängigkeit von der relativen Feuchte (r. F.) 100 bis 1% untersucht. Auch der Zusammenhang von r. F. und Wassergehalt der Zellen wurde dabei durch Wägung geklärt. Für die durch UV ausgelösten Farbmutationen sowie die Zelltötung ergab sich auch beim Stamm CV (für Tötung beim Stamm HY) die von uns am Stamm K gefundene, zunächst geringe Zunahme beider Effekte von 98% r. F. an bis etwa 50% r. F. sowie der dann steile Aufstieg bis etwa 30% r. F. (50% r. F. entspricht etwa 8% Wassergehalt, 30% r. F. etwa 5,5% Wassergehalt). Nach noch niederen Feuchtigkeiten hin fallen dann die Werte für die Strahleneffekte wieder ab. Ein gesicherter Unterschied im Verhalten von Mutation und Tötung war nicht bemerkbar; auch die drei Mutationstypen reagierten ähnlich. Durch spezielle Versuche fand Metzger von 100% r. F. nach etwa 95% (40% Wassergehalt) einen kleinen, aber gesicherten Abfall der zwei UV-Sensibilitäten. Mit Röntgenstrahlen war dieser relativ erheblich stärker. Bei höheren Trocknungsgraden unterschieden sich die Kurven mit dieser Strahlung von denen mit UV jedoch nicht wesentlich, auch die Röntgen-Effekte waren unterhalb 40 bis 50% r. F. um ein Vielfaches gegenüber höherer Feuchtigkeit verstärkt.

Wir zogen daraus u. a. folgende Schlüsse: 1. Die Ähnlichkeit des H_2O-Einflusses auf die Effekte beider Strahlenarten deutet auf ähnliche Wirkungsarten; vielleicht sind auch bei den Röntgenstrahlen vorwiegend die Anregungen wirksam. 2. Die räumlich indirekte Wirkung spielt wohl nur bei sehr wasserhaltigen Zellen eine Rolle, bei trockenen entscheidet die vom H_2O-Gehalt beeinflußte Struktur des DNS über die Effektivität der Treffer in ihr. Da Mutation und Tötung sehr wahrscheinlich (s. u.) auf verschiedenen Läsionen beruhen, dürfte ihr gleichartiges Verhalten gegen die Feuchte durch einen von der Läsionsart unabhängigen Parameter verursacht sein. Ich denke z. B. an Energieleitung längs des DNS-Molekels, die von der Hydratation stark abhängen könnte.

II. Temperatur während der UV-Bestrahlung (H. Hainz)

Da über den Einfluß insbesondere tiefer Temperaturen auf die UV-Effekte bisher nur wenig bekannt ist, wurden an *Serratia* CV die Tötung sowie die Farbmutationen, beim Phagen $\varkappa$ die Inaktivierung sowie die *c*-Mutation untersucht. Während der Bestrahlung befanden sich die noch voll wasserhaltigen Zellen bzw. Phagen auf dem Gelatinefilm eines Glasplättchens, welches auf einem gekühlten bzw. angewärmten (—55° bis +40°C) Cu-Block in einer Kammer mit Quarzfenster lag. Sofort nach der UV-Bestrahlung wurden die Zellen bzw. Phagen in Saline suspendiert und ausgeplattet. Die Manipulationen außer der Bestrahlung änderten den Gehalt an Spontanmutanten nicht, das Überleben nur wenig. Obwohl die Methodik möglichst sorgfältig konstant gehalten wurde, zeigten die Ergebnisse ziemliche Schwankungen. Daher wurden die Versuche vielfach wiederholt und die Mittelwerte durch Fehlerrechnungen analysiert. Sie ergaben, daß die UV-Tötung der Bakterien nach niederen Temperaturen hin zunimmt, besonders steil zwischen —40° und —55°. Um 0° herum verläuft die Kurve etwa konstant und steigt von da wieder bis 30°, dann fällt sie nach 40°. Die Mutabilität der Farbgene ist viel weniger temperaturabhängig, es ließ sich nur ein geringer Anstieg von den tiefen zu den höheren Temperaturen um etwa den Faktor 2 sichern. Einen ähnlichen nach höheren Temperaturen ansteigenden Trend zeigte die Inaktivierung des Phagen sowie dessen *c*-Mutation. Wir schließen daraus, daß die Bakterientötung wohl auf anderen UV-Läsionen beruht als die Bakterien- und die Phagenmutation sowie die Inaktivierung des Phagen.

Beukers, IJlstra und Berends (1958) fanden, daß die zur Veränderung des Absorptionsspektrums führende, durch UV hervorgerufene Photoreaktion (Dimerisierung) von Thymin und Thymidylsäure bei —45° in gefrorener Lösung sehr viel stärker ist als bei Zimmertemperatur, während die Reaktion von Cytosin (Wasseraddition) bei Kälte etwas geringer ist als bei Wärme. Man könnte demnach vermuten, daß die Bakterientötung vorwiegend durch UV-Wirkung auf die Thymingruppen in der DNS, die anderen biologischen Effekte hauptsächlich durch Änderung an den Cytosingruppen entstehen.

III. Nachbehandlung mit Biochemikalien nach UV bei Serratia CV (E. Goes)

Außer Licht und Temperatur beeinflussen auch Chemikalien die nach der UV-Bestrahlung ablaufenden Vorgänge, welche die Mutation fertigstellen und also den Ertrag an Mutanten bestimmen. Goes prüfte in den letzten Jahren in breitem Umfang viele biochemisch wichtige Stoffe auf Beeinflussung dieser UV-Folgeprozesse bei *Serratia* CV. Es sollte Anhalt dafür gefunden werden, welche metabolischen Vorgänge für die Perfektierung der Mutationen bedeutsam sind, und ferner, ob verschiedene Gensorten desselben Genoms unterschiedlich beeinflußt werden. Bei diesem Bakterium können ja drei Farbmutationstypen getrennt in der gleichen Zelle studiert werden, nämlich zu weißen (w), zu hellroten (h) sowie zu sektorierten (s) Kolonien führende.

Auf Grund ihrer Studien an den Prototrophiemutationen einer try-Mutante von *Salmonella* tm. bzw. einer ähnlichen von *E. coli* hatten Witkin sowie weiterhin Haas und Doudney annehmen können, daß für die Mutationsperfektierung Proteinsynthese förderlich oder nötig ist und daß die DNS-Verdopplung den Vorgang abschließt. Die beiden letzten Autoren stellten ferner die Hypothese auf, daß das UV zunächst RNS-Vorstufen ändert, die mittels Proteinsynthese stabilisiert und dann in RNS inkorporiert werden, und daß die geänderte RNS schließlich die Synthese mutierter DNS verursacht.

Tabelle 1. Wirkung von Nachbehandlung (2 h, 25 °C) mit verschiedenen Chemikalien auf die UV-induzierten *h*-, *s*- und *w*-Mutationen von *Serratia* CV (nach Goes, 1961)

Stoff	Mutantentyp *h*	*s*	*w*
Casaminosäuren (CAS)	−	+	O
Hefeextrakt	−	+	O
Glutamat	−	(O, −)	(−)
Aminobuttersäure	+	−	O
Tyroxin	+	+	O
DNS-Hydrolysat (DN-Hy)	−	−	O
RNS-Hydrolysat (RN-Hy)	+	(+)	O
Adenin (A), Uracil (U) oder Thymin (T)	O	O	O
Guanin (G)	+	(−)	O
Cytosin (C)	+	O	O
A + G + C + U (je 1/4)	+	O	O
A + G + C + T (je 1/4)	+	O	O
Hypoxanthin, Xanthin	−	O	O
Chloramphenicol (CHL)	+	−	O
Streptomycin (Str.)	+	+	O
Na-Azid	+	+	O
Sulfathiazol	+	−	O
PAB	(−)	+	O
DN-Hy + RN-Hy	+ D	O D?	O
DN-Hy + Str.	+ D	+ D	O
DN-Hy + CHL	− A	− A	O
DN-Hy + CAS	− A	O A	O
CAS + CHL	+	(+) A	O
CAS + Str.	(+) D	+ A	O
Str. + CHL	+ S	+ A	O
% Mutationen in Pufferkontrolle (Durchschnitt)	21	5,3	4,3

+ = Zunahme, − = Abnahme, O = ohne Veränderung gegenüber Pufferkontrolle, ( ) = statistisch schwach gesichert. In den Stoffgemischen: A = additive oder subtraktive Wirkung, S = gegenseitige Schwächung gleichsinnig wirkender Stoffe, D = Dominanz der Wirkung des zweiten Stoffes über die des ersten.

In unseren Versuchen wurden die Stoffkonzentrationen so gering gewählt, daß die UV-Tötung nicht oder kaum beeinflußt wurde, sondern nur die Mutationen. Beteiligung von selektiver Tötung oder Reaktivierung an den beobachteten Effekten wurde ferner durch spezielle Versuche ausgeschlossen. Es wurden sowohl Verminderungen (—) und Zunahmen (+) des Mutationsertrags wie auch fehlender Einfluß (○) festgestellt. Die wichtigsten Befunde sind in Tab. 1 zusammengestellt.

Zunächst fällt auf, daß die drei Mutationsgruppen sehr individuell auf die Nachbehandlungen reagieren. Die *w*-Mutationen sind von kaum einem Stoff beeinflußbar, nur Glutamat wirkt schwach und unsicher. Die *h*- und *s-Mutationen* werden bisweilen gleichsinnig, oft aber gegensinnig oder unabhängig beeinflußt. Diese Elektivität der Nachbehandlungen zeigt an, daß es wenigstens drei Sorten von Genen hinsichtlich ihrer prämutativen Vorgänge gibt. Falls Indirektheit der UV-Wirkung vorliegt, müßten mindestens drei Mechanismen, z. B. „Strahlengifte", existieren, die gen-elektiv wirken müssen. Dies macht direkte Erzeugung prämutativer UV-Läsionen in den Genen (DNS) wahrscheinlicher, wobei diese prämutierten Gene wegen ihrer unterschiedlichen Beschaffenheit elektiv von den Nachbehandlungen angegriffen werden. Die oben angeführten Hypothesen WITKINS sowie HAAS' und DOUDNEYS können höchstens für manche Gene zutreffen, keineswegs für alle. Die Vorstellung WITKINS, daß die Stoffe durch Förderung oder Verzögerung des Eintritts der ersten Zellteilung wirken und also die Perfektierung der Mutationen proportional dem Zeitintervall bis zur DNS-Replikation ist, wird zweifelhaft; denn bei Dehnung oder Kürzung dieses Intervalls durch einen Stoff sollten alle Mutationen im gleichen Sinne reagieren, was nicht immer zutrifft.

Die Reziprozität der Wirkung von CAS und CHL sowie DNS-Hy und Str. sowie die Dominanz der „Gifte" über die „Baustoffe" der Synthesen deuten auch in unseren Versuchen auf Beteiligung der Protein- sowie der DNS-Synthese an der Mutationsperfektierung. Jedoch erzeugt eine Förderung der Proteinsynthese (CAS) nur bei den *s*-Mutationen eine Zunahme analog wie bei den try$^+$-Mutationen von *E. coli*; die *h*-Mutationen werden dadurch sogar vermindert. Der Mechanismus des Eingriffs der Proteinsynthese ins Mutationsgeschehen ist also Gen(gruppen) spezifisch. Da wir drei verschieden reagierende Mutationssorten vor uns haben, kann die Unterschiedlichkeit nicht nur auf der Differenz zwischen intragenischen (Gen-) und intergenischen (Bruch + Umbau-)Mutationen beruhen, wie in den Fällen, wo nur zwei Mutationstypen in derselben Zelle studiert werden.

Literatur

BEUKERS, R., J. IJLSTRA, und W. BERENDS, 1958: The effect of UV-light on some components of the nucleic acids. Recueil Trav. chim. Pays-Bas **77**, 729.

GOES, E., und R. W. KAPLAN, 1959: Elektive Wirkung von Nachbehandlungen mit Chemikalien auf die UV-induzierten Farbmutationen von *Serratia* CV. Naturwissenschaften **36**, 335.

HAAS, F. L., und C. O. DOUDNEY, 1959: Mutation induction and expression in bacteria. Proc. nat. Acad. Sci. U. S. **45**, 1620 (dort weitere Literatur).

KAPLAN, R. W., und CH. KAPLAN, 1956: Influence of water content on UV-induced *s*-mutation and killing in *Serratia*. Exper. Cell Res. **11**, 378.

KAPLAN, R. W., 1959: Studien über die UV-Mutabilität verschiedener Stämme von *Serratia* und über die Natur der Färbungsmutationen. Arch. Mikrobiol. **32**, 138.

METZGER, K., 1961: Der Einfluß des Wassergehaltes auf Inaktivierung und Mutabilität von *Serratia marcescens* durch UV- und Röntgenbestrahlung. Z. allg. Mikrobiol. **1**, 29—45.

WINKLER, U., 1960: Genetische Untersuchungen an farbsektorierten Mutantenkolonien von *Serratia* CV unter Ausnutzung des Pigment-Syntrophie-Phänomens. Z. allg. Mikrobiol. **1**, 83—85.

WITKIN, E. M., 1956. Time, temperature, and protein synthesis. A study of ultraviolet-induced mutation in bacteria. Cold Spring Harbor Symp. **21**, 123.

Strahleninaktivierung von *Saccharomyces* in Abhängigkeit von Ploidiegrad und Genotyp

W. LASKOWSKI

I. Physikalisches Institut der Freien Universität Berlin, Abt. für Biophysik

Daß die Inaktivierung von *Saccharomyces* durch Röntgenstrahlen u. a. auf der Induktion rezessiver und dominanter Letalmutationen beruht, wurde von MORTIMER (1955) sowie OWEN und MORTIMER (1956) in Kreuzungs- und Sporulationsversuchen nachgewiesen. Ein Modell, das eine quantitative Abschätzung der Anteile strahleninduzierter chromosomaler rezessiver und dominanter Letalmutationen sowie plasmatischer Letaleffekte gestattet, wurde kürzlich entwickelt (STEIN und LASKOWSKI, 1958) und an einer für eine derartige Analyse besonders gezüchteten Serie weitestgehend homozygoter, isogener *Saccharomyces*-Stämme unterschiedlichen Ploidiegrades (haploid bis tetraploid) angewendet (LASKOWSKI, 1960; LASKOWSKI und STEIN, 1960; STEIN und LASKOWSKI, 1960). Die mit dieser indirekten Methode gewonnenen Ergebnisse sind mit den aus den direkten Kreuzungs- und Sporulationsversuchen gewonnenen Resultaten recht gut verträglich. Als wesentliches Ergebnis stellte sich heraus, daß die Strahleninaktivierung von Haplonten vorwiegend durch die Induktion rezessiver Letalmutationen erfolgt, während die Inaktivierung polyploider Stämme vorwiegend auf die Induktion dominanter Letalmutationen zurückzuführen ist.

Um die Strahlenresistenz in Abhängigkeit von Ploidiegrad noch eingehender zu analysieren, wurden, ausgehend von der bereits erwähnten Serie haploider bis tetraploider *Saccharomyces*-Stämme (LASKOWSKI, 1960), zwei weitere weitestgehend homozygote, isogene Stammserien gezüchtet, die die Voraussetzung zur Gewinnung weitestgehend homozygoter und isogener penta- und hexaploider Stämme boten. Der Stammbaum dieser Stammserien ist in Abb. 1 wiedergegeben.

Sämtliche Stämme stammen von dem haploiden Ausgangsstamm α-41 ab. Durch Isolierung von spontan in haploiden Kulturen auftretenden diploiden Zellen, die für die Paarungstypallele entweder homozygot oder heterozygot sind (ROMAN und SANDS, 1953), sowie durch geeignete Sporulations- und Kreuzungsschritte wurden mehrere diploide bis tetraploide Stämme hergestellt, die bis auf die Paarungstypallele vollkommen homozygot und isogen sind (Zentrum der Abb. 1). Weiterhin wurden mit Hilfe der Stempeltechnik (LEDERBERG und LEDERBERG, 1952) aus zwei haploiden Stämmen (211–1a und 211–1b) spontan aufgetretene auxotrophe Mutanten isoliert (211–1aM und 211–1bN), die wiederum den Ausgangspunkt für den Aufbau von homozygoten isogenen Di-, Tri- und Tetraplonten darstellten. Eine dieser auxotrophen Mutanten ist bedürftig für die Aminosäure Serin (211–1bN), während die andere für die Aminosäuren Valin und Isoleucin bedürftig ist. Kreuzungsexperimente und Ascusanalysen ergaben für beide Mutanten, daß jeweils ein mutiertes Gen für die Bedürftigkeit verantwortlich ist. Mit Hilfe der Prototrophtechnik (POMPER und BURKHOLDER, 1949) wurden durch Verwendung geeigneter Di- und Triplonten die penta- bzw. hexa-

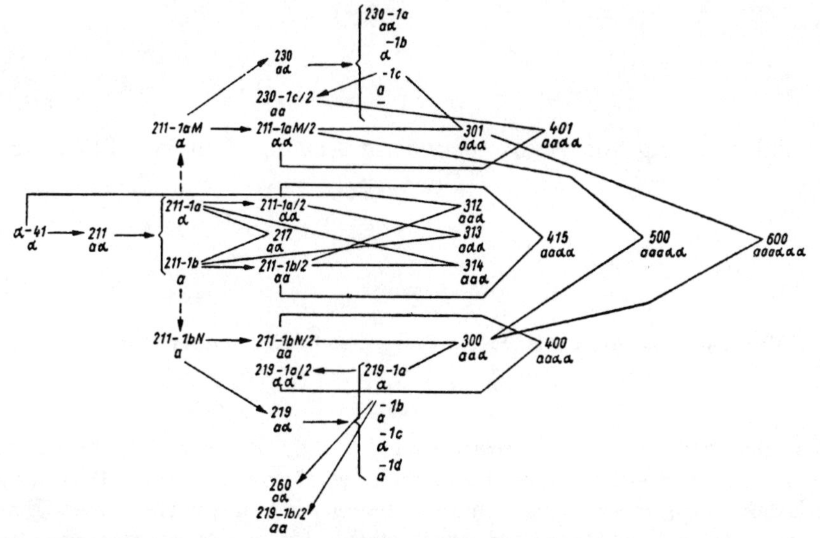

Abb. 1. Stammbaum dreier Serien weitestgehend homozygoter, isogener haploider bis tetraploider *Saccharomyces*-Stämme sowie je eines penta- und hexaploiden Stammes. Der Ploidiegrad der polyploiden Stämme ist jeweils aus der ersten Ziffer der dreistelligen Stammnummer erkennbar. Die gleiche Charakterisierung wurde für die *αα*-Diplonten gewählt. Durch Selbstdiploidisierung hervorgegangene *aa*- bzw. *αα*-Diplonten sind durch die Anfügung von „/2" an die Nummer des jeweiligen haploiden Ausgangsstammes gekennzeichnet. Die Paarungstypkonstitution ist unter jeder Stammnummer angegeben (Einzelheiten im Text)

ploiden Stämme 500 und 600 isoliert, deren Ploidiegrad durch genetische Analysen bestätigt wurde (Laskowski, 1962a).

Infolge der angewendeten Züchtungsmethode sind die haploiden bis tetraploiden Stämme jeder Serie mit Ausnahme der Paarungstypallele völlig homozygot und isogen. Untereinander unterscheiden sich die Stämme, der einzelnen Serien nur durch ein jeweils homozygot vorhandenes mutiertes Allél, durch das in einem Falle die Synthesefähigkeit von Serin blockiert wird (Abkömmlinge von 211–1bN, unterer Teil der Abb. 1), im anderen die Bedürftigkeit für Valin und Isoleucin bedingt ist (Abkömmlinge von 211–1aM, oberer Teil der Abb. 1).

Bei einer derart weitgehenden Homozygotie und Isogenie der Erbfaktoren konnte eine ebenfalls weitgehend gleiche Strahlenresistenz der Stämme, die dem gleichen Ploidiegrad angehören, erwartet werden.[1]) Tatsächlich wurde diese Erwartung auch bei Inaktivierungsexperimenten mit ultraviolettem Licht (UV) erfüllt. Sämtliche auf der Abb. 1 eingetragenen haploiden Stämme zeigen die gleiche Strahlenempfindlichkeit, das gleiche gilt jeweils für die di-, tri- bzw. tetraploiden Stämme; die Dosiseffektkurven sind also mit den bereits früher veröffentlichten Kurven der zentralen Stammserie identisch (Laskowski, 1960). Da nur jeweils ein penta- und hexaploider Stamm vorhanden ist, konnte ein solcher Vergleich in diesen Ploidiegraden nicht vorgenommen werden. Generell ist festzustellen, daß die Resistenz gegenüber UV von den Haplonten über die Diplonten bis zu den Triplonten ansteigt; bei Tetraplonten fällt sie dagegen wieder ab.

Eine derartige identische Empfindlichkeit aller Stämme einer Ploidiestufe ist dagegen bei den Röntgenstrahlen-Experimenten nicht vorhanden. In Abb. 2 sind die Dosiseffektkurven einer Serie haploider bis tetraploider Stämme sowie die Kurven für den penta- und hexa-

---

[1]) Die bei den Bestrahlungsversuchen verwendeten Methoden sind die gleichen wie bei Laskowski, 1960.

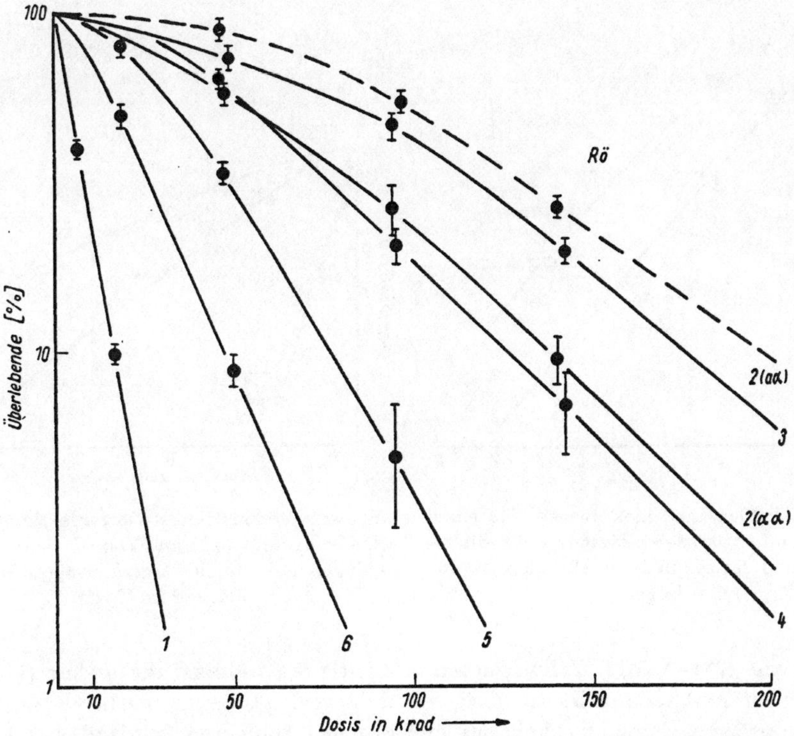

Abb. 2. Röntgenstrahlen-Dosiseffektkurven für weitestgehend homozygote, isogene haploidebis hexaploide *Saccharomyces*-Stämme. Folgende Stämme wurden verwendet (vgl. hierzu Abb. 1):

1	(haploid)	Stamm Nr. 211-1bN	4	(tetraploid)	„ „	400
2 (αα)	(diploid)	„ „ 211-1bN/2	5	(pentaploid)	„ „	500
2 (*a*α)	(diploid)	„ „ 219	6	(hexaploid)	„ „	600
3	(triploid)	„ „ 300				

Die Senkrechten geben die halbe maximale Schwankungsbreite an

ploiden Stamm wiedergegeben. Deutlich ist eine Abnahme der Strahlenresistenz von Ploidiestufen > 3 bzw. > 2 an zu bemerken. Diese bei den polyploiden Stämmen mit zunehmendem Ploidiegrad festzustellende Zunahme der Strahlenempfindlichkeit kann als Anzeichen für die Induktion dominanter Letalmutationen als vorwiegende Inaktivierungsursache gewertet werden. Auffällig ist eine unterschiedliche Resistenz innerhalb der Gruppe der Diplonten. Stämme, die in allen Allelenpaaren einschließlich der Paarungstypallele homozygot sind (*aa*- bzw. αα-Stämme), sind röntgenstrahlen-empfindlicher als Stämme, die bei sonst völliger Homozygotie für die Paarungstypallele *a* und α heterozygot sind (*a*α-Stämme). Die *a*α-Diplonten sind sogar resistenter als die Triplonten. Dieser „*a*α-Effekt" tritt bei allen geprüften Diplontenpaaren (siehe in Abb. 1 z. B. 211 und 211–1a/2, 219 und 211–1bN/2, 260 und 219–1b/2, 230 und 211–1*a*M/2 usw.) nach Bestrahlung mit Röntgen- oder Po-α-Strahlen auf (s. Abb. 3) und ist möglicherweise charakteristisch für ionisierende Strahlen.

Außer den Paarungstypallelen zeigen aber auch Allele, die eine Aminosäurebedürftigkeit verursachen, einen Einfluß auf die Resistenz gegenüber Röntgen- und Po-α-Strahlen. Dieser Einfluß wirkt sich allerdings nur in Ploidiestufen > 1 aus. Während die Haplonten aller Stammserien aus Abb. 1 die gleiche Röntgenstrahlenempfindlichkeit zeigen, sind die auxotrophen Diplonten wesentlich strahlenresistenter als die prototrophen Diplonten. In Abb. 4 sind als Beispiel die Inaktivierungskurven für zwei diploide αα-Stämme wiedergegeben, von

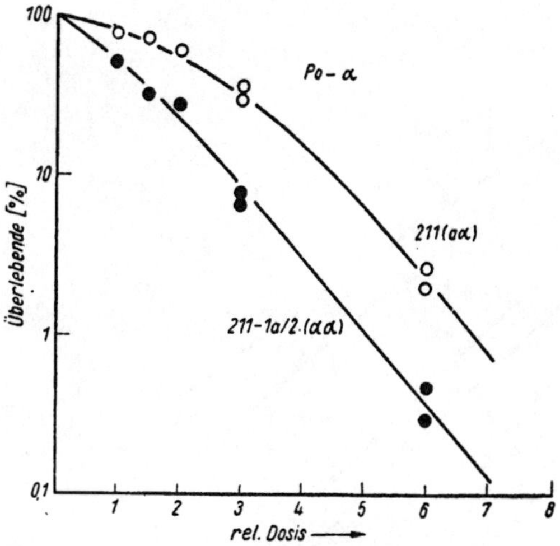

Abb. 3. Po-α-Strahlen-Dosiseffektkurven für einen *aa*- und einen αα-*Saccharomyces*-Stamm. Beide diploiden Stämme sind für die übrigen Allele homozygot und isogen

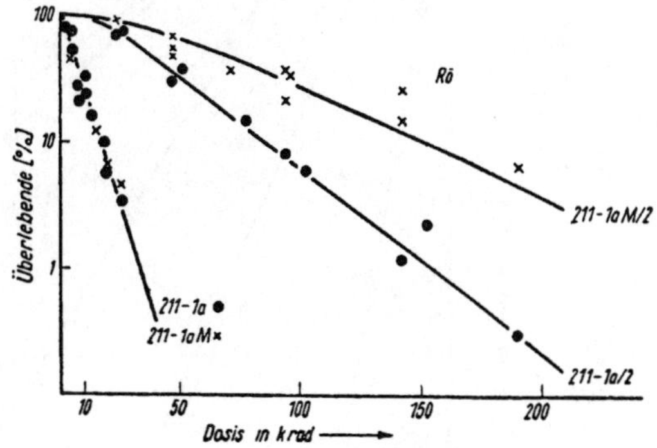

Abb. 4. Röntgenstrahlen-Dosiseffektkurven für verschiedene prototrophe und Valin-Isoleucin-bedürftige haploide und diploide *Saccharomyces*-Stämme (Erklärung im Text)

denen der eine (211–1aM/2) Valin-Isoleucin-bedürftig ist, während der andere (211–1a/2) unabhängig von Aminosäuren ist. Im übrigen sind die beiden Stämme, wie bereits ausgeführt, isogen. Die größere Empfindlichkeit des prototrophen Diplonten ist deutlich erkennbar, während die haploiden auxo- und prototrophen Ausgangsstämme 211–1aM und 211–1a die gleiche Empfindlichkeit zeigen. Ein gleichartiger Resistenzunterschied ist nach Po-α-Bestrahlung zu beobachten. Als Beispiel sind in Abb. 5 die Po-α-Inaktivierungskurven für den Serin-bedürftigen *aa*-Stamm 211–1bN/2 und den prototrophen sonst isogenen Stamm 211–1b/2 gezeigt.

Da zwischen *aa*- und αα-Diplonten der Valin-Isoleucin-bedürftigen Stammserie und entsprechenden Diplonten der Serin-bedürftigen Stammserie kein wesentlicher Unterschied im Ausmaß des Resistenzgewinns gegenüber Röntgen- bzw. Po-α-Strahlen besteht, sei dieser mit einer Aminosäurebedürftigkeit korrelierte Resistenzeffekt im folgenden allgemein als „AS-Effekt“ bezeichnet. Einige für den AS-Effekt charakteristische Merkmale seien zusammenfassend angeführt.

1. Außer bei den erwähnten diploiden Stämmen wirkt sich der AS-Effekt auch bei allen entsprechenden tri- und tetraploiden Stämmen des hier behandelten Stammsystems aus. Er tritt jedoch nicht – wie bereits erwähnt – bei haploiden Stämmen auf.

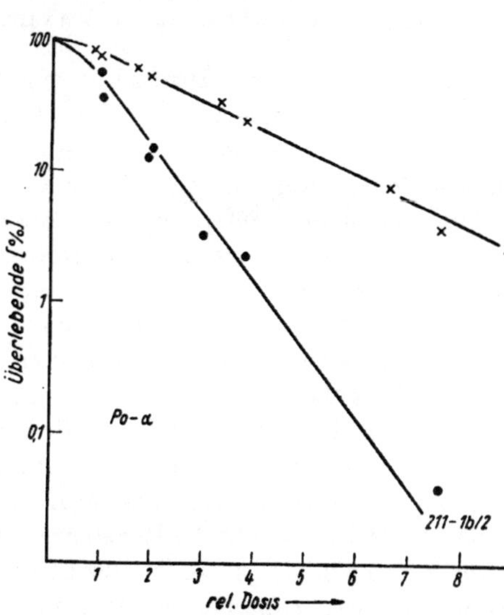

Abb. 5. Po-α-Strahlen-Dosiseffektkurven für einen prototrophen (211-1b/2) und einen Serin-bedürftigen (211-1bN/2) diploiden *Saccharomyces*-Stamm

2. Der AS-Effekt ersetzt nicht den vorher erwähnten $a\alpha$-Effekt, sondern letzterer überlagert sich noch dem AS-Effekt (Abb. 6).

3. Werden aus Aminosäure-bedürftigen diploiden Stämmen prototrophe Mutanten isoliert, so zeigen letztere etwa die gleiche Strahlenempfindlichkeit wie die prototrophen Ausgangsstämme (Abb. 6). Der AS-Effekt ist also in Zellen, die mit hoher Wahrscheinlichkeit für die Aminosäurebedürftigkeit heterozygot sind, nicht ausgeprägt. (Dabei ist bisher unentschieden, ob die isolierten Mutanten auf echte Rückmutationen oder auf Suppressormutationen zurückzuführen sind.)

4. Der AS-Effekt tritt bereits maximal in Erscheinung, wenn ein diploider Stamm für eine Aminosäure bedürftig ist. Mehrfach bedürftige Mutanten, wie z. B. ein Valin-Isoleucin- und Methioninbedürftiger Stamm, zeigen keine größere Resistenz als einfach bedürftige Stämme.[1])

5. Der AS-Effekt tritt – wie bereits erwähnt – ebenso wie der $a\alpha$-Effekt nur in Inaktivierungsversuchen mit ionisierenden Strahlen (Röntgen- und α-Strahlen), nicht aber bei UV-Bestrahlung auf (Abb. 7).

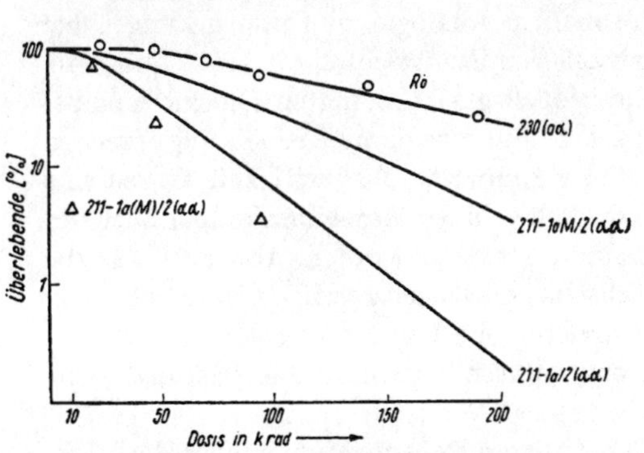

Abb. 6. Röntgenstrahlen-Dosiseffektkurven der αα-Stämme 211-1aM/2 und 211-1a/2 aus Abb. 4. Dazu die Dosiseffektkurve für den entsprechenden $a\alpha$-Stamm 230 sowie Inaktivierungsraten der prototrophen „Rückmutanten" 211-1a(M)/2

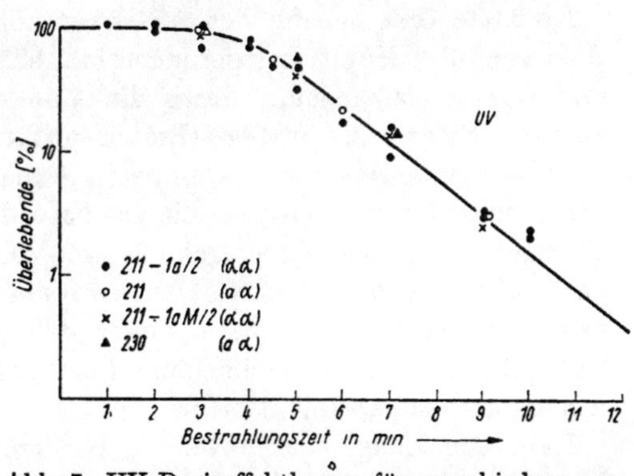

Abb. 7. UV-Dosiseffektkurve für verschiedene prototrophe und Valin-Isoleucin-bedürftige diploide *Saccharomyces*-Stämme (Erklärung im Text)

Diese Befunde sind in mehrfacher Hinsicht nicht uninteressant. Allgemein zeigen beide Resistenzeffekte, der $a\alpha$- sowohl als der AS-Effekt, daß durch bestimmte, von einzelnen Genen gesteuerte Änderungen der Zellphysiologie die inaktivierende Wirkung ionisierender Strahlen reduziert wird. Sieht man von der wenig plausiblen Annahme ab, daß durch die infolge einer geänderten Paarungstypkonstitution oder Aminosäurebedürftigkeit geänderten zellphysiologischen Zustände spezifische Modifikationen an den für die Inaktivierung kritischen Treffbereichen erfolgen, so weist die Existenz der beiden Effekte auf ein Vorkommen indirekter Treffermechanismen hin. Diese Folgerung steht in gewissem Gegensatz zu der gerade in jüngster Zeit wieder häufiger diskutierten Annahme einer direkten Trefferwirkung bei der Inaktivierung von Hefezellen durch ionisierende Strahlen (Sommermeyer und Magnus, 1960). Letztere Annahme wurde aus der Modifizierbarkeit der inaktivierenden Wirkung ionisierender Strahlen durch äußere (extrazelluläre) Änderung bestimmter Bestrahlungsbedingungen (wie z. B. Temperatur, O_2-Druck) abgeleitet. Im Gegensatz hierzu werden die hier diskutierten Effekte durch intrazellulär genetisch bedingte Milieuänderungen verursacht. Eine Unterscheidung zwischen räumlich-zeitlicher und rein zeitlicher Indirektheit (s. z. B.

[1]) Anmerkung bei Korrektur: Neuere Versuche haben gezeigt, daß Diplonten, die nur Methioninbedürftig sind, keine höhere Strahlenresistenz als verwandte prototrophe Diplonten haben.

Kaplan, 1959) der zur Inaktivierung führenden strahleninduzierten Prozesse ist allerdings in diesem System zur Zeit nicht möglich und dürfte, wenn überhaupt, nur schwierig herbeizuführen sein. Immerhin mögen Untersuchungen über das Verhalten beider Effekte in Abhängigkeit von bestimmten modifizierenden Mitteln Aufschlüsse über beeinflußbare Parameter geben. Derartige Versuche sind geplant.

Über obige grundsätzliche Aussagen hinaus ist über die konkreten, die beiden Effekte verursachenden biochemischen Veränderungen innerhalb der Zelle noch wenig bekannt. Für den *aα*-Effekt mag eine kürzlich nachgewiesene erhöhte Katalasekonzentration der *aα*-Diplonten im Vergleich zu *aa*- oder *αα*-Diplonten verantwortlich gemacht werden (Laskowski, Lochmann, Wacker und Stein, 1960; Laskowski, 1962b). Dieser biochemische Unterschied zwischen den bis auf die Paarungstypallele isogenen *aα*- bzw. *aa*- und *αα*-Stämmen erscheint im Hinblick auf die unterschiedliche Sporulations- und Kopulationsfähigkeit der aα-Zellen (kopulationsunfähige Sporenbildner) und *aa*- oder *αα*-Zellen (sporulationsunfähige Kopulationspartner) nicht unplausibel.

Konkrete Ursachen für den AS-Effekt könnte man in Analogie zu Untersuchungen bei Bakterien über Mutationsfixierungen in Abhängigkeit von Stoffwechselprozessen (Doudney und Haas, 1959) in einer durch die Aminosäurebedürftigkeit bedingten Blockierung bestimmter Stoffwechselprozesse (Proteinsynthese) sehen. Für eine derartige Deutung sprechen der gleich große Resistenzgewinn bei unterschiedlicher Aminosäurebedürftigkeit. Gegen eine derartige Deutung spricht aber die Tatsache, daß die Zellen in der Regel unmittelbar nach der Bestrahlung auf ein Hefeextrakt-Pepton-Vollmedium gebracht wurden, also auch für die auxotrophen Zellen alle zum Ablauf von Stoffwechselprozessen notwendigen Nährstoffe vorhanden waren. Dagegen spricht weiter, daß eine dreistündige Verzögerung der Aufbringung bestrahlter Zellen auf Vollmedium (Aufbewahrung währenddessen in Pufferlösung) ohne Einfluß auf den AS-Effekt bleibt.

Kann demzufolge eine mittelbare Wirkung der Aminosäurebedürftigkeit über die Blokkierung bestimmter Stoffwechselprozesse als Ursache des AS-Effekte kaum in Frage kommen, so wird man dazu geführt, an eine mehr unmittelbare Schutzwirkung der betreffenden Aminosäuren oder deren Vorstufen zu denken. Eine derartige Hypothese wird bestärkt durch die eindeutige Schutzwirkung, die von bestimmten Aminosäuren (u. a. Valin, Isoleucin, Serin und Methionin) auf Desoxyribonucleinsäure bei Röntgenbestrahlung in vitro ausgeübt wird (Lochmann, persönliche Mitteilung). Weitere Experimente zur Klärung der biochemischen Ursachen des AS-Effektes sind in Vorbereitung.

Abschließend sei noch auf die Besonderheit des AS-Effektes hingewiesen, sich nur in di- und polyploiden Zellen auszuwirken. Sucht man zur Erklärung dieser Erscheinung nach möglichen Unterschieden im Inaktivierungsmechanismus zwischen haploiden und höherploiden *Saccharomyces*-Zellen, so fallen die eingangs erwähnten Unterschiede in den jeweiligen Anteilen rezessiver und dominanter Letalmutationen bei der Inaktivierung haploider bzw. di- und polyploider Zellen auf. Während die Inaktivierung von Haplonten vorwiegend auf induzierten rezessiven Letalmutationen beruht, wird die Inaktivierung höher-ploider Zellen vorwiegend durch induzierte dominante Letalmutationen verursacht. Im Hinblick auf diesen Tatbestand wäre die Ploidiegradabhängigkeit des AS-Effektes durch die Annahme erklärbar, daß dieser Effekt vornehmlich die Induktionswahrscheinlichkeit dominanter Letalmutationen erniedrigt, sich deswegen also nur in Ploidiestufen > 1 merklich auswirkt. Da nach allgemeiner Auffassung dominante Letalmutationen vorwiegend durch Chromosomenbrüche und daraus folgende Aberrationen verursacht werden, würde das bedeuten, daß durch den AS-Effekt besonders die Wahrscheinlichkeit für die Strahleninduktion von Chromosomenbrüchen reduziert wird. Da eine frühere Analyse gezeigt hat, daß der Anteil dominanter Letalmutationen bei der UV-Inaktivierung diploider bis tetraploider Zellen wesentlich geringer als bei Inakti vierung mit ionisierenden Strahlen ist (STEIN und LASKOWSKI, 1960), wird das Fehlen des AS-Effektes in UV-Inaktivierungsexperimenten verständlich.

Zusammenfassend sei noch einmal die Notwendigkeit hervorgehoben, bei der Analyse strahleninduzierter Inaktivierungen in Abhängigkeit vom Ploidiegrad mögliche Beeinflussungen der ploidiegradbedingten Strahlenempfindlichkeit durch unvollständige Homozygotie oder Isogenie der untersuchten Stämme zu beachten. Erst bei Verwendung streng homozygoter, isogener Stämme unterschiedlichen Ploidiegrades ist eine Beurteilung der allein durch eine Vermehrung des Chromosomensatzes pro Zellkern bedingten Strahlenempfindlichkeit möglich. Diese an einem speziellen Objekt, den Hefen, gewonnenen Erkenntnisse, dürften auch für Strahlenversuche mit anderen Organismen nicht ohne Bedeutung sein.

Für verantwortungsvolle technische Assistenz bei der Durchführung der Versuche bin ich Fräulein CHRISTA UMLAUF und Fräulein GABRIELE RIEBKE zu Dank verpflichtet.

Die Versuche wurden mit Unterstützung der Deutschen Forschungsgemeinschaft und des Bundesministeriums für Atomkernenergie und Wasserwirtschaft durchgeführt.

Literatur

DOUDNEY, C. O., and F. L. HAAS, 1959: Mutation induction and macromolecular synthesis in bacteria. Proc. nat. Acad. Sci. U. S. **45**, 709–722.

KAPLAN, R. W., 1959: Strahlengenetik der Mikroorganismen. In: Strahlenbiologie, Strahlentherapie, Nuklearmedizin und Krebsforschung. Stuttgart.

LASKOWSKI, W., 1960: Inaktivierungsversuche mit homozygoten Hefestämmen verschiedenen Ploidiegrades. I. Aufbau homozygoter Stämme und Dosiseffektkurven für ionisierende Strahlen, UV und organische Peroxyde. Z. Naturforsch. **15b**, 495–506.

LASKOWSKI, W., 1962a: VI. Über den Aufbau weitgehend isogener, homozygoter penta- und hexaploider Stämme sowie den Einfluß bestimmter mutierter Allele auf die Strahlenresistenz. Z. f. Naturforsch. **17b**, 93–108.

LASKOWSKI, W., 1962b: Der *aα*-Effekt, eine Korrelation zwischen Paarungstypkonstitution und Strahlenresistenz bei Hefen. Zbl. Bakt. Orig. **184**, 251–258.

LASKOWSKI, W., and W. STEIN, 1960: Inaktivierungsversuche mit homozygoten Hefestämmen verschiedenen Ploidiegrades. II. Mikroskopische Beobachtungen nach Inaktivierung mit ionisierenden Strahlen, UV und organischen Peroxyden. Z. Naturforsch. **15b**, 604–612.

LASKOWSKI, W., E.-R. LOCHMANN, A. WACKER und W. STEIN, 1960: Inaktivierungsversuche mit Hefestämmen verschiedenen Ploidiegrades. III. Biochemische Eigenschaften in Abhängigkeit von Ploidiegrad und Kombination der Paarungstypallele. Z. Naturforsch. **15b**, 730–734.

LEDERBERG, J., and E. M. LEDERBERG, 1952: Replica plating and indirect selection of bacterial mutants. J. Bact. **63**, 399–406.

MORTIMER, R. K., 1955: Evidence for two types of X-ray-induced lethal damage in *Saccharomyces cerevisiae*. Radiation Research **2**, 361–368.

OWEN, M. E., and R. K. MORTIMER, 1956: Dominant lethality induced by X-rays in haploid *Saccharomyces cerevisiae*. Nature (London) **177**, 625–626.

POMPER, S., and P. R. BURKHOLDER, 1949: Studies on the biochemical genetics of yeast. Proc. nat. Acad. Sci. U. S. **35**, 456–464.

ROMAN, H., and S. M. SANDS, 1953: Heterogeneity of clones of *Saccharomyces* derived from haploid ascospores. Proc. nat. Acad. Sci. U. S. **39**, 171–179.

SOMMERMEYER, K., und H. MAGNUS, 1960: Vergleichende Untersuchungen über die Strahlenempfindlichkeit von *Bacterium coli* und von Hefe, insbesondere bei Anwendung von energiereichen Strahlen. Z. Naturforsch. **15b**, 770–778.

STEIN, W., und W. LASKOWSKI, 1958: Zur mathematischen Analyse der Strahleninaktivierung mikrobiologischer Objekte verschiedenen Ploidiegrades unter Berücksichtigung genetischer und nichtgenetischer Anteile. Z. Naturforsch. **13b**, 651–657.

STEIN, W., und W. LASKOWSKI, 1960: Inaktivierungsversuche mit homozygoten Hefestämmen verschiedenen Ploidiegrades. IV. Quantitative Deutung unter Berücksichtigung genetischer und nichtgenetischer Anteile. Z. Naturforsch. **15b**, 734–743.

UV and Ionizing Irradiation Inactivation of the Streptomycin Resistance Markers in Transforming DNA of Different Species of Bacteria

R. PAKULA[1])

State Institute of Hygiene, Warsaw

Data concerning inactivation of some genetic markers in transforming DNA have been reported by several workers. ZAMENHOF et al. (1957) revealed that different transforming activities differ in their sensitivity to ultraviolet irradiation. They found that up to a 60 times higher UV energy dose is required to exert the same amount of destruction for the streptomycin resistance marker in DNA of type b *Hemophilus influenzae* than for capsule formation in DNA of the types a, b, or c. Also the ultraviolet sensitivities of the analogous streptomycin resistance markers in DNA of types a, b, and d are different.

RUPERT and GOODGAL (1960) found that the cathomycin, viomycin and erytromycin resistance markers of *H. influenzae* are equally susceptile to UV irradiation but that the streptomycin resistance activity is less sensitive.

From the studies of STUY (1961) with the transforming principle of *H. influenzae* extracted from irradiated bacteria it became evident that linked markers, like streptomycin-cathomycin resistance and kanamycin-cathomycin resistance, are more readily inactivated by UV irradiation than the individual ones. The sensitivity of the large complexes is about equal to the sum of the sensitivities of the single markers.

In transforming DNA of pneumococci several markers differ in their sensitivity to UV irradiation. LERMAN and TOLMACH (1959) found that destruction of 90% activity of the streptomycin resistance marker of pneumococci requires an energy dose about 8 times higher than for the same amount of destruction of optochin resistance activity. At lower survival levels there is a striking increase of the difference in the sensitivities.

The effects of ionizing radiation on DNA have been investigated mainly with the purpose of obtaining information on the size of the genetic markers active in bacterial transformation. MARMUR and FLUKE (1955) using dried preparations of pneumococcal DNA exposed to an electron beam did not find any differences in sensitivity of four markers. However linked marker were more sensitive than single activities.

Similar results were obtained by STUY (1961) on intracellular DNA of *H. influenzae* after irradiation with X-rays. It is evident from all the data available that while different markers of *H. influenzae* or pneumococci are usually differentially inactivated by UV irradiation, their discrimination by means of ionizing irradiation inactivation has not been demonstrated.

---

[1]) The second part of this communication on the effects of ionizing radiation was done in collaboration with D. SHUGAR, State Institute of Hygiene, Warsaw.

It seemed therefore of interest to make a comparative study on the sensitivities of the analogous streptomycin resistance markers of different species of bacteria to UV and ionizing irradiation.

Materials and Methods

The recipients used for transformation were the capsuleless Rd strain of *Hemophilus influenzae*, the pneumococcal rough strain R36A, and, for streptococci, the strains *Streptococcus sanguis*, type I/II, and the strain Challis for the group H hemolytic streptococci. As donors were used mutants of the above mentioned strains resistant to 2 mg/ml of streptomycin.

Preparation of transforming DNA and transformation techniques have been previously described (Pakula et al., 1959).

UV irradiation of DNA solutions were carried out with a Westinghouse germicidal lamp the energy output of which, at a proper distance, was calculated according to the graphs of Buttolph (1955). DNA solutions in 0,3 ml portions at a concentration of 100 μg/ml of DNA were subjected to UV irradiation in an open glass dish 18 mm in diameter. The dish was mounted on a electromagnetic vibrator for constant agitation.

The source of ionizing irradiation was a 22 mm disc of 50 curies of ^{60}Co mounted on a rod which could be displaced vertically. DNA samples were irradiated both in solutions and in the form of dried films. The solvent medium was 0.15 M NaCl. The dried preparations were formed as follows: Microscope cover glasses, cut to size 10 $\times$ 20 mm, were carefully degreased, washed, and 10 μl of a solution of DNA deposited on each slide; if the slides have been properly cleaned, the entire surface is instantly covered by this volume. They were then dried for 30 mins. at 30° in the presence of $CaCl_2$ and in vacuum over P_2O_5 prior to irradiation. Such preparations do not lose more than 25–30% of their transforming activity during a period of 10–20 hours at room temperature.

For direct comparisons of the sizes of the sensitive volume of the streptomycin resistance markers of DNA of the different bacterial species, the dried "films" were prepared from solutions containing a 1:1 mixture of two DNA samples in 10% yeast extract. The purpose of addition of the yeast extract was to provide additional protection against the indirect effect of free radicals. Irradiations were performed on mixtures of *H. influenzae* and streptococcal DNA and on *H. influenzae* and pneumococcal DNA. A combination of streptococcal and pneumococcal DNA could not be applied because of reciprocal transformation reactions.

The dried preparations on cover glasses were stacked parallel to each other on a specially constructed "scaffold" two cover glasses being located on each level and subjected to the same dose, the mean of two being used to estimate percentage of inactivation. The scaffold was inserted into a metal cylinder which was then closed with a lid containing a rubber gasket. The cylinder contained a receptacle with P_2O_5 and was exhausted to a pressure of a couple of mm Hg. The roof of the cylinder consisted of a thin layer of aluminium foil. The source was brought down until it made contact with the roof of the cylinder containing the samples.

Solutions were irradiated in 5 mm diameter test tubes supported at various positions on a plastic stand symmetrically with respect to the source. Duplicate samples were irradiated at any distance from the source. The dose rate at various distances from the source was obtained from a calibration curve prepared by means of a ferrous sulphate dosimeter.

Transformation reactions with competent cells preserved in the frozen state were carried out at concentrations of 0.01, 0.1 or 1 μg/ml. of undamaged or irradiated DNA.

Results and Discussion

Ultraviolet irradiation

The dose-survival curves for UV irradiation are of complex character, as has been shown by many workers. This was also shown in our own studies.

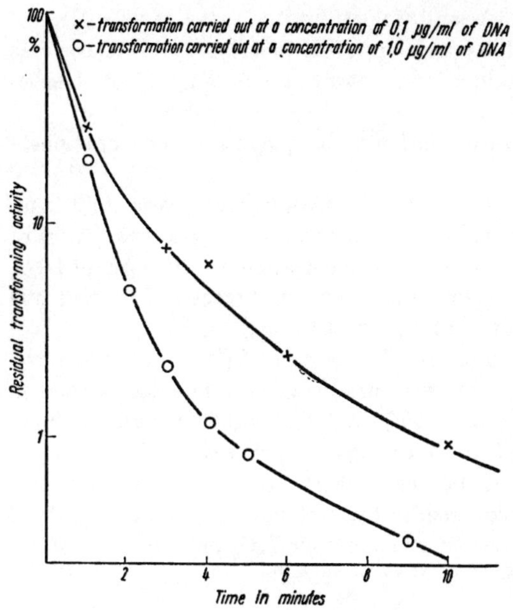

Fig. 1 represents a semilogarithmic plot of UV inactivation of transforming DNA of *Hemophilus influenzae*. The activity of the irradiated DNA to streptomycin resistance was tested at two concentrations of 1.0 and 0.1 μg/ml.

As may be seen the relative survival is found to be significantly higher when the irradiated DNA is assayed at the lower concentration. This is true not only for the streptomycin resistance activity of *H. influenzae* but also for the analogous activities of DNA of the pneumococcus and streptococcus inactivated to different extents by UV irradiation.

Fig. 1. UV inactivation of DNA of *H. influenzae*. Dose rate 11.6 ergs. $mm^{-2} \cdot sec^{-1}$; × = transformation carried out at a concentration of 0.1 μg/ml of DNA; ○ = transformation carried out at a concentration of 1.0 μg/ml of DNA

Table 1

Relative survival of streptomycin resistance activities in UV partially inactivated DNA's tested at various concentrations

Concentrations of DNA's used in transformation [μg/ml of competent cells]	Activity of non damaged DNA [%]	Activity of partially damaged DNA's [%]	
		Pneumococcus	*Streptococcus*
5	100	1.3	24.5
1	100	1.5	32.4
0.1	100	2.3	59.5
0.01	100	2.6	62.0
0.001	100	2.4	52.9
0.0001	100	—	42.2

Table 1 represents data for partially inactivated DNA of the pneumococcus strain R 36 A, sm and of the streptococcal strain Challis. The DNA's of both organisms were inactivated to different residual activities and then tested in transformation reactions at different concentrations.

The differences in survival after UV irradiation when low and high concentrations are used may be responsible for the discrepancies in survival curves represented by different workers, because not all workers are using the same concentrations of DNA for assay.

A possible explanation for the observed phenomenon may be competition for competent cells of damaged and non-damaged DNA molecules. Dilution of partially inactivated DNA does not change the ratio of the numbers of active and non-active molecules. However, dilution causes a decrease in the ratio of the number of damaged molecules to competent cells. This means that inhibition of transformation exerted by UV damaged molecules should be higher at a saturating dose than at a low, non saturating dose.

To test if this supposition is true, a solution of DNA from the strain *Str. sanguis*, sm was inactivated by UV to 2,4% residual activity and mixed at a concentration of 1 μg/ml. with an equal part of the same non-irradiated DNA at the same concentration. From this mixture proper dilutions were prepared for assay of activity in transformation reactions. Controls were analogous dilutions of non-irradiated DNA in 0.15 M NaCl.

Table 2

Transforming activities of different concentrations of a mixture of UV irradiated and non irradiated DNA in per cent of activities of non-irradiated DNA only. Recipient: *Str. sanguis*; donor: *Str. sanguis*, sm)

Concentrations of non-irradiated DNA only [μg/ml]	Number of transformants in 1 ml	Activity [%]	Concentrations of each kind of DNA in a mixture of irradiated and non-irradiated DNA [μg/ml]	Number of transformants in 1 ml	Relative transforming activity to corresponding concentrations of active DNA only
0.5	190×10^3	100	0.5	408×10^2	21.5
0.05	40.8×10^3	100	0.05	298×10^2	73.0
0.005	6.16×10^3	100	0.005	32.6×10^2	106.0
0.0005	0.68×10^3	100	0.0005	6.9×10^2	101.2

Results presented in Table 2 show an increase in relative transforming activity with diminution of concentration of the mixture of active and UV irradiated DNA. These results may only be due to inhibition of transformation by the inactivated DNA molecules and they agree with our previous data on inhibition of transformation by genetically non active DNA (Pakula et al., 1959).

Fig. 2 represents dose survival curves of pneumococcal and streptococcal streptomycin resistance activities assayed in homospecific transformation reactions, e. g., in reactions carried out with wild type host cells of the strain analogous to the donor strain.

It may be seen that there are no differences in survival of the pneumococ-

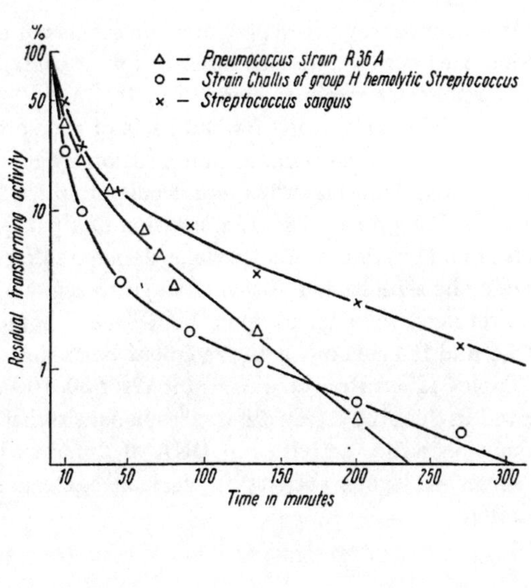

Fig. 2. UV inactivation of transforming activities to streptomycin resistence of DNA extracted from: △ = the pneumococcus strain R36A; ○ = the strain Challis of group H hemolytic *streptococcus*; × = *Streptococcus sanguis*; Transformation carried out at concentration of 1.0 μg/ml of DNA; Dose rate 11.6 ergs.mm^{-2} · sec^{-1}

cal DNA and of the DNA from *Str. sanguis*, sm at the initial part of the curves. At higher energy doses the streptomycin resistance marker of the latter strain is significantly less sensitive.

The marker under test of the strain Challis, sm is more sensitive at lower energy doses but more resistant at the terminal part of the curve.

An interesting phenomenon previously observed in this laboratory (Pakula et al., 1959) and also by Green (1959) is represented in Fig. 3. A solution of UV irradiated DNA appears to be inactivated to different extents when different recipient bacteria are employed to assay the surviving activity. This was found also true with the UV irradiated DNA of the pneumococcal strain R 36 A assayed with the homologous strain and the streptococcus strain Challis as hosts.

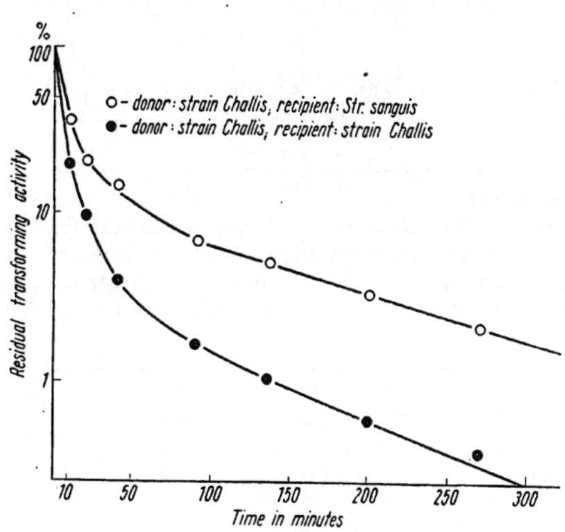

Fig. 3. Survival of UV irradiated DNA tested in homospecific and heterospecific transformation; Dose rate 11.6 erg.mm^{-2} · sec^{-1}; Concentration of DNA in transformation 1 μg/ml of DNA; ○ = donor: strain Challis; recipient *Str. sanguis*; ● = donor: strain Challis; recipient strain Challis

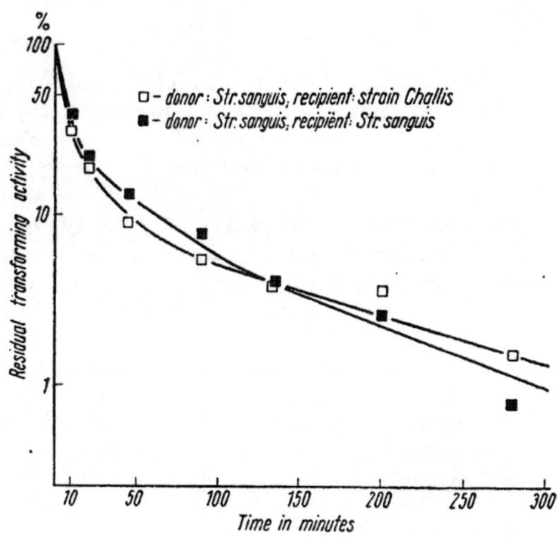

Fig. 4. Survival of UV irradiated DNA tested in homospecific and heterospecific transformation; Dose rate 11.6 ergs.mm^{-2} · sec^{-1}; Concentration of DNA in transformation 1 μg/ml of DNA; □ = donor: *Str. sanguis*; recipient strain Challis; ■ = donor: *Str. sanguis*; recipient *Str. sanguis*

However the described phenomenon seems not to be of general character since no significant differences were found with DNA of *Str. sanguis*, sm tested with the strain Challis, and with the homologous strain as hosts (Fig. 4).

A possible explanation for the basis of the so called "instability" of the irradiated DNA, as suggested in a personal communication by Ravin, might be the nature of the intracellular reactions which occur when a molecule of transforming DNA is integrated into the bacterial genome. The probability of integration of a paticular genetic marker might depend upon the extent and proximity of ultraviolet damage suffered in regions adjacent to the genetic marker itself. The significance of this damage to regions of the molecule adjacent to the site of the marker itself will depend upon the degree of homology between the molecule of transforming DNA and the genome of the recipient bacterium.

Table 3 illustrates energies required for 50, 10 and 1% survival of DNA's of all the organism tested in this study. The differences in sensitivity to UV irradiation of the analogous streptomycin resistance activities in DNA of different bacteria are striking. This finding suggests that an analogous activity in various bacteria may be under control of different genetic markers.

Table 3

UV energies in ergs.mm^{-2} required for 50, 90 and 99% destruction of the transforming activities to streptomycin resistance of DNA of *H. influenzae*, of the *Pneumococcus* strain R36A, of *Str. sanguis* and strain Challis in homospecific and heterospecific transformations. Dose rate 11.6 erg.mm^{-1} sec^{-1}; Concentration of DNA in transformation 1.0 μg/ml.

Residual activity	Homospecific transformation				Heterospecific transformation	
	H. influenzae	*Pneumococcus*	*Streptococcus* Challis	*Streptococcus Str. sanguis*	Donor: Challis Recipient: *Str. sanguis*	Donor: *Str. sanguis* Recipient: Challis
50%	300	5500	3500	5500	5000	5500
10%	1000	31500	14500	44000	38500	33000
1%	3000	110000	105000	220000	273000	240000

Gamma ray inactivation

Irradiation in aqueous solutions. Fig. 5 presents the dose survival curves for DNA samples irradiated in 0.15 M saline at neutral pH. Each point is the mean of at least three independent experiments on duplicate samples. It will be seen that for each DNA the survival curve exhibits a break at approximately 1% residual activity.

Of the three different transforming factors, it will be noted that the sensitivity of *H. influenzae* DNA is slightly smaller for small doses and least for the larger doses. For streptococcal DNA the situation is reversed; it is more sensitive for small doses and least sensitive for larger doses (not included in the figure). On the whole, however, the differences between inactivation of the three transforming factors are not appreciable. For the DNA concentration used, the D37% dose is about 900 r for pneumococcal DNA as compared to a value of 750 r reported by LATARJET et al. (1959) for the same marker in liquid solution under analogous conditions.

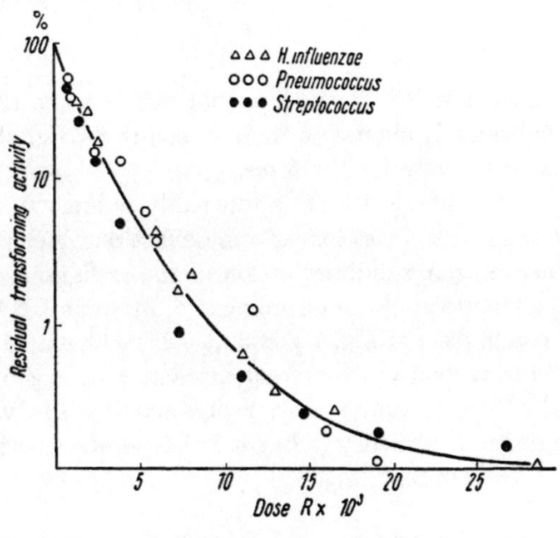

Fig. 5. Dose survival curves in saline solutions of the following transforming DNA samples: △△△ *H. influenzae*; ○○○ *Pneumococcus*; ●●● *Streptococcus*. Concentration of DNA in irradiated solutions 1 μg/ml; 0.1 μg/ml used for transformations

As might be expected the influence of DNA concentration is quite marked. Fig. 6 exhibits the inactivation curves for concentrations of 1, 10 and 100 μg/ml of streptococcal DNA. Only initial portions of the survival curves are shown, but the lower segments exhibit the usual break at about 2–3% residual activity.

Irradiation of dried preparations. The survival curves for dried films are represented in Fig. 7 and are considerably more complex than the analogous curves obtained by irradiation in aqueous medium. The curves in Fig. 7 may be resolved into three, or even more, straight line components. The initial components, to about 10–30% residual activity, are approximately linear, bearing in mind the experimental error in the individual determinations

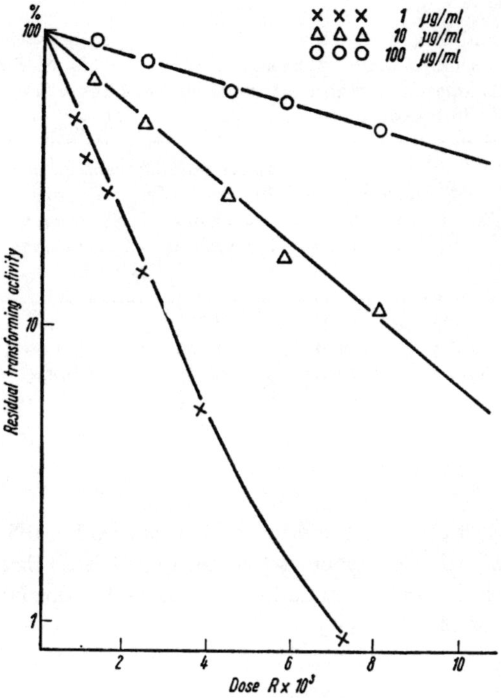

Fig. 6. Dose-survival curves for streptococcal DNA in saline solution at concentrations of: ××× 1 μg/ml; △△△ 10 μg/ml; ○○○ 100 μg/ml with 0.1 μg/ml for transformation

of transforming activity. It should, nonetheless, be emphasized that the breaks shown in the figure for the invidual DNA preparations were found to be reproducible in repeated experiments.

Furthermore, as for DNA in solution, *H. influenzae* DNA exhibits intitially a higher radiation resistance, while streptococcal DNA is most sensitive. For the lower portions of the curves, at residual activities at about 1–3%, the sensitivity of *H. influenzae* DNA is increased with respect to that for pneumococci. On the other hand, at very high doses, not represented in the figure, it becomes more sensitive than streptococcal DNA.

Summarizing, it appears that the relative radiation sensitivities for the DNA films are qualitatively similar to those of solutions, but the differences are considerably more marked quantitatively for the films.

Those differences are appreciably enhanced for DNA films prepared from solutions containing 10% yeast extract, the effect of which is presumably to reduce even further the influence of any indirect action of the radiation. All of these experiments were conducted on films containing a mixture of two different DNA samples as described above (Fig. 8). From a comparison of Fig. 7 and 8, it will be observed that there is a marked protective effect due to the presence of the yeast extract. This is shown also in Table 4. The figures given in the table are, of course, only representative and will vary considerably for different residual activities, particularly beyond the points where the curves exhibit pronounced breaks, as can be seen from Fig. 8.

Table 4

Doses, in roentgens, required to reduce activity of transforming DNA films to 50% in presence and absence of yeast extract

Transforming DNA	Without yeast extract	With yeast extract
H. influenzae	1.2×10^5	2.8×10^6
D. pneumoniae	0.6×10^5	2.3×10^6
Streptococcus	0.5×10^5	16×10^6

To a residual activity of 70% *H. influenzae* DNA is more resistant than streptococcal DNA; below 70% residual activity the relationship is reversed.

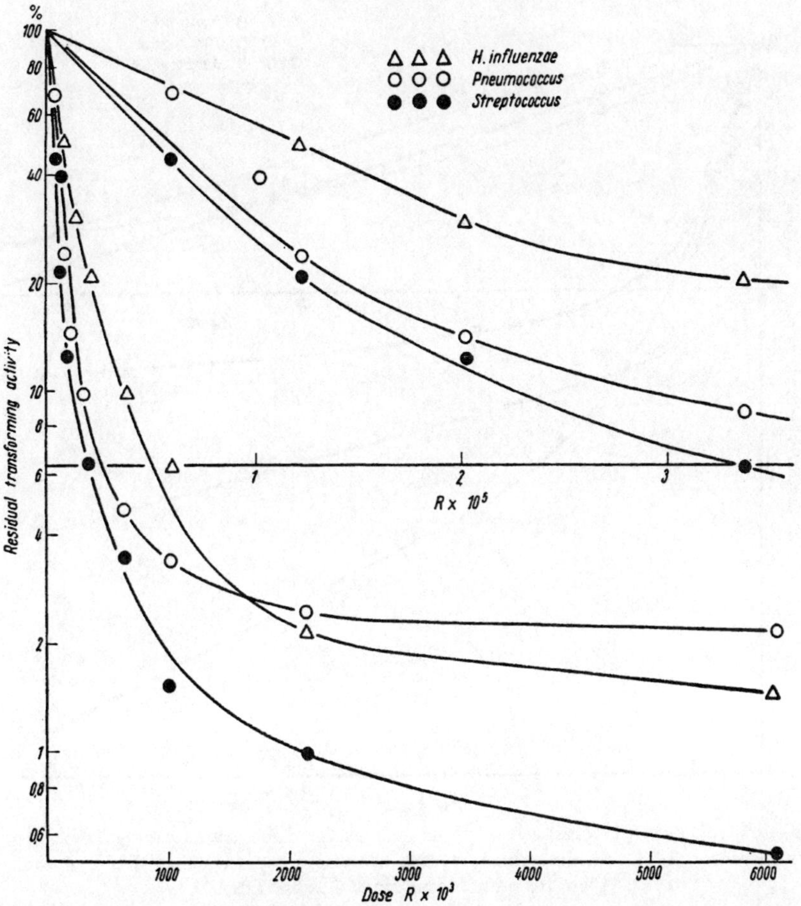

Fig. 7. Dose-survival curves of preparations of transforming DNA dried from saline solutions; 2 μg DNA on cover slip; 0.5 μg/ml used for transformation; △△△ *H. influenzae*; ○○○ *Pneumococcus*; ●●● *Streptococcus*

The rather complex form of the survival curves in the presence of yeast extract raises considerable doubts as to the validity of simple target theory for estimation of the size of the units responsible for biological activity. To a rough approximation, each of the curves in Fig. 8 may be regarded as consisting of two components, with the break occuring for each sample at residual activities well above 37%. If we extrapolate the initial portions of the curves to 1/e or 37% residual activity, and apply the usual procedure for calculation of the equivalent molecular weights of those portions of the DNA molecules responsible for transforming activity, we obtain the following results (Table 5).

Table 5
Calculated streptomycin resistance marker sizes in transforming DNA

Transforming factor	D_{37} [r]	Mol. wt.
Pneumococcal, R36A, sm	2.4×10^6	1.2×10^5
H. influenzae, Rds	3.2×10^6	0.9×10^5
Streptococcal, Challis, sm	4.1×10^6	0.75×10^5

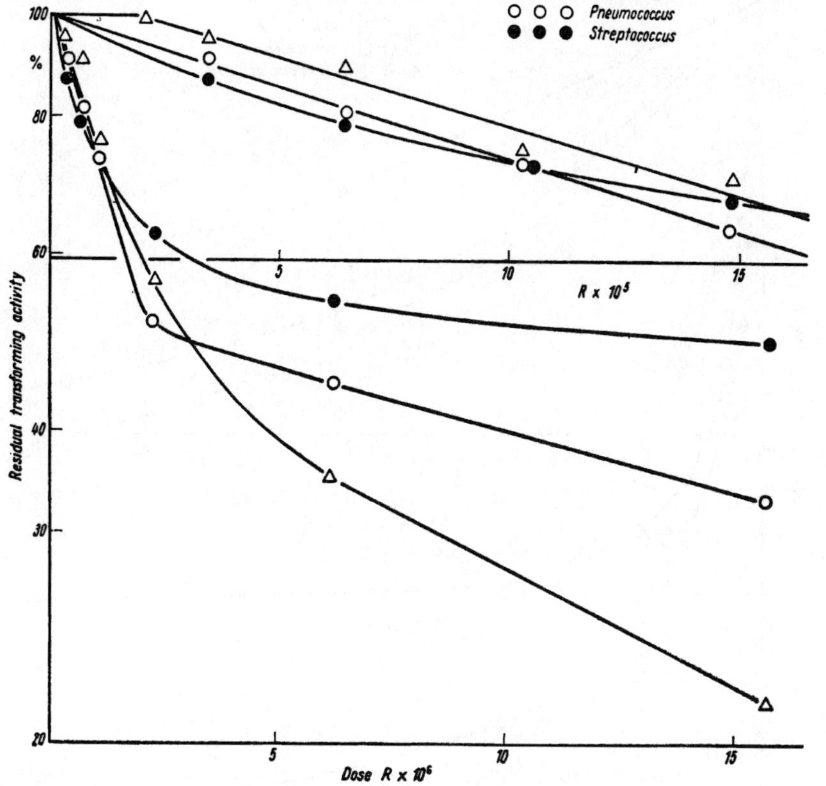

Fig. 8. Dose-survival curves in presence of yeast extract for dried transforming DNA of: △△△ *H. influenzae*; ○○○ *Pneumococcus*; ●●● *Streptococcus*; with 2 μg DNA of each on cover glass for transformation 0.5 μg/ml DNA

If, as has been done by others, we use the slopes of the lower components of the curves to calculate marker sizes, the results fall well below the values of 10^5 in molecular weight units for the streptomycin resistance activities in all three DNA samples, a result the validity of which is most improbable.

More significant, however, is the reversal in relative sensitivities of the streptococcal and *H. influenzae* markers at the higher radiation dose (Fig. 8). In fact if any doubts existed as to the reality of the differences in sensitivity between the different markers at low radiation doses, these are at once dispelled by the data for the higher doses.

The magnitudes of the various streptomycin resistance markers calculated from the initial slopes of the inactivation curves obtained in the presence of yeast extract are approximately of the same order of magnitude as that reported by Latarjet et al. (1959) for pneumococcal DNA irradiated in the presence of 10% yeast extract plus catalase in frozen solution. However, in view of the influence of yeast extract in our experiments on the shapes of the inactivation curves, the validity of calculations of marker size from simple target theory remains in our opinion, at least questionable. More significant, in this study is the demonstration of differences in apparent size if the streptomycin resistance markers in DNA from different species of bacteria. This finding is quite unequivocal, the more so in that the differences, if any, are so slight in aqueous medium where the inactivation process is due entirely to the action of free radicals in the solvent medium.

Summary

1. The relative survival of UV inactivated streptomycin resistance activities in transforming DNA of various bacteria is significantly higher when assayed at low concentrations. The reason for this is inhibition of transformation by UV inactivated DNA molecules at saturating concentrations used for assay.

2. The extent of UV inactivation of transforming DNA from some bacteria, but not from others, appears to be lower when assayed in heterospecific than in homospecific transformation reactions.

3. Striking differences were found in energy doses (especially in the terminal parts of the inactivation curves) required to provoke a certain amount of damage in the streptomycin resistance activities of various species of bacteria.

4. Slight but repeatable differences were found in the inactivation for the streptomycin resistance markers in DNA of different bacteria irradiated by gamma rays in aqueous solution. These differences were more distinct when dried preparations of DNA were irradiated and especially when these preparations were protected by yeast extract.

5. Both UV and gamma ray inactivation suggest that the streptomycin resistance activities of various species of bacteria are under control of different markers.

References

BUTTOLPH, L. J., 1955: Practical applications and sources of ultraviolet energy. In: Radiation biology, Vol. II. McGraw-Hill Book Comp. Inc., pp. 41—93.

GREEN, M. D., 1959: A host-specific variation affecting relative frequency of transformation of two markers in pneumococcus. Exper. Cell Res. 18, 466—480.

LATARJET, R., H. EPHRUSSI-TAYLOR, and H. REBEYROTTE, 1959: On the target size of a transforming factor based on X-ray inactivation. Radiation Research 1, 417—430.

LERMAN, L. S., and L. J. TOLMACH, 1959: Genetic transformation. II. The significance of damage to the DNA molecule. Biochim. biophysica Acta 33, 371—387.

MARMUR, J., and J. FLUKE, 1955: Uniformity of ionizing radiation action on several transforming factors of pneumococcus. Arch. Biochem. Biophys. 57, 506—514.

PAKULA, R., E. HULANICKA, and W. WALCZAK, 1959: Inhibition of transformation in *Streptococcus* sbe by desoxyribonucleic acids of different origin. Bull. Acad. Polon. Sci., serie sci. biol. VIII, 49—55.

PAKULA, R., E. HULANICKA, and W. WALCZAK, 1959: Sensitivity to ultraviolet irradiation of the streptomycin resistance activity in transforming DNA of different bacteria. Bull. Acad. Polon. Sci., serie sci. biol. VI, 217—222.

RUPERT, C. S., and S. H. GOODGAL, 1960: Shape of ultraviolet inactivation curves of transforming deoxyribonucleic acids. Nature (London) 185, 556—557.

STUY, J., 1961: Radiation inactivation of intracellular transforming deoxyribonucleic acid (thesis). Philips Research Laboratory, Eindhoven, Netherlands, pp. 48—58.

ZAMENHOF, S., G. LEIDY, S. GREER, and E. HAHN, 1957: Differential stabilities of individual heredity determinants in transforming principle. J. Bact. 74, 194—199.

Zur Reaktivierung UV-inaktivierter T1-Bakteriophagen

W. ECKART und W. ESCHKE

Physikalisch-Technisches Institut der Deutschen Akademie der Wissenschaften zu Berlin

Die Ergebnisse, über die nachstehend berichtet wird, sind z. T. zusammen mit Dr. WETZEL erarbeitet worden.

Der Reaktivierung muß per definitionem eine Inaktivierung vorausgehen. Die Kurve der UV-Inaktivierung des Coliphagen T1 besitzt bekanntlich zwei Abschnitte unterschiedlicher Neigung. Die Hypothese von GAREN und ZINDER [1], die durch Experimente von TESSMAN [2, 3] unterstützt wird, postuliert zwei Treffbereiche oder zwei Arten von Treffern, von denen der strahlenempfindlichere durch das Bakterium mit einer gewissen Wahrscheinlichkeit reaktiviert werden kann. Er bedingt den steileren Anfangsast der Inaktivierungskurve.

Dies betrifft Experimente mit der Wellenlänge 254 nm. Wie Prof. SHUGAR hier vor zwei Tagen dargelegt hat, ist es nun nützlich und notwendig, den Einfluß der Wellenlänge auf die verschiedenen Effekte zu untersuchen.

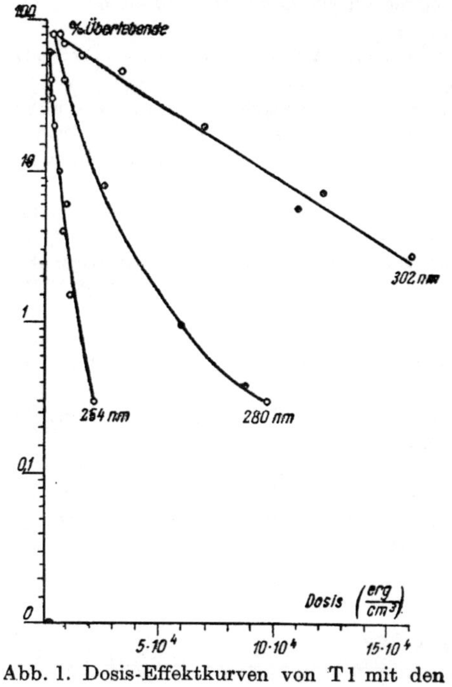

Abb. 1. Dosis-Effektkurven von T1 mit den Wellenlängen 254, 280 und 302 nm

Der Phage ist eine aus Nucleoprotein aufgebaute Partikel. Wir kennen die Absorptionsspektren beider Komponenten. Die DNS besitzt ihr breites Maximum zwischen 250 und 270 nm, das Protein zeigt ein Nebenmaximum bei 280 nm und eine steigende Absorption für Wellenlängen über 300 nm.

Wir haben einen Quarzmonochromator verwendet, um T1 in Pufferlösung mit 254, 280 bzw. 302 nm zu bestrahlen [4]. Abb. 1 zeigt die Ergebnisse dieser Versuche. Zur Deutung der Befunde möchten wir vorschlagen, daß die DNS an der stärker strahlenempfindlichen Reaktion (254 nm) beteiligt ist und daß die Reaktion geringerer Strahlenempfindlichkeit über eine Denaturierung des Proteins abläuft (302 nm). Die Kurve bei 280 nm ist das Ergebnis der Wirkung auf beide Komponenten (stark ausgeprägte Krümmung der Kurve).

Wenden wir uns nun der Frage zu, ob eine anschließende Photoreaktivierung ebenfalls zu unterschiedlichen Effekten führt. Bei diesen Untersuchungen ließen wir die UV-bestrahlten Phagen an Bakterien adsorbieren und brachten sie anschließend auf Platten. Die Platten wurden 90 min lang mit dem Spektrum eines Quecksilberbogens oberhalb 330 nm bestrahlt [5].

Als Maß der Reaktivierung diente das Verhältnis der Lochzahl auf den Reaktivierungsplatten zu derjenigen auf den im Dunkeln bebrüteten Platten:

$$\frac{\text{Überlebende} + \text{Reaktivierte}}{\text{Überlebende}} = 1 + \frac{\text{Reaktivierte}}{\text{Überlebende}}$$

Abb. 2 zeigt dieses Verhältnis als Ordinate gegen den log der Überlebenden in % (Abszisse) aufgetragen. Wir haben diese Form der Darstellung gewählt, um die Unterschiede der Photoreaktivierung für die mit den verschiedenen Wellenlängen inaktivierten Phagen besonders deutlich zu machen. Üblicherweise trägt man die Kurven ohne und mit Reaktivierung in das gleiche Koordinatenkreuz (Dosis-Effekt-Skala) ein und bestimmt hieraus den Dosisreduktionsfaktor. Eine einfache geometrische Überlegung zeigt, daß beide Darstellungsarten gleichbedeutend sind.

Wir sehen, daß der größte Anstieg der Photoreaktivierung für die mit 302 nm inaktivierten Phagen (Protein) erfolgt. Dieser Strahlenschaden ist wahrscheinlich weniger schwerwiegend als der mit 254 nm gesetzte (DNS). Ein sehr interessantes Verhalten bei der Reaktivierung weisen die mit 280 nm inaktivierten Phagen auf. Bis zu einer Überlebensrate von 2% herab ist die Reaktivierung normal. Über diesen Punkt hinaus nimmt sie jedoch wieder ab. Wahrscheinlich wirken hier die höheren Dosen durch eine irreversible Denaturierung des Proteins, ein Effekt, der bei geringeren Dosen entweder gar nicht auftritt oder innerhalb dieses Dosisbereiches ebenfalls photoreversibel ist.

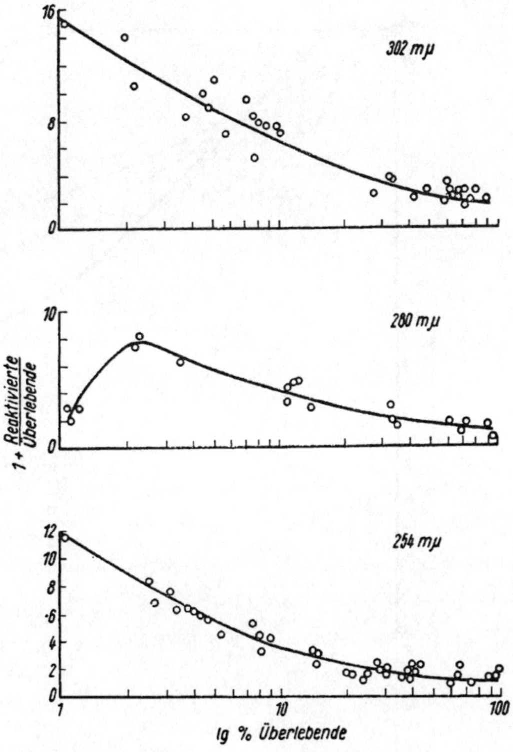

Abb. 2. Photoreaktivierung von T1-Phagen nach Inaktivierung mit 254, 280 bzw. 302 nm. Abszisse: % Überlebende, Ordinate: 1 + (Reaktivierte/Überlebende)

Die Rolle der Wirtszellen bei diesen Reaktionen wird durch die Untersuchung der Strahlenempfindlichkeit der Kapazität demonstriert, d. h. der Fähigkeit der Zellen, Phagennachkommen zu erzeugen. Wir benutzten zu diesen Versuchen *E. coli* B/r, weil mit diesem Stamm die Ergebnisse viel besser reproduzierbar waren als mit *E. coli* B.

T1-Phagen wurden mit 254, 280 bzw. 302 nm bis zu einer Überlebensrate von 1% bestrahlt. Gleiche Mengen dieser inaktivierten Phagen wurden 30 min zwecks Adsorption zu Bakterien gegeben, die zuvor in Puffer suspendiert und mit steigenden UV-Dosen von 254 nm bestrahlt worden waren. Dann kamen sie mit frischen Bakterien versetzt zur Lochzählung auf die Platte. Abb. 3 zeigt die Ergebnisse [6].

Kurve B/r stellt die Inaktivierungskurve der Bakterien dar (Koloniebildungsvermögen). Wir sehen, daß die Kapazitätskurven für Phagen, die mit 280 und 302 nm inaktiviert wurden, nicht wesentlich voneinander abweichen. Ihre Endneigung ist annähernd parallel zu der der Kapazität für unbestrahlte T1 (nach TESSMAN [2]). Dies könnte Zufall sein.

Die Inaktivierung mit 254 nm führt zu einer Kapazitätskurve ganz anderen Charakters. Niedrige Dosen stimulieren (wie dies auch schon von TESSMANN [2] gefunden wurde) eine Reaktion analog zu der erstmals für *E. coli* B und T3-Phagen von WEIGLE und DULBECCO 1953 beschriebenen Wirtsreizungsreaktivierung [7]. Man könnte meinen, daß diese Reaktion

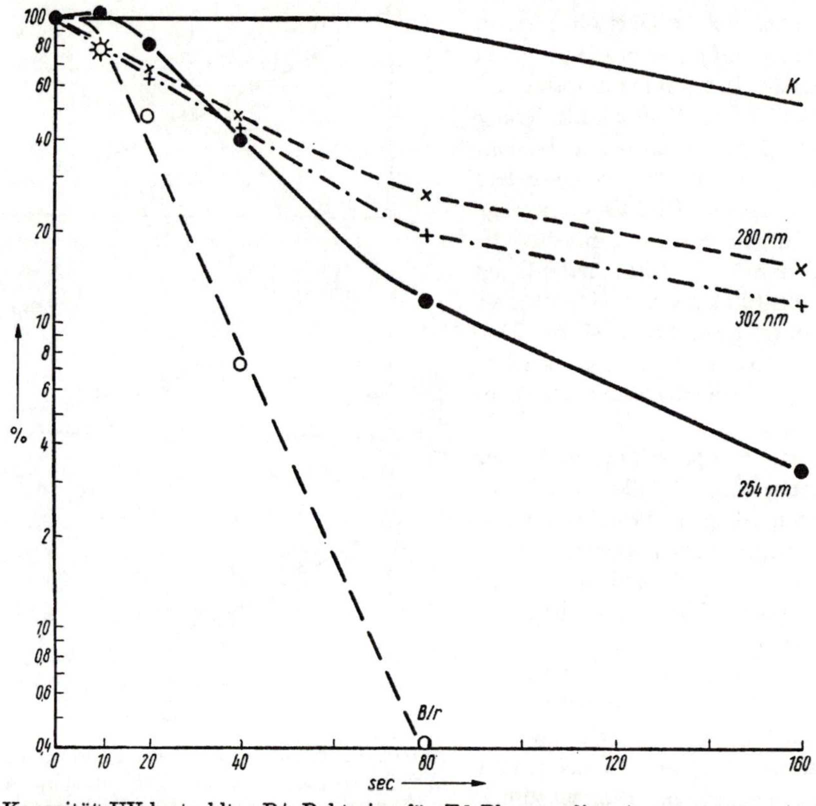

Abb. 3. Kapazität UV-bestrahlter B/r-Bakterien für T1-Phagen, die mit 254, 280 bzw. 302 nm auf 1% inaktiviert wurden. Die Abszisse gibt die UV-Dosis auf die Bakterien an. Kurve B/r zeigt die Inaktivierung der Bakterien. Kurve K zeigt nach TESSMAN [2] die Kapazität von B/r für unbestrahlte T1

durch eine vorhergehende Zerschlagung der Zellen-DNS in passende Stücke in Gang gesetzt wird, welche die geschädigten Strukturen im Phagengenom leichter zu ersetzen vermögen als bei ausschließlichem Vorhandensein völlig intakter DNS-Moleküle in der Zelle.

Abb. 3 macht deutlich, daß eine Bestrahlung der Zellen mit höheren Dosen die Kapazität der Zellen für mit 254 nm geschädigte Phagen schneller vermindert als für solche, die mit 280 oder 302 nm inaktiviert wurden. Wahrscheinlich kann diese Frage etwas klarer (oder noch komplizierter?) werden, wenn man auch die Kapazität von Bakterien, die mit 280 bzw. 302 nm vorbestrahlt wurden, für mit 254, 280 bzw. 302 nm inaktivierte Phagen untersucht.

Schließlich möchten wir noch einige vorläufige Ergebnisse über die Wellenlängenabhängigkeit der Mehrfachreaktivierung (multiplicity reactivation) anführen. Bekanntlich funktioniert die Mehrfachreaktivierung mit den geradzahligen T-Phagen am besten. Daher begannen wir mit der MR von T2-Phagen, die mit verschiedenen Dosen von 254 nm inaktiviert worden waren. Man sollte einen deutlichen Einfluß der Infektionszahl (multiplicity of infection) erwarten. Aber dieser kann tatsächlich innerhalb eines weiten Bereiches (1 bis 9) vernachlässigt werden.

Abb. 4 zeigt unsere ersten Ergebnisse. In das doppelt logarithmische Achsenkreuz ist die Überlebensrate in % als Ordinate eingetragen und als Abszisse der Reaktivierungsfaktor, der in folgender Weise einfach definiert ist:

$$\frac{\text{Lochzahl mit Mehrfachinfektion}}{\text{Lochzahl mit Einzelinfektion}} = R.$$

Für eine Phagenpopulation überwiegend aktiver Partikeln muß der Faktor kleiner als 1 sein; denn n Phagen, die ein Bakterium infizieren, werden nur ein Loch erzeugen, während n Phagen, die an n Bakterien adsorbiert sind, n Löcher hervorrufen. Abb. 4 zeigt die Zunahme des Reaktivierungsfaktors mit steigender Inaktivierungsdosis. Diese Zunahme erfolgt für Phagen, die mit 280 nm inaktiviert wurden, etwas langsamer.

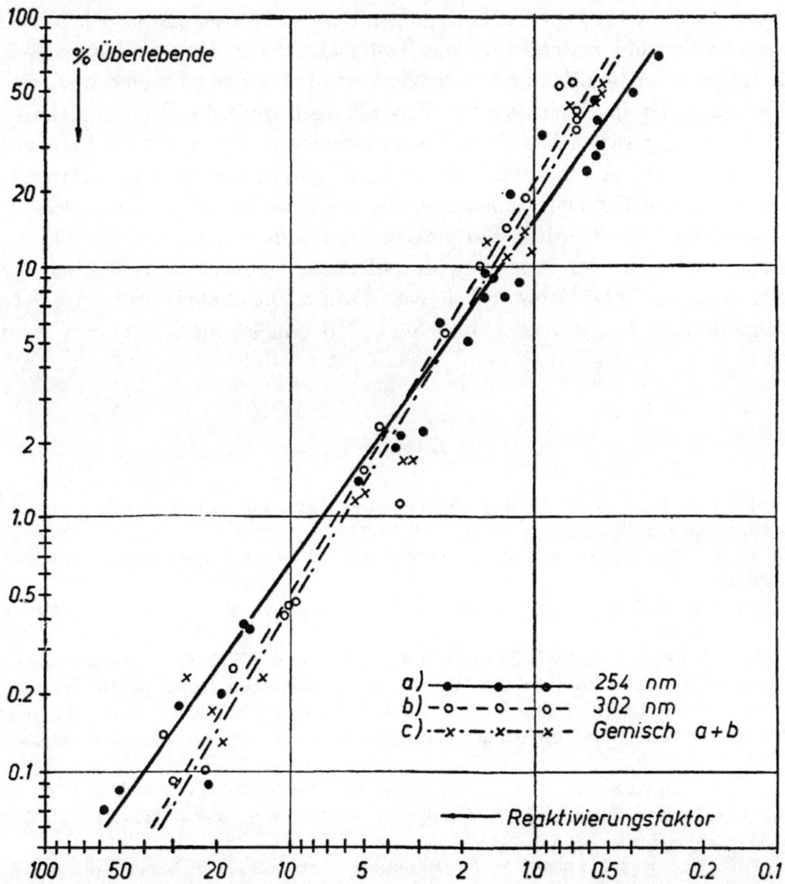

Abb. 4. Mehrfachreaktivierung (multiplicity reactivation) von T2-Phagen, die mit 254 nm (●——●) bzw. 280 nm (○——○) inaktiviert wurden. Die dritte Kurve (×·—·—×·) zeigt die Reaktivierung für ein Gemisch aus beiden Inaktivierungslösungen.

Nun erwarten wir eine deutlich größere Mehrfachreaktivierung für eine gemischte Suspension aus Phagen, die mit 254 nm, mit solchen, die mit 280 nm bestrahlt worden waren. Wir meinten, daß der unterschiedliche Strahlenschaden, den diese beiden Wellenlängen hervorrufen, die gegenseitige Kreuzungsaktivierung innerhalb der Wirtszelle sehr erleichtern müßte.

Wie man aus Abb. 4 ersehen kann, weicht die Kurve für das Gemisch praktisch nicht von der für 280 nm ab, ein Ergebnis, das unserer Annahme widerspricht.

Das gilt jedoch für T2, während unsere sonstigen Experimente mit T1 ausgeführt wurden. Leider fanden wir bisher nur eine sehr kleine Mehrfachreaktivierung bei T1, die zu klein ist, um zur Entscheidung des Problems beitragen zu können. Wir hoffen jedoch, bald neue Daten gewonnen zu haben.

Zusammenfassend möchten wir drei Punkte zur Deutung der Befunde vorschlagen:

1. Die Inaktivierung des Phagen T1 mit UV der Wellenlänge 254 nm erfolgt hauptsächlich über die DNS, mit UV von 280 nm sowohl über die DNS als auch über das Protein und mit UV von 302 nm im wesentlichen, falls nicht ausschließlich, über das Protein.
2. Die UV-Schädigung des Proteins geht entsprechend der absorbierten Energie in zwei Stufen vor sich, von denen die eine photoreaktivierbar, die andere irreversibel ist.
3. Die Photoreaktivierung betrifft entweder nur den Schaden am Protein, oder sie funktioniert auf jeden Fall für Proteinschädigungen sehr viel besser als für solche an der DNS.

Für eine solche Deutung spricht auch die Beobachtung von HILL und ROSSI [8], daß T1-Phagen, die in getrocknetem Zustand inaktiviert wurden, keine oder eine stark verminderte Photoreaktivierung zeigen. Nach neueren Vorstellungen über den Denaturationsprozeß ist dieser mit einer Konfigurationsänderung der Proteine verbunden. Es ist anzunehmen, daß bei der UV-Denaturation stabilisierende Bindungen zerstört werden und im Anschluß daran als Sekundäreffekt eine Konfigurationsänderung des Proteins durch thermische Stöße von Flüssigkeitsmolekeln bewirkt wird. Bei getrocknetem Protein ist die Ausbildung des Sekundäreffektes erschwert. Zur Erzielung des gleichen Denaturationseffektes werden also höhere Dosen benötigt als bei Anwesenheit von Wasser. Die Inaktivierung des getrockneten Phagen erfolgt deshalb hauptsächlich über die DNS und ist nicht oder nur wenig reaktivierbar.

Literatur

[1] GAREN, A., and N. D., ZINDER, 1955: Radiological evidence for partial genetic homology between bacteriophage and host bacteria. Virology **1**, 347–376.

[2] TESSMAN, E. S., 1956: Growth and mutation of phage T1 on ultraviolet-irradiated host cells. Virology **2**, 679–688.

[3] TESSMAN, I., and T. OZAKI, 1957: Multiplicity reactivation of bacteriophage T1. Virology **4**, 315–327.

[4] WETZEL, R., W. ESCHKE und W. ECKART, 1960: Über den UV-Inaktivierungsmechanismus des Bakteriophagen T1 bei verschiedenen Wellenlängen. Monatber. Dtsch. Akad. Wiss. **2**, 697–700.

[5] ESCHKE, W., W. ECKART und R. WETZEL, 1960: Über den Einfluß der inaktivierenden Wellenlänge der UV-Strahlung auf die Reaktivierung von Bakteriophagen. Naturwissenschaften **47**, 521.

[6] ECKART, W., W. ESCHKE und R. WETZEL, 1960: Zur Kapazität UV-bestrahlter *E. coli* B/r-Bakterien für T1-Phagen, die mit UV verschiedener Wellenlänge inaktiviert wurden. Naturwissenschaften **47**, 606–607.

[7] WEIGLE, J. J., and R. DULBECCO, 1953: Induction of mutations in bacteriophage T3 by ultraviolet light. Experientia (Basel) **9**, 372–373.

[8] HILL, R. F., and H. H. ROSSI, 1954: The ultraviolet sensitivity and photoreactivability of T1 bacteriophage. I. Effect of irradiation conditions upon survival curves. Radiation Research **1**, 282–293.

Étude Quantitative de Réversions Biochimiques Induites chez la Levure *Schizosaccharomyces Pombe* par des Radiations et des Substances Radiomimétiques

H. Heslot

Laboratoire de Génétique, Institut National Agronomique, Paris

1. Matériel et méthodes

Dans une récente publication [14], nous avons exposé pour quelles raisons nous avions choisi la levure *Schizosaccharomyces pombe* pour l'étude quantitative des mutations induites par les rayonnements et diverses substances radiomimétiques.

Rappelons brièvement que cet organisme est constitué de cellules uninucléées haploïdes se divisant par fission. Il représente donc un système plus simple que les conidies plurinucléées de *Neurospora*. On peut le cultiver sans difficulté sur un milieu minimum comportant du glucose, des sels minéraux, de l'azote inorganique et quelques substances de croissance (acide pantothénique, acide nicotinique, méso-inositol, biotine).

Sous l'action du rayonnement ultra-violet, U. Leupold [23 à 27] a obtenu un grand nombre de mutants biochimiques à déterminisme monogénique.

Enfin, il existe des cellules de plusieurs sortes: h^+, h^-, h^{90}. Les types h^+ et h^- sont hétérothalliques et fusionnent en donnant un zygote diploïde qui subit la méiose avec formation de quatre ascospores. Au bout de quelques jours, ces spores sont libérées spontanément par lyse de la paroi de l'asque. U. Leupold a mis au point un traitement par l'alcool qui permet de tuer sélectivement les cellules végétatives, en ne conservant que les ascospores, nettement plus résistantes. Il est donc possible de procéder à des analyses génétiques en utilisant ces suspensions d'ascopores.

Parmi les mutants biochimiques, induits par l'ultra-violet, qui nous ont été obligeamment fournis par U. Leupold, nous avons recherché ceux qui paraissaient susceptibles d'utilisation pour un test quantitatif. Les trois souches suivantes ont finalement été retenues:

arg-1 (975/230 × 972) 1a h^-: arginineless
leuc-3 (975/241 × 972) 2a h^-: leucineless
Ur-1 (975/171 × 972) 1c h^-: uracileless

Ces souches sont cultivées sur un milieu complet solide contenant:

Bacto Yeast extract Difco .	5 g
Glucose	30 g
Bacto Agar Difco	15 g
Eau distillée	1000 g

D'autre part, le milieu minimum – qui permet la croissance des souches «sauvages» – contient du glucose, des sels minéraux, du sulfate d'ammoniaque et quelques vitamines (mentionnées plus haut). Ce milieu permet aussi la croissance des mutants biochimiques, à condition d'y ajouter le métabolite dont la souche considérée a besoin.

Dans toutes nos expériences d'induction de réversions, des cellules du mutant biochimique choisi sont ensemencées dans des tubes sur le milieu complet et incubées à 30 °C pendant 4 jours. On introduit ensuite dans chaque tube 5 cc d'eau distillée stérile, et on met les cellules en suspension par pipetages répétés. Les suspensions cellulaires provenant des différents tubes sont ensuite extraites et mélangées dans un Erlenmeyer stérile.

Un examen microscopique révèle que la suspension contient une majorité de cellules isolées, mais aussi des amas de 2, 3, 4 . . . cellules, restées attachées les unes aux autres par leurs membranes. Pour dissocier ces amas, nous soumettons la suspension à un traitement de 5–6 minutes par les ultra-sons (20000 cycles). Ce procédé, très efficace, permet de diminuer considérablement la fréquence des amas: celle-ci se trouve ramenée à 1% au maximum.

A l'aide d'un hématimètre, on détermine alors le nombre de cellules par cm^3 et l'addition d'eau stérile permet d'ajuster ce nombre à 10^8/cc environ.

La suspension cellulaire est alors répartie dans une série de tubes stériles, à raison de 5 cc par tube. On centrifuge, on élimine le liquide surnageant et on le remplace par 5 cc d'eau stérile. On remet les cellules en suspension afin de les laver et de les débarrasser des substances nutritives qu'elles contiennent. On centrifuge une deuxième fois et on jette le liquide surnageant.

On notera que ces lavages sont effectués avec de l'eau (et non avec une solution isotonique). Il s'agit donc d'un traitement assez brutal qui doit sans doute, par choc osmotique, permettre d'éliminer certains métabolites solubles contenus à l'intérieur des cellules (en particulier les précurseurs de l'A.D.N.).

Si l'on désire étudier l'activité mutagène d'une substance alkylante, on prépare des solutions aqueuses de ce corps, en concentrations croissantes, et on verse 5 cc de chaque solution dans chaque tube. Les cellules sont remises en suspension, et les tubes agités mécaniquement pendant toute la durée de l'expérience dans un bain-marie maintenu à 25° C par un thermostat.

Le traitement est arrêté par centrifugation, élimination de la solution surnageante et son remplacement par de l'eau (deux centrifugations, deux lavages).

A partir de chaque tube, on procède alors à une série de dilutions successives 10^{-1}, 10^{-2}, . . . 10^{-5}. Sur milieu complet (ou sur milieu minimum additionné du métabolite dont la souche a besoin), on ensemence 1 cc de la dilution 10^{-5}, ce qui permet de compter le nombre de cellules vivantes dans la suspension témoin et dans les suspensions traitées. On détermine ainsi la courbe de survie en fonction de la concentration (ou de la dose) de l'agent mutagène.

Sur le milieu minimum, on ensemence 1 cc des dilutions 10^0 et 10^{-1}, ou 10^{-1} et 10^{-2}, selon la souche considérée. Ceci permet de déterminer le nombre des réversions spontanées et induites. Nous employons à dessein le terme réversion pour qualifier les cellules mutantes capables de former des colonies sur le milieu minimum. Il peut s'agir de mutations réverses vraies, ou de la mutation de gènes suppresseurs. C'est un point que nous préciserons ultérieurement.

En pratique, on prépare des tubes contenant 20 cc de milieu (complet ou minimum). On liquéfie le milieu par chauffage et on le maintient liquide en plaçant les tubes dans un bain-marie à 40° C. C'est dans ces tubes que l'on ajoute 1 cc des différentes solutions. On les agite pour homogénéiser et on verse le contenu de chaque tube dans une boîte de Pétri. Le milieu gélosé se solidifie par refroidissement.

Les boîtes sont alors incubées dans une étuve à 30 °C. Au bout de 8 jours, on peut procéder au comptage des colonies apparues sur milieu complet. Au contraire, on attend 15 jours avant de dénombrer les colonies mutantes sur milieu minimum: certaines poussent en effet très lentement.

Fig. 1. Action de la 1,4-Butane sultone sur arg-1

Milieu	Dilution	Témoin	Concentration de la 1,4-Butane sultone (Trait^t = 1 heure) 3/500	6/500	9/500	12/500	15/500	18/500	21/500	24/500
Complet	10^{-5}	662–682–653– (M = 665,7)	639–626–652– (M = 639)	623–581–572– (M = 592)	510–481–469– (M = 486,7)	500–459–473– (M = 477,3)	439–448–458– (M = 448,3)	361–355–328– (M = 348)	314–310–278– (M = 300,7)	231–242–246– (M = 239,7)
	% survie	100	96,0	88,9	73,1	71,7	67,3	52,3	45,2	36,0
Minimum + arginine	10^{-5}	684–639–668– (M = 663,7)	627–605– (M = 616)	620–642–626– (M = 629,3)	475–478–480– (M = 477,7)	492–498–463– (M = 484,3)	446–454–423– (M = 441)	356–364–376– (M = 365,3)	324–312–324– (M = 320)	277–272–272– (M = 273,7)
100 mg/litre	% survie	100	92,8	94,8	71,9	72,6	66,4	55,0	48,1	41,2
Minimum	10^{-1}	74–65–72–69–86– (M = 73,2)	166–174–158–194–205– (M = 179,4)	376–363–316–340–382– (M = 355,4)	497–503–522–482–505– (M = 501,8)	581–560–600–586–540– (M = 573,4)	725–760–692–714–756– (M = 719,6)	725–711–714–754–778– (M = 736,4)	791–802–814– (M = 802,3)	923–897–939– (M = 919,7)
	10^{-2}	5–5 (M = 5)	13–19–14–18– (M = 16)	28–39–34–50– (M = 37,75)	39–42–46–40– (M = 41,75)	63–49–53–60– (M = 56,25)	76–76–75–83– (M = 77,5)	72–79–69–70– (M = 72,5)	76–82–81–73– (M = 78)	86–85–83– (M = 84,7)

Nombre N de réversions pour 10^7 survivants:

			N
Témoin	110,0	pour 10^7	0
3/500	280,8	,, ,,	170,8
6/500	600,3	,, ,,	490,3
9/500	1031,0	,, ,,	921,0
12/500	1201,3	,, ,,	1091,3
15/500	1605,2	,, ,,	1495,2
18/500	2116,1	,, ,,	2006,1
21/500	2668,1	,, ,,	2558,1
24/500	3836,9	,, ,,	3726,9

Fig. 2. Action du sulfate neutre d'éthyle sur ur–1

Milieu	Dilution	Témoin	Concentrations du sulfate d'éthyle (Trait^t = 1 heure) 0,5/500	1/500	1,5/500	2/500	2,5/500	3/500
Complet	10^{-5}	600–593–589 561– (M = 585,75)	529–546–535 572– (M = 545,5)	464–431–462 468– (M = 456,25)	433–414–441 428– (M = 429)	441–452–461 385– (M = 434,75)	366–391–377 414– (M = 387)	341–355– 389–366– (M = 362,75)
	% survie	100	93,1	77,9	73,2	74,2	66,1	61,9
Minimum	10^{0}	43–42–29– 36– (M = 37,5)	32–40–30– 36–43– (M = 36,2)	51–54–54– 64–65– (M = 57,6)	101–79–81– 87–82– (M = 86)	101–104–90– 102– (M = 99,25)	104–108–94– 83–101– (M = 98)	95–98–93– 98–110– (M = 98,8)
	10^{-1}	1–2– (M = 1,5)	5–2–2– (M = 3)	1–3–6– (M = 3,3)	7–6–6– (M = 6,3)	12–9–6– (M = 9)	7–8–7– (M = 7,3)	11–2–8– (M = 7)

Nombre N de réversions pour 10^7 survivants:

		N
Témoin	6,40 pour 10^7	0
0,5/500	6,64 „ „	0,24
1/500	12,62 „ „	6,22
1,5/500	20,05 „ „	13,65
2/500	22,83 „ „	16,43
2,5/500	25,32 „ „	18,92
3/500	27,24 „ „	20,84

Fig. 3. Action des rayons γ (Co^{60}) sur leuc-3

Milieu	Dilution	Témoin	Dose de rayons γ							
			10000 r	20000 r	40000 r	60000 r	80000 r	100000 r	120000 r	140000 r
Complet	10^{-5}	616–607– 630–679– (M = 633)	550–558– 575–582– (M = 566,25)	534–567– 551–524– (M = 544)	433–402– 386–422– (M = 410,75)	381–320– 340–327– (M = 342)	282–291– 279–275– (M = 281,75)	190–170– 153–200– (M = 178,25)	110–140– 125–145– (M = 130)	102–93– 85–91– (M = 92,75)
	% survie	100	89,5	85,9	64,9	54,0	44,5	28,2	20,5	14,7
Minimum	10^{0}	0–4–2–0– (M = 1,5)	9–7–15– 13–8–7– (M = 9,8)	9–9–19– 14–15–15– (M = 13,5)	28–19–24– 24–25– (M = 24)	26–34–24– 38–37–31– (M = 31,7)	35–31–36– 34–29–24– (M = 31,5)	35–44–30– 28–41–36– (M = 35,7)	38–39–37– 44–40–41– (M = 39,8)	41–46–48 41–42– (M = 43,6)
	10^{-1}	1–0–1– (M = 0,3)	3–3–4– (M = 3,3)	1–2–2– (M = 1,7)	2–1–1– (M = 1,3)	2–5–3– (M = 3,3)	4–3–4– (M = 3,7)	2–2–0– (M = 1,3)	6–3–7– (M = 5,3)	3–8–1– (M = 4)

Nombre N de réversions pour 10^7 survivants:

		N
Témoin	0,24 pour 10^7	0
10000 r	1,74 ,, ,,	1,50
20000 r	2,48 ,, ,,	2,24
40000 r	5,84 ,, ,,	5,60
60000 r	9,26 ,, ,,	9,02
80000 r	11,18 ,, ,,	10,94
100000 r	20,01 ,, ,,	19,77
120000 r	30,64 ,, ,,	30,20
140000 r	47,01 ,, ,,	46,77

Avant chaque expérience, il est nécessaire de procéder à une purification de la souche. Les réversions spontanées, favorisées par la sélection naturelle, risquent en effet de s'y accumuler et de constituer une gène. Il suffit de faire une dilution et d'ensemencer les tubes destinés à l'expérience avec des cellules provenant d'une seule colonie. Dans ces conditions, la fréquence des réversions est à peu près constante pour chaque souche et a les valeurs moyennes suivantes:

arg-1 : 50 pour 10^7 cellules vivantes
ur-1 : 2 pour 10^7 ,, ,,
leuc-3: 0,3 pour 10^7 ,, ,,

Nous avons réalisé un grand nombre d'expériences avec les souches arg-1, ur-1, leuc-3 et nos résultats seront présentés sous forme graphique. Toutefois, pour plus de clarté, nous indiquons ici les chiffres obtenus dans trois expériences type (fig. 1, 2 et 3).

En ce qui concerne la souche arg-1 (fig. 1) on voit que le nombre des colonies sur le milieu complet est du même ordre que celui observé sur le milieu minimum auquel on a ajouté de l'arginine. Ces deux milieux peuvent donc être utilisés indifféremment, à condition que le taux de survie des cellules soit suffisamment élevé. Bien entendu, pour chaque concentration et chaque dilution, il faut ensemencer plusieurs boîtes de Pétri; on fait ensuite la moyenne M du nombre de colonies par boîte.

Avec la souche arg-1, on observe qu'il y a une correspondance satisfaisante entre le nombre de colonies sur le milieu minimum aux dilutions 10^{-1} et 10^{-2} (rapport de 10 à 1). C'est là un point très important dans ce type d'expérience. Grigg [6] a en effet montré que, sur milieu minimum, des concentrations élevées de cellules auxotrophes étaient capables de gêner, ou même d'empêcher complètement la croissance de mutants prototrophes. Ce phénomène, qui a reçu le nom d'«effet Grigg», résulte de la consommation du sucre présent dans le milieu par les cellules auxotrophes. La quantité de sucre restante est alors insuffisante pour permettre la croissance des prototrophes. Il est clair que, dans le cas de la souche arg-1, il n'y a pas d'effet Grigg. La souche peut donc être utilisée pour des études quantitatives de réversions induites.

Avec les souches ur-1 (fig. 2) et leuc-3 (fig. 3), l'effet Grigg n'intervient pas non plus, mais il apparaît moins clairement que le rapport des comptages dans les dilutions 10^0 et 10^{-1} est égal à 10:1. Ceci résulte simplement de ce que, pour un même agent mutagène, utilisé dans les mêmes conditions de concentration et de durée, le nombre moyen de réversions est beaucoup plus faible avec ces deux souches. En conséquence, les fluctuations individuelles du nombre des colonies sur les boîtes de Pétri à la dilution 10^{-1} deviennent importantes. Par exemple, avec le sulfate d'éthyle (3/500, pendant 1 heure) on a trouvé:

arg-1 : 7643,7 pour 10^7 cellules survivantes
leuc-3: 187,2 pour 10^7 ,, ,,
ur-1 : 20,8 pour 10^7 ,, ,,

(soustraction faite du taux de réversion spontané)

Les chiffres précédents montrent que la souche arg-1 donne beaucoup plus de réversions que la souche leuc-3 et surtout ur-1. C'est là une propriété générale, comme l'indiquent les données suivantes, relatives à d'autres agents mutagènes:

$CH_3SO_2OC_2H_5$ (20/500 – 1 heure)
- arg-1 : 4092 pour 10^7 cellules survivantes
- leuc-3: 57,8 pour 10^7 ,, ,,
- ur-1 : 11,0 pour 10^7 ,, ,,

CH_2	arg-1 :	1842,7 pour 10^7 cellules survivantes
| $>NCH_2CH_2C{\equiv}N$	leuc-3 :	77,5 pour 10^7 ,, ,,
CH_2	ur-1 :	13,8 pour 10^7 ,, ,,
(4/500– 1 heure)		

CH_2–CH–CH_2Cl	arg-1 :	3417,2 pour 10^7 cellules survivantes
\O/	leuc-3 :	244,3 pour 10^7 ,, ,,
(0,8/500 – 5 heures)	ur-1 :	28,6 pour 10^7 ,, ,,

(soustraction faite du taux des réversions spontanées)

Il en résulte donc que la souche arg-1 permet d'obtenir des indications numériques beaucoup plus précises que leuc-3 et surtout ur-1. Pour cette dernière, il existe en plus le handicap d'un taux de réversion spontané relativement élevé (2 pour 10^7). Pour toutes ces raisons, nous avons utilisé de préférence la souche arg-1 dans nos expériences.

Pour ce qui est des substances mutagènes, les corps que nous avons étudiés sont presque tous instables en présence d'eau. Les solutions aqueuses destinées aux traitements ont donc été préparées immédiatement avant l'emploi. Lorsqu'il s'agissait de substances solides, nous les avons pesées et dissoutes dans un volume déterminé d'eau. Avec les corps liquides, les quantités nécessaires ont été prélevées à l'aide de propipettes, la densité de ces substances étant ultérieurement mesurée. Il a donc été possible, dans tous les cas, de calculer les concentrations moléculaires correspondantes, et par conséquent de comparer entre elles les activités de diverses substances.

Bien entendu, toute substance nouvelle fait d'abord l'objet d'expériences préliminaires visant à déterminer la concentration correspondant à la dose létale 50.

2. *Action mutagène des rayonnements*

Nous avons étudié à la fois l'action des radiations ionisantes (rayons γ émis par le Cobalt 60) et celle du rayonnement ultra-violet.

a) Rayons γ

Les irradiations ont été réalisées avec une source de Cobalt 60 donnant une intensité de 2600 r/minute. Les levures ont été irradiées dans des tubes de verre à des doses s'échelonnant de 20000 r à 140000 r (la DL 50 se situe entre 40000 et 50000 r).

Avec la souche arg-1 (fig. 4), deux expériences ont donné des résultats très comparables. La relation entre la dose et le nombre N de réversions pour 10^7 cellules survivantes paraît être linéaire (courbe à 1 coup), ce qui est en bon accord avec les théories généralement admises pour une action directe des radiations ionisantes (théorie de la cible).

Par contre, pour la souche leuc-3 (fig. 4), deux expériences ont également donné des résultats cohérents, mais cette fois, la courbe est du type quadratique (à deux coups).

Enfin, une seule expérience a été faite avec la souche ur-1. Bien que la dose ait été poussée jusqu'à 140000 r, aucune activité mutagène n'a pu être détectée. Il est vrai que le taux des réversions spontanées dans le témoin non irradié, s'est trouvé à être élevé, ce qui a pu masquer l'action mutagène des rayons γ. Mais, de toute façon, cette action est certainement très faible.

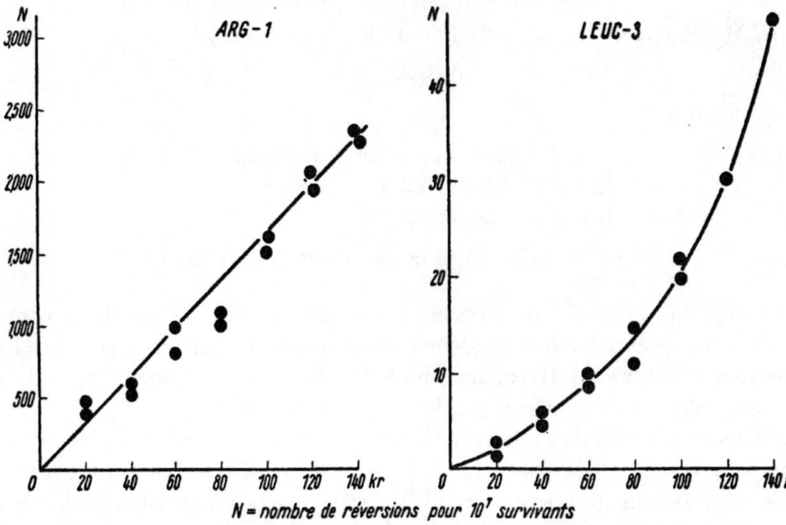

Fig. 4. Induction de réversions dans les souches arg-1 et leuc-3 action des rayons ∅ (^{60}Co)

La relation linéaire, trouvée avec arg-1, ne constitue pas un sujet d'étonnement; elle semble signifier que la mutation exige la production d'une ionisation à l'intérieur d'un volume sensible. Par contre, la loi quadratique que donne leuc-3 pose un problème: elle pourrait s'interprêter formellement en supposant que la mutation demande la production de deux ionisations indépendantes au sein d'un volume sensible. Nous ne pensons pas que ce soit là une explication satisfaisante et nous suggérons que cette loi résulte des propriétés physiologiques particulières de la souche leuc-3. Nous nous proposons de soumettre cette hypothèse à une vérification expérimentale, ce qui exige d'obtenir par croisement une souche double mutante arg-1 leuc-3, de l'irradier par les rayons γ, et de déterminer parallèlement les lois de réversion pour arg-1 et pour leuc-3 en ensemençant les cellules sur:

(Milieu minimum + leucine) → réversions pour arg-1

(Milieu minimum + arginine) → réversions pour leuc-3

le nombre de survivants étant mesuré par ensemencement sur

(Milieu minimum + arginine + leucine).

Ainsi, les réversions de arg-1 et leuc-3 pourront être étudiées dans des conditions physiologiques identiques.

b) ultra-violet

Nous avons utilisé comme source un tube germicide émettant un rayonnement ultra-violet vers 2500 Å. L'irradiation a été faite sur des volumes de 5 cc de suspension cellulaire (10^8 cellules/cm^3) dans des boites de Pétri de 10 cm de diamètre, constamment agitées à la main. La distance des boîtes au tube était de 37 cm. Les temps d'irradiation ont varié entre 20 et 160 secondes. Les énergies incidentes, exprimées en ergs/cm^2 ne sont pas connues. La dose létale 50 correspond à une irradiation de l'ordre de 50 secondes.

La fig. 5 montre que la relation entre la dose et le nombre N de réversions pour 10^7 survivants n'est pas linéaire, mais affecte une forme parabolique. WITKIN [39] a récemment fait

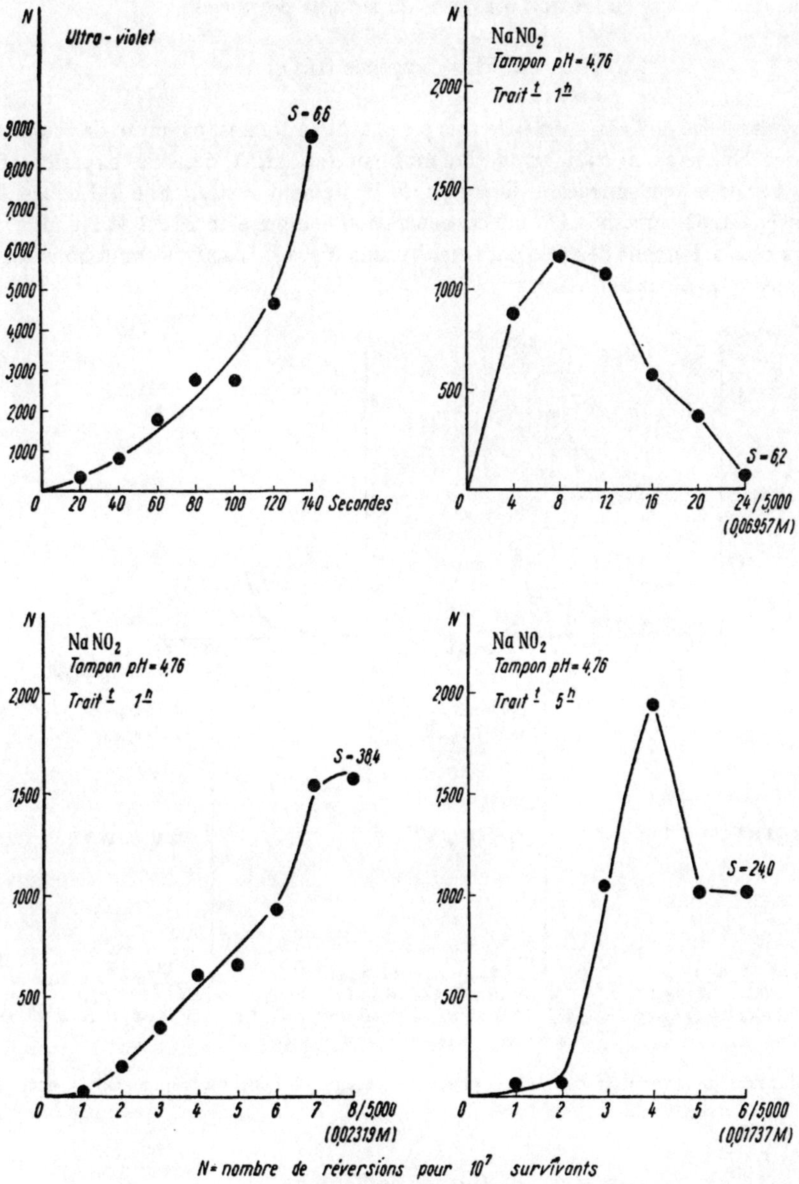

Fig. 5. Induction de réversions dans la souche arg-1; action de l'ultra-violet et de $NaNO_2$

une étude critique de ce type de réponse à l'irradiation par l'ultra-violet. En effet, lorsque l'on étudie, chez les bactéries, la réversion de mutants biochimiques sous l'action de ce rayonnement, on trouve systématiquement une courbe qui diffère fortement de la relation linéaire à laquelle on s'attendait. Nous discuterons plus tard de son interprétation.

3. Action mutagène de certains péroxydes

a) Eau oxygénée (H_2O_2)

Nous avons utilisé de l'eau oxygénée très pure, à 110 volumes, exempte de toute substance stabilisatrice. Nous avons étudié son action mutagène sur arg-1, dans des expériences où nous avons fait varier la concentration, le temps de traitement restant fixé à 1 heure. Dans ces conditions, la DL 50 correspond à une concentration comprise entre 2,1 et 2,4 M.

Les deux essais donnent des résultats très voisins (fig. 6). L'activité mutagène est faible et ne se manifeste qu'à haute dose.

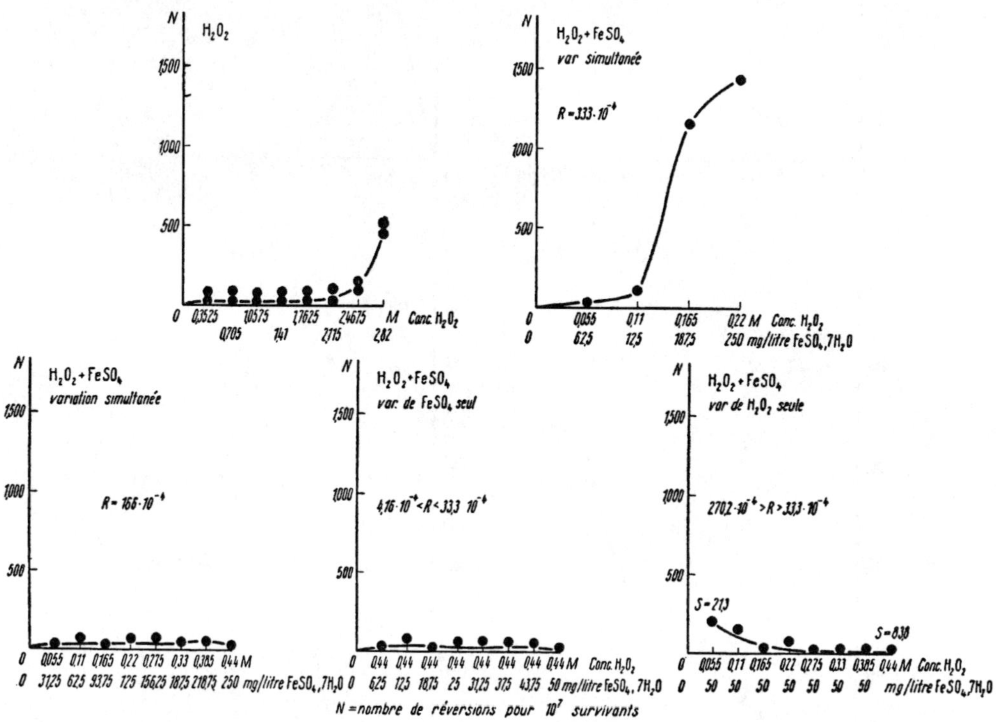

Fig. 6. Induction de réversions dans la souche arg-1; action de l'eau oxygénée et du réactif de Fenton

b) Réactif de Fenton

Il s'agit d'un mélange de H_2O_2 et de $FeSO_4$. On sait depuis longtemps que les ions Fe^{++} entraînent une décomposition catalytique de H_2O_2. La cinétique du phénomène a été étudiée par Haber et Weiss [7]. Ces auteurs ont proposé le mécanisme de réaction suivant:

$$Fe^{++} + H_2O_2 \rightarrow Fe^{+++} + OH^- + HO^{\cdot} \quad (1)$$

$$HO^{\cdot} + H_2O_2 \rightarrow H_2O + HO_2^{\cdot} \quad (2)$$

Il y a par conséquent formation des radicaux libres $HO^{\cdot}$ et $HO_2^{\cdot}$ Or ces mêmes radicaux sont également produits au sein de la matière vivante soumise aux rayonnements ionisants et l'on considère qu'ils sont, au moins pour une part, responsables des effets génétiques observés. Le réactif de Fenton permet de produire ces radicaux par une voie purement chimique. Il était donc intéressant d'étudier ses propriétés mutagènes. Ce système a été déjà examiné par DUBININ, SIDOROV et SOKOLOV [4], qui ont conclu qu'il était capable de provoquer des cassures et remaniements chromosomiques.

Il s'agit d'un système dans lequel on peut faire varier séparément ou simultanément les concentrations de H_2O_2 et de $FeSO_4$. La fig. 6 montre les résultats que nous avons obtenus et indique que – dans certaines conditions – l'activité mutagène de H_2O_2 peut être considérablement renforcée par la présence de l'ion Fe^{++}. Toutefois l'activité dépend à la fois de la concentration absolue de H_2O_2 et $FeSO_4$ et du rapport

$$R = \frac{\text{nombre de mg de } FeSO_4,\ 7\ H_2O}{\text{nombre de mg de } H_2O}$$

Pour un rapport $R = 333.\ 10^{-4}$, un effet mutagène très net est observé. Par contre pour $R \leq 166 \cdot 10^{-4}$, aucun effet génétique n'est détectable. Ceci se traduit par des conséquences à première vue paradoxales: dans une expérience où la concentration de $FeSO_4$, $7\ H_2O$ est maintenue fixe à 50 mg/litre, mais où la concentration de H_2O_2 est progressivement diminuée de 0,44 M à 0,55 M, on observe que l'effet toxique et l'action mutagène sont plus élevés pour les faibles concentrations:

H_2O_2 0,44 M: % de survivants = 83,8 ($R = 33{,}3 \cdot 10^{-4}$)

H_2O_2 0,055 M: % de survivants = 21,3 ($R = 270{,}2 \cdot 10^{-4}$)

Dans une expérience ultérieure, la concentration en H_2O_2 étant maintenue fixe à 0,44 M, nous avons fait varier par palliers la concentration de $FeSO_4$, $7\ H_2O$ de 500 mg/litre à 7000 mg/litre. Le rapport R varie donc de $333 \cdot 10^{-4}$ à $4667.\ 10^{-4}$. Or, on constate que la courbe de mutation a une forme en cloche et présente un maximum correspondant à $R = 2667 \cdot 10^{-4}$, soit un mélange réactif comportant environ 30 molécules H_2O_2 pour un ion Fe^{++}.

On sait que le radical $HO^{\cdot}$ est beaucoup plus réactif que le radical $HO_2^{\cdot}$. Il semble donc possible d'interpréter nos résultats expérimentaux à l'aide des équations [1] et [2] de HABER et WEISS si l'on suppose que l'induction de réversions biochimiques dépend essentiellement de la concentration en radical $HO^{\cdot}$.

L'équation [1] montre en effet que la vitesse de formation de $HO^{\cdot}$ augmente lorsque les concentrations de Fe^{++} et H_2O_2 croissent. L'équation [2] indique toutefois que le radical $HO^{\cdot}$ est transformé en $HO_2^{\cdot}$ (moins actif) s'il existe un excès de H_2O_2 relativement à Fe^{++}.

L'existence d'un maximum d'activité mutagène, lorsque R augmente paraît être plus difficile à expliquer. Il peut s'agir toutefois de phénomènes secondaires (déjà suggérés par HABER et WEISS) tels que:

$$HO^{\cdot} + Fe^{++} \rightarrow Fe^{+++} + OH^{-}$$

l'ion Fe^{++}, aux hautes concentrations, réagissant avec $HO^{\cdot}$ pour former l'ion OH^{-} inactif.

Ces interprétations ne sont suggérées que sous réserve et nous souhaitons que des chimistes compétents nous donnent leur avis.

c) Acide periodique (HIO_4)

Ce réactif a été employé dissous dans un tampon de pH = 4,76. C'est en effet en milieu acide que ses propriétés oxydantes sont les plus marquées.

On sait que cette substance est capable de dégrader la molécule de ribose par oxydation, en raison de la présence dans cette molécule de deux groupes–OH en cis (positions 2′ et 3′). Ainsi [3] pour un nucléotide 5′ – phosphate:

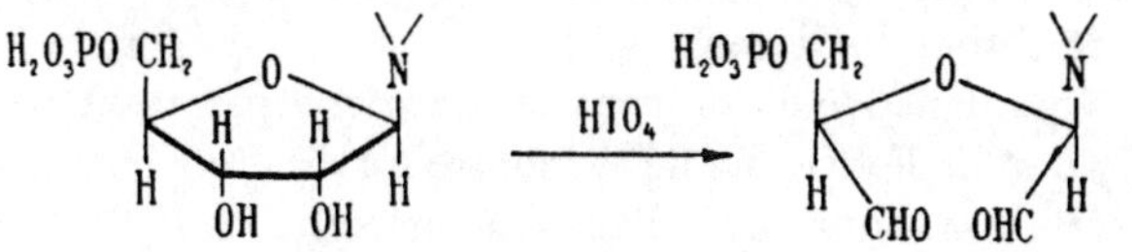

Par contre, le désoxyribose n'est pas attaqué, puisqu'il ne possède pas deux groupes –OH en cis.

Il résulte de ces propriétés que l'acide periodique est susceptible de dégrader les acides ribonucléiques, mais qu'il est inactif sur les acides désoxyribonucléiques. On notera toutefois que si les liaisons phosphoriques entre nucléotides dans les acides ribonucléiques se réalisent entre les positions 3′ et 5′ de molécules successives de ribose, l'acide periodique s'avérera inactif puisqu'il n'existe pas alors de groupes –OH libres en cis. Cependant, l'acide periodique conserve ses possibilités de dégradation des ribonucléosides non encore intégrés dans les acides ribonucléiques.

Dans ces conditions nous avons voulu savoir si l'acide periodique avait une action mutagène. Nous l'avons testé sur la souche arg-1, mais les résultats ont été complètement négatifs. C'est ce à quoi on doit s'attendre si l'induction de mutations exige de provoquer des modifications chimiques dans l'ADN.

4. Action mutagène du nitrite de sodium ($NaNO_2$)

Le nitrite de sodium a été dissous dans un tampon acétate de sodium/HCl de pH = 4,76. On sait depuis longtemps que ce réactif est capable de désaminer les groupes $-NH_2$ pour y substituer des groupes –OH. En particulier, il réagit avec les groupes $-NH_2$ des bases puriques et pyrimidiques, que celles-ci soient libres ou combinées dans des nucléosides, nucléotides ou polynucléotides.

Schuster et Schramm [34, 35] ont établi que $NaNO_2$ désaminait les bases puriques et pyrimidiques de l'acide ribonucléique du virus de la mosaïque du tabac. Le virus est inactivé sans qu'il se produise de dégradation de la chaîne de polynucléotides. Gierer et Mundry [5, 29] ont démontré que ce traitement provoque l'apparition de mutants chez le virus.

Vielmetter et Wieder [36, 37] ont traité le bactériophage T_2 par $NaNO_2$ et obtenu un accroissement net du nombre des mutants *r*. Le pourcentage de mutants *r* dans les particules ayant survécu est une fonction linéaire du temps de traitement (courbe à 1 coup).

Kaudewitz [15, 16, 17] est parvenu à induire des mutations biochimiques chez les bactéries (*Escherichia coli*, *Salmonella typhimurium*) sous l'effet de $NaNO_2$. Toutefois, le pourcentage de mutants parmi les survivants est une fonction quadratique du temps de traitement (courbe à deux coups). De plus, les colonies mutantes, détectées par la technique des copies sur milieu minimum, sont constituées uniquement de cellules mutantes (on n'y recontre pas de prototrophes).

Kaudewitz remarque que la technique de détection des colonies mutantes ne permet pas d'identifier celles d'entre-elles qui comporteraient à la fois des auxotrophes et des prototrophes: ces colonies, qui sont classées comme non mutantes, résultent en fait d'une mutation induite dans une seulement des chaînes de la spirale de Watson-Crick, l'autre étant inaffectée.

Pour qu'une colonie soit classée comme mutante, KAUDEWITZ considère qu'il faut deux désaminations, une dans chaque chaîne: l'une donnerait une mutation dans une chaîne et l'autre inactiverait l'autre chaîne. Ainsi, la colonie ne comporterait que des cellules mutantes auxotrophes issues d'une seule chaîne.

La courbe à deux coups observée dans ces expériences serait une simple conséquence de la technique utilisée pour la détection des mutants, la mutation proprement dite étant un processus à un coup.

Nous avons étudié l'action mutagène de $NaNO_2$ sur *Schizosaccharomyces pombe* en traitant pendant un temps fixe (1 heure ou 5 heures) avec des concentrations croissantes du réactif (dans un tampon pH = 4,76). Les trois souches arg-1, leuc-3 et ur-1 ont été utilisées. Avec arg-1 (fig. 5), la relation entre N (nombre de réversions pour 10^7 survivants) et la concentration, a la forme d'une courbe en cloche: N passe donc par un maximum pour diminuer ensuite. Ce type de courbe se rencontre à la fois dans les expériences de durée 1 heure ou 5 heures. Si, à plus faibles concentrations de $NaNO_2$, on étudie la partie initiale de la courbe, on s'aperçoit qu'il s'agit en gros d'une sigmoïde qui pourrait facilement, si on se limitait à des concentrations plus faibles encore, être confondue avec une courbe quadratique (à deux coups).

Avec leuc-3, dans la gamme de concentrations étudiées, les courbes sont précisément du type quadratique (fig. 7). Enfin avec ur-1, nous retrouvons une sigmoïde (Fig. 8).

Nous discuterons plus tard de l'interprétation que l'on peut donner de ces courbes.

5. *Action mutagène des substances alkylantes*

a) Méthodes

Nous avons étudié des substances alkylantes appartenant aux groupes chimiques suivants: esters sulfoniques et sulfuriques (auxquels nous joindrons les sultones), β-halogéno-éthyle amines, époxydes, éthylène-imines.

Il s'agit de corps qui s'hydrolysent généralement dans l'eau, aussi est-il nécessaire de ne préparer les solutions aqueuses que juste avant l'emploi.

On peut envisager deux types principaux d'expériences: soit faire varier la concentration de l'agent mutagène en laissant fixe le temps de traitement (1 heure ou 5 heures dans nos essais), soit choisir une concentration fixe et faire varier le temps de traitement (de 1 heure à 5 heures dans nos essais).

Ce deuxième type d'expérience est manifestement moins satisfaisant que le premier. En effet, en solution aqueuse, une hydrolyse se produit, qui augmente en fonction du temps. La concentration en substance active diminue donc progressivement et la comparaison des activités mutagènes d'une substance après 1 h et 5 h doit évidemment tenir compte de ce fait important. A titre d'exemple, la fig. 9 donne les variations de N en fonction du temps de traitement, pour la souche arg-1, avec une solution aqueuse de sulfate neutre d'éthyle dont la concentration initiale est de 2/500. On obtient une courbe en cloche dont la signification sera discutée plus tard.

Par contre, dans les expériences où la concentration est variable, le temps de traitement restant fixe, l'hydrolyse se produit aussi, mais affecte dans chaque cas la même fraction de substance active initialement présente. En conséquence, les rapports mutuels des concentrations en substance mutagène ne varient pas au cours de l'expérience. Ainsi, si l'on réalise au départ les rapports suivants des concentrations: 1, 2, 3, 4, 5, 6, 7, 8, ces rapports sont

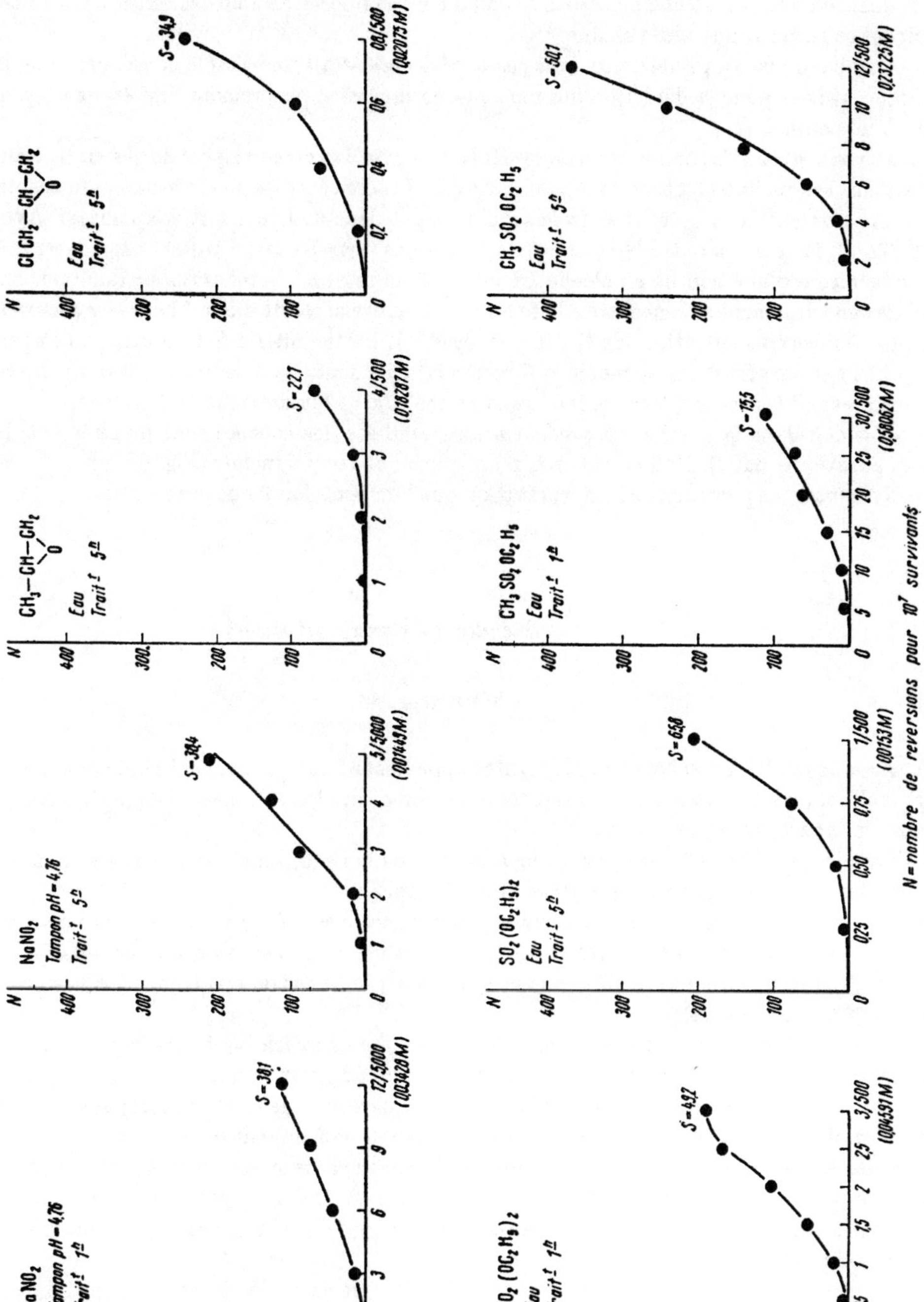

Fig. 7. Induction de réversions dans la souche leuc-3

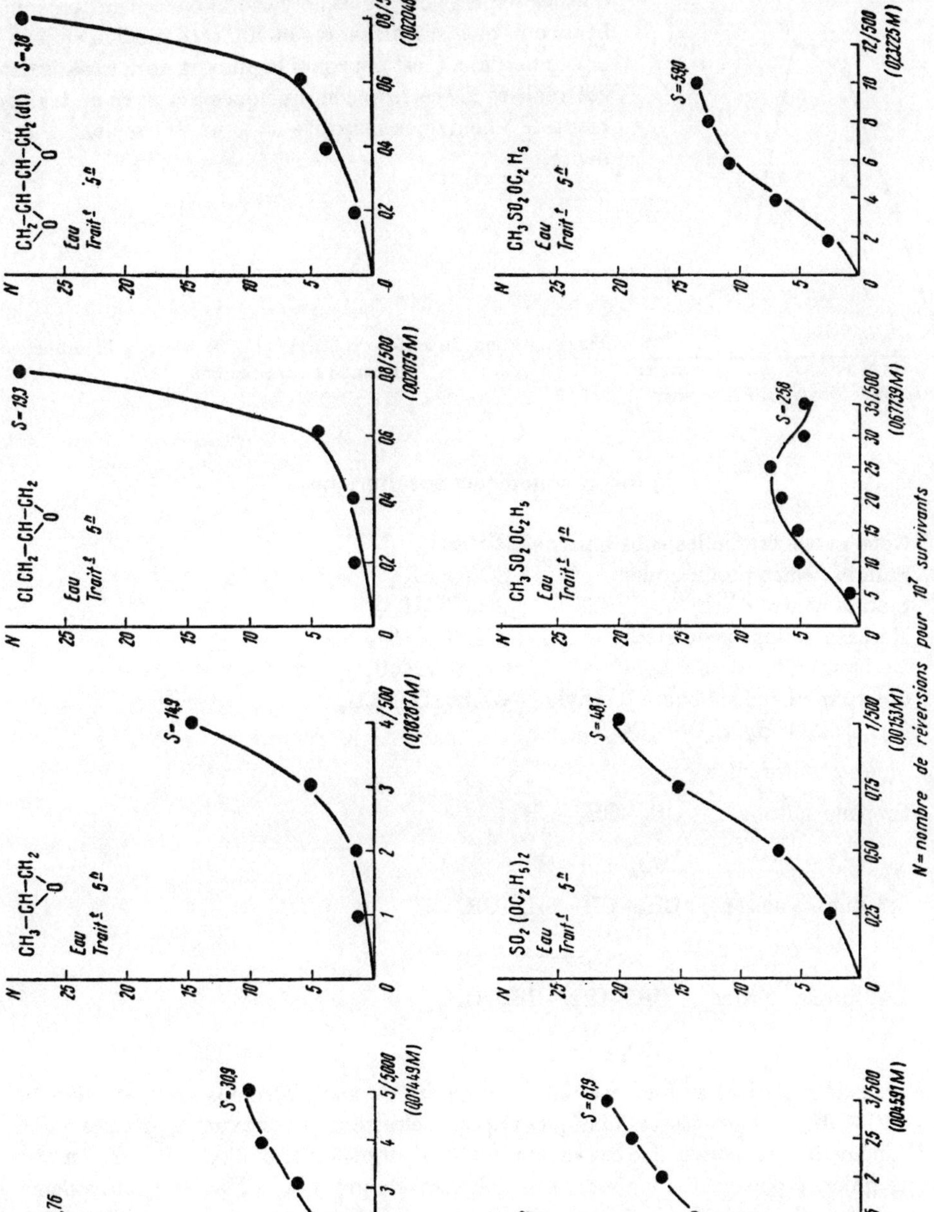

Fig. 8. Induction de réversions dans la souche ur-1

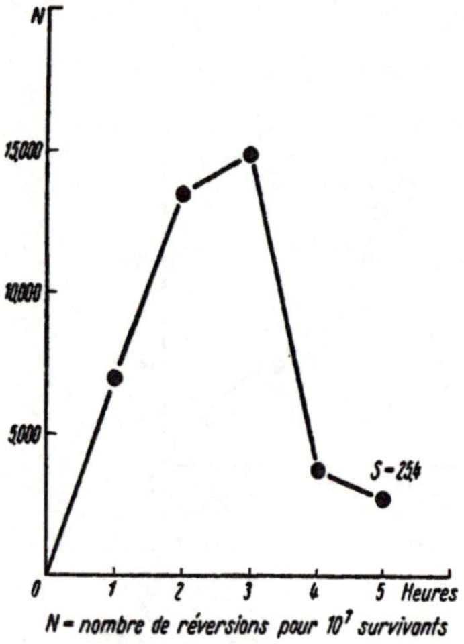

inaltérés au bout d'une heure (ou de 5 heures) de traitement, bien qu'une hydrolyse se soit produite. Il en résulte que les expériences où seule varie la concentration sont beaucoup plus satisfaisantes et d'interprétation en principe plus aisée. C'est pourquoi la plupart de nos essais ont été réalisés de cette façon, quelques uns avec un traitement de 5 heures, la majorité avec un traitement d'une heure.

Fig. 9. Action du sulfate d'éthyle (2/500) sur arg-1; influence du temps de traitement

b) Esters sulfoniques et sulfuriques

Nous avons étudié les substances suivantes:
Sulfate neutre de méthyle: $SO_2(OCH_3)_2$
Sulfate neutre d'éthyle: $SO_2(OC_2H_5)_2$
Méthane sulfonate de méthyle: $CH_3SO_2OCH_3$
Méthane sulfonate d'éthyle: $CH_3SO_2OC_2H_5$
Chloro méthane sulfonate d'éthyle: $ClCH_2SO_2OC_2H_5$
$C_2H_5OSO_2OCH_2CH_2OSO_2OC_2H_5$

ainsi que certaines sultones:

Propane sultone: CH_2—CH_2—CH_2 (cycle: SO_2——O)

1,3 Butane sultone: CH_3—CH—CH_2—CH_2 (cycle: SO_2——O)

1,4 Butane sultone: CH_2—CH_2—CH_2—CH_2 (cycle: SO_2——O)

Les esters sulfoniques et sulfuriques ont la propriété de transférer des groupes alkyl (ici: —CH_3 ou —C_2H_5) aux structures chimiques riches en électrons (anions variés, groupes —NH_2, etc.). On pouvait donc penser que ces substances seraient capables d'alkyler l'ADN, en particulier les groupes phosphates ionisés, ainsi que certains groupes —NH_2 et éventuellement certains atomes d'azote inclus dans les cycles des bases puriques et pyrimidiques. Ceci laissait donc prévoir que ces corps seraient doués d'activité mutagène. Au surplus, KØLMARK et WESTERGAARD [18 à 22, 38] ont déjà établi que certains de ces composés sont actifs sur *Neurospora*, et RAPOPORT [30] a montré qu'ils provoquaient des changements génétiques chez *Drosophila*.

Par contre, à notre connaissance, les sultones n'ont pas encore attiré l'attention des généticiens. Ces corps ont été synthétisés par des chimistes allemands, en particulier HELBERGER, MANECKE, HEYDEN, FISCHER, LAUTERMANN, WINTER [9 à 13] qui ont signalé leurs propriétés alkylantes. Informé de ces caractéristiques, nous avons pensé que les sultones devaient être mutagènes. C'est ce que l'expérience a confirmé. Nous nous trouvons donc en présence d'une nouvelle classe d'agents mutagènes alkylants particulièrement actifs.

Seuls le sulfate neutre d'éthyle et le méthane sulfonate d'éthyle ont été étudiés sur les souches leuc-3 et ur-1 (fig. 7 et 8) avec des temps de traitement de 1 heure et de 5 heures. La relation entre N (Nombre de réversions pour 10^7 cellules survivantes) et la concentration, affecte la forme d'une parabole dans sa première partie, s'infléchit ensuite et passe par un maximum. Selon l'échelle des concentrations étudiées, la courbe peut donc avoir l'aspect d'une parabole, ou d'une sigmoïde, ou encore d'une sigmoïde suivie d'un maximum.

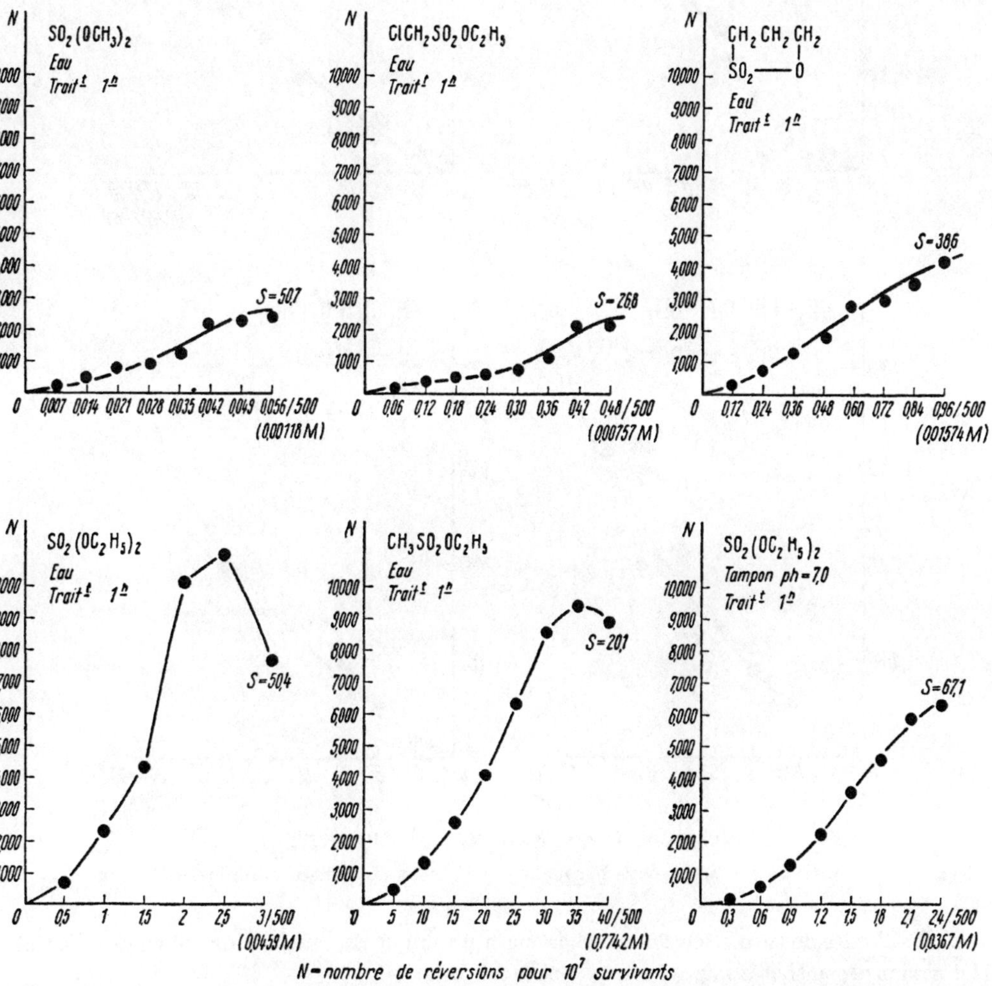

Fig. 10. Induction de réversions dans la souche arg-1; action d'esters sulfoniques et sulfuriques

Les neufs corps mentionnés ci-dessus ont été essayés sur arg-1, avec des temps de traitement de 1 heure (parfois aussi de 5 heures). Les résultats sont résumés dans les fig. 10 et 11. On voit que les courbes obtenues sont aussi des sigmoïdes suivies d'un maximum, pour peu que la concentration soit assez élevée. La figure 12 compare les courbes obtenues avec ces neuf composés (temps de traitement 1 heure), l'abcisse étant graduée en concentrations molé-

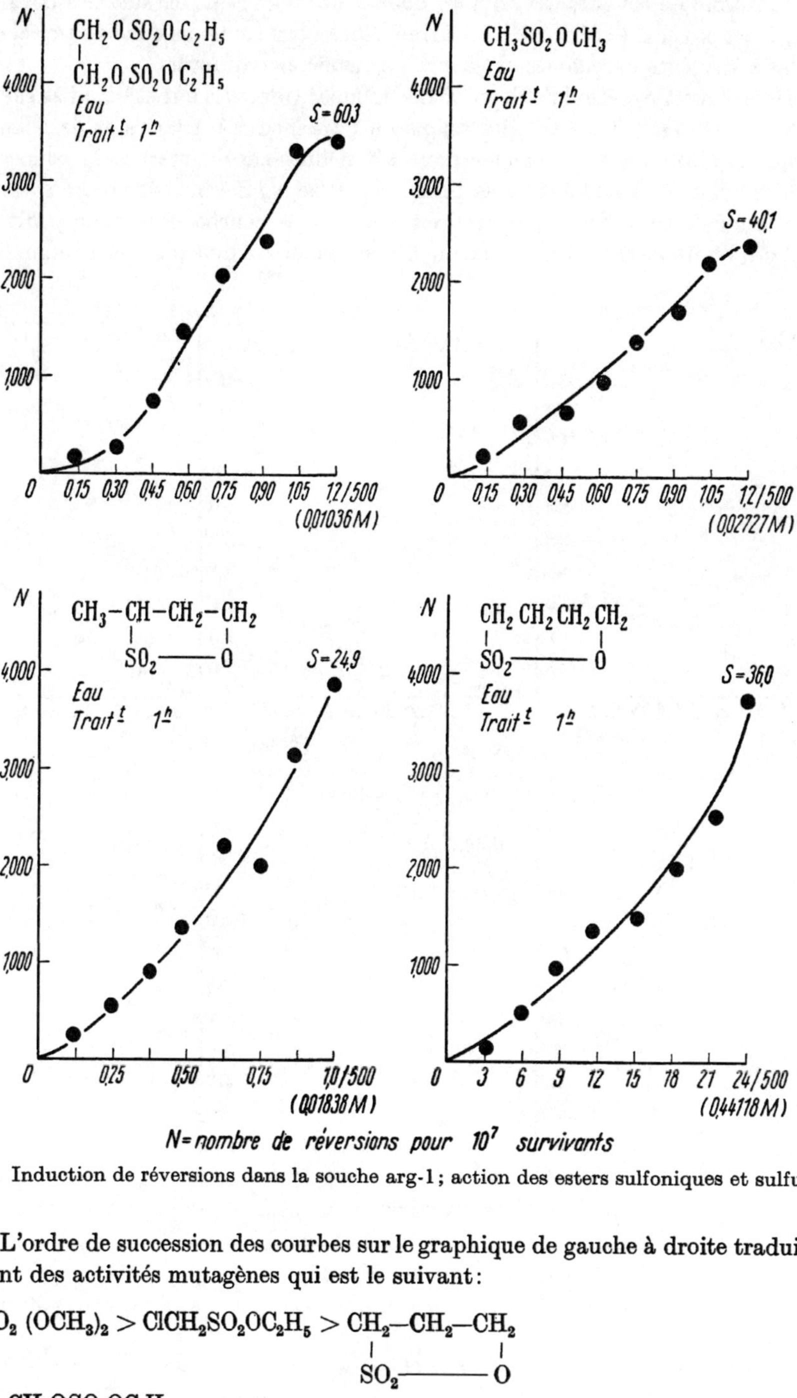

Fig. 11. Induction de réversions dans la souche arg-1; action des esters sulfoniques et sulfuriques

culaires. L'ordre de succession des courbes sur le graphique de gauche à droite traduit l'ordre décroissant des activités mutagènes qui est le suivant:

$$SO_2\,(OCH_3)_2 > ClCH_2SO_2OC_2H_5 > \underset{SO_2\text{———}O}{CH_2\text{—}CH_2\text{—}CH_2}$$

$$> \begin{matrix} CH_2OSO_2OC_2H_5 \\ | \\ CH_2OSO_2OC_2H_5 \end{matrix} > \text{1,3 Butane sultone} > SO_2(OC_2H_5)_2$$

$$> CH_3SO_2OCH_3 > CH_3SO_2OC_2H_5 > \text{1,4 Butane sultone}$$

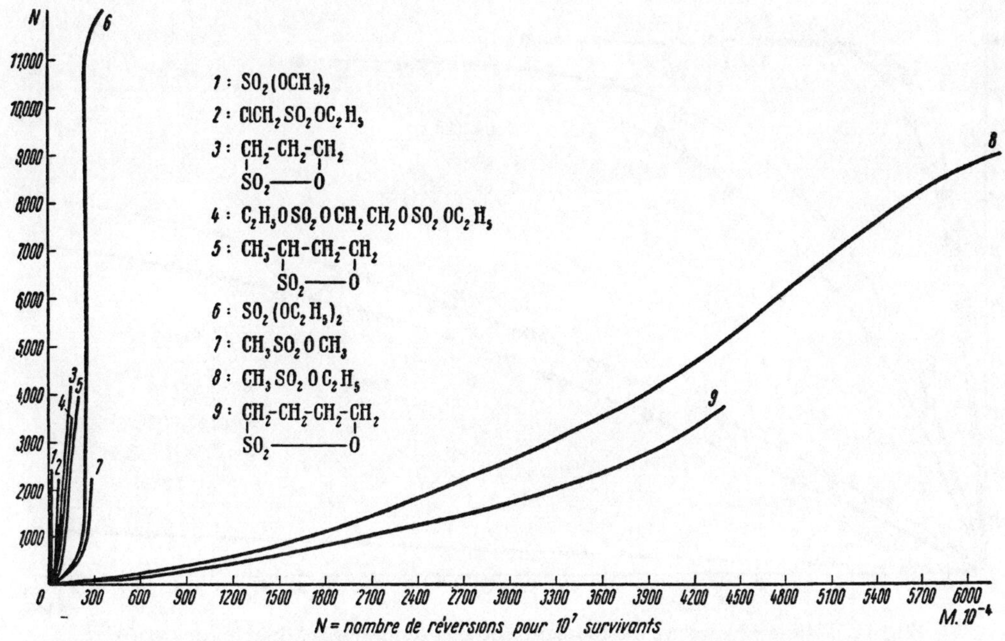

Fig. 12. Induction de réversions dans la souche arg-1; action d'esters sulfoniques et sulfuriques

Quel sens doit-on attribuer à cette séquence particulière? Il semble exister une corrélation, dans cette série, entre la réactivité chimique d'un composé et son activité mutagène. A défaut de pouvoir mesurer aisément la vitesse de réaction de ces corps avec l'ADN, nous avons déterminé quelle était la vitesse de réaction avec l'eau (vitesse d'hydrolyse). La fig. 13 montre les courbes obtenues à 50 °C. Les périodes correspondantes sont les suivantes:

Composé	Période à 50 °C	Période à 21 °C
$SO_2(OCH_3)_2$	5–6 minutes	–
$SO_2(OC_2H_5)_2$	7–8 ,,	183 minutes
1,3-Butane sultone	27 ,,	–
Propane sultone	36 ,,	–
$CH_3SO_2OCH_3$	140 ,,	–
$CH_3SO_2OC_2H_5$	168 ,,	–
1,4-Butane sultone	27 heures	–

La période de certains composés n'a pu être déterminée, faute d'en posséder une quantité suffisante.

Il est clair qu'il existe une certaine corrélation entre la valeur de la période et l'activité mutagène. Cette corrélation n'est toutefois pas absolue, mais il faut remarquer que – pour des raisons de facilité technique – les hydrolyses ont été conduites à 50°, alors que toutes les expériences génétiques ont été faites à 25°. D'autre part, et ceci est fort important, lavitesse d'hydrolyse n'est qu'un aspect restreint de la réactivité chimique des substances en question. Il n'est nullement certain que l'ordre subsisterait inchangé pour la réaction avec les anions et les amines. Or, l'action sur l'ADN consiste sans doute en réactions de ce dernier type.

Aucune conclusion définitive ne pourra donc être formulée tant que des études précises de cinétique chimique n'auront pas été conduites avec une série d'anions et d'amines, voire même avec l'ADN in vitro.

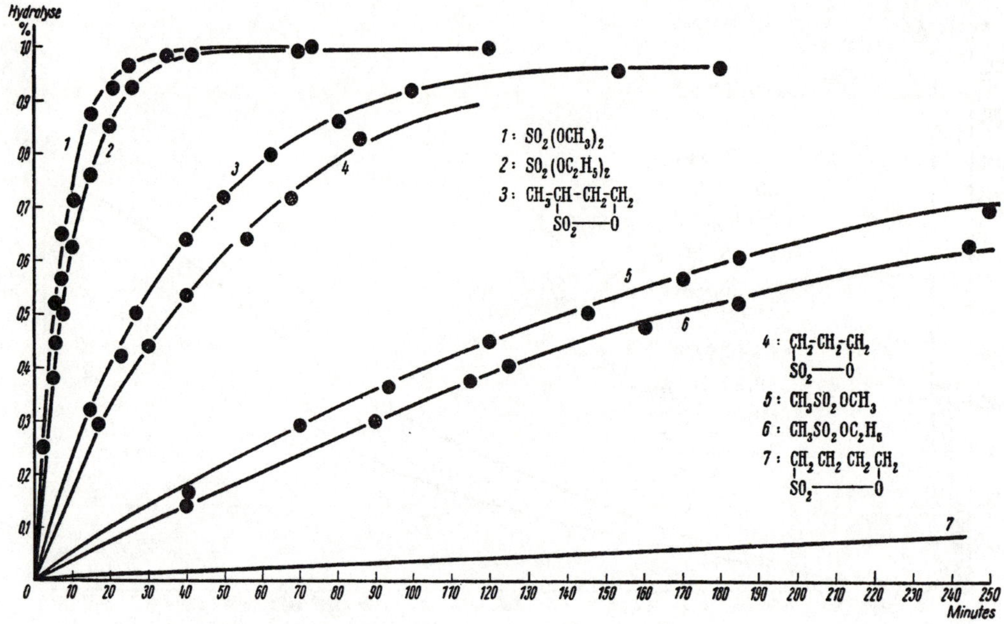

Fig. 13. Vitesse d'hydrolyse d'esters sulfoniques et sulfuriques et de sultones (50° C)

On sait que les esters sulfoniques et sulfuriques réagissent généralement par un mécanisme bimoléculaire ($S_N 2$) qui, selon Ross [32, 33] peut être schématisé de la façon suivante, en explicitant l'état de transition:

$$R'SO_2OR + A^- \rightarrow R'SO_2O \ldots\ldots R \ldots\ldots A$$
$$\rightarrow R'SO_2\bar{O} + RA$$

Dès lors, des considérations électroniques permettent de prévoir quelle sera l'action (positive ou négative) de divers substituants sur la vitesse de réaction. Ainsi, pour les méthane sulfonates:

$$\left.\begin{array}{l} CH_3SO_2OCH_3 \\ CH_3SO_2OC_2H_5 \end{array}\right\downarrow$$

on peut prévoir que la réactivité diminuera dans le sens indiqué par la flèche, parce que les substituants $-CH_3$ et $-C_2H_5$ se classent dans l'ordre suivant en ce qui concerne l'intensité avec laquelle ils repoussent les électrons:

$$-CH_3 < -C_2H_5$$

De même le remplacement d'un hydrogène par un chlore (qui attire fortement les électrons) dans la molécule $CH_3SO_2OC_2H_5$, conduit à $ClCH_2SO_2OC_2H_5$ dont la réactivité doit être notablement accrue. C'est bien ce que l'on observe.

Par ailleurs, le passage de $CH_3SO_2OCH_3$ à $SO_2(OCH_3)_2$ ou de $CH_3SO_2OC_2H_5$ à $SO_2(OC_2H_5)_2$ conduit à remplacer $-CH_3$ respectivement par $-OCH_3$ ou $-OC_2H_5$ dont le caractère électronique doit entraîner un accroissement de la réactivité.

Pour les sultones enfin, il est normal que la réactivité diminue lorsque l'on passe de la propane sultone à la 1,4 butane sultone dont le cycle comporte un atome de carbone de plus, ce qui accroît la stabilité.

Ainsi, les théories électroniques permettent de prévoir les relations entre la structure moléculaire et la réactivité et, par voie de conséquence, d'estimer dans une certaine mesure les variations de l'activité mutagène.

On notera, dans la fig. 12, que le nombre N de réversions pour 10^7 survivants atteint un maximum variable selon les composés. Ce maximum dépend, entre autres facteurs, de la toxicité du corps considéré. On voit qu'ici le sulfate neutre d'éthyle et le méthane sulfonate d'éthyle ont permis d'atteindre les valeurs N les plus élevées. Ce sont tous deux des agents d'éthylation, mais le transfert d'un groupe $-C_2H_5$ ne présente pas cependant un caractère particulier et unique comme cela a été suggéré par Loveless [28]. En effet $ClCH_2SO_2OC_2H_5$, qui est lui aussi un agent d'éthylation, conduit à des valeurs de N beaucoup plus faibles.

Il apparaît nettement qu'il n'y a pas de relation entre le maximum de N et l'activité mutagène exprimée en fonction de la concentration moléculaire.

Enfin, nous noterons que les maxima de N atteints ici (11000 et 9500 respectivement pour $SO_2(OC_2H_5)_2$ et $CH_3SO_2OC_2H_5$ avec des survies de 47,5% et 40%) sont bien plus élevés que ceux obtenus avec les radiations ionisantes (2300 pour 140000 r avec une survie de 10% environ). Par conséquent, du point de vue pratique, si l'on cherche à obtenir le plus grand nombre possible de mutations avec une survie élevée, il vaudra mieux utiliser ces agents mutagènes chimiques de préférence aux radiations ionisantes.

c) β-halogéno-éthyle amines

Nous avons étudié les substances suivantes:

$H_2NCH_2CH_2Br \cdot HBr$
$(C_2H_5)_2NCH_2CH_2Cl \,.\, HCl$
$CH_3N(CH_2CH_2Cl)_2 \cdot HCl$
$N(CH_2CH_2Cl)_3 \,.\, HCl$

utilisées sous forme de chlorhydrates (ou de bromhydrates). Des solutions aqueuses ont été préparées au moment de l'emploi et la base a été libérée par addition de la quantité théorique de NaOH (1 molécule par molécule de chlorhydrate ou de bromhydrate).

On notera que les corps étudiés appartiennent à trois catégories: mono-, bi- et trifonctionnels.

Les résultats de nos expériences sont consignés dans les fig. 14 et 15 et concernent la souche arg-1. Les courbes obtenues sont des sigmoïdes. La fig. 15 compare les activités mutagènes des quatre composés étudiés, les concentrations moléculaires figurant en abcisse.

On constate qu'il n'y a pas de différence fondamentale de nature entre les composés mono- et polyfonctionnels: c'est ainsi que $H_2NCH_2CH_2Br$ et $CH_3N(CH_2CH_2Cl)_2$ ont des activités du même ordre.

De gauche à droite, les courbes se succèdent dans l'ordre des activités mutagènes décroissantes:

$$N(CH_2CH_2Cl)_3 > H_2NCH_2CH_2Br > CH_3N(CH_3CH_2Cl)_2 > (C_2H_5)_2NCH_2CH_2Cl$$

Afin d'interpréter ces résultats, il est nécessaire d'examiner le mode de réaction des β-halogéno-éthyle amines en solution aqueuse.

Hanby et al. [8], Cohen et al. [2] ont montré que ces substances réagissent par un mécanisme monomoléculaire (S_N 1), la première étape consistant en la formation d'un ion cyclique éthylène-imonium. Cet ion réagit ensuite soit avec l'eau, soit avec les anions (ou plus généralement les groupes riches en électrons) présents dans le milieu:

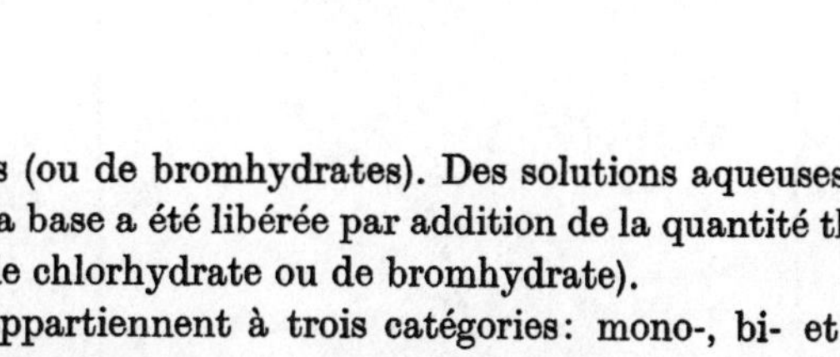

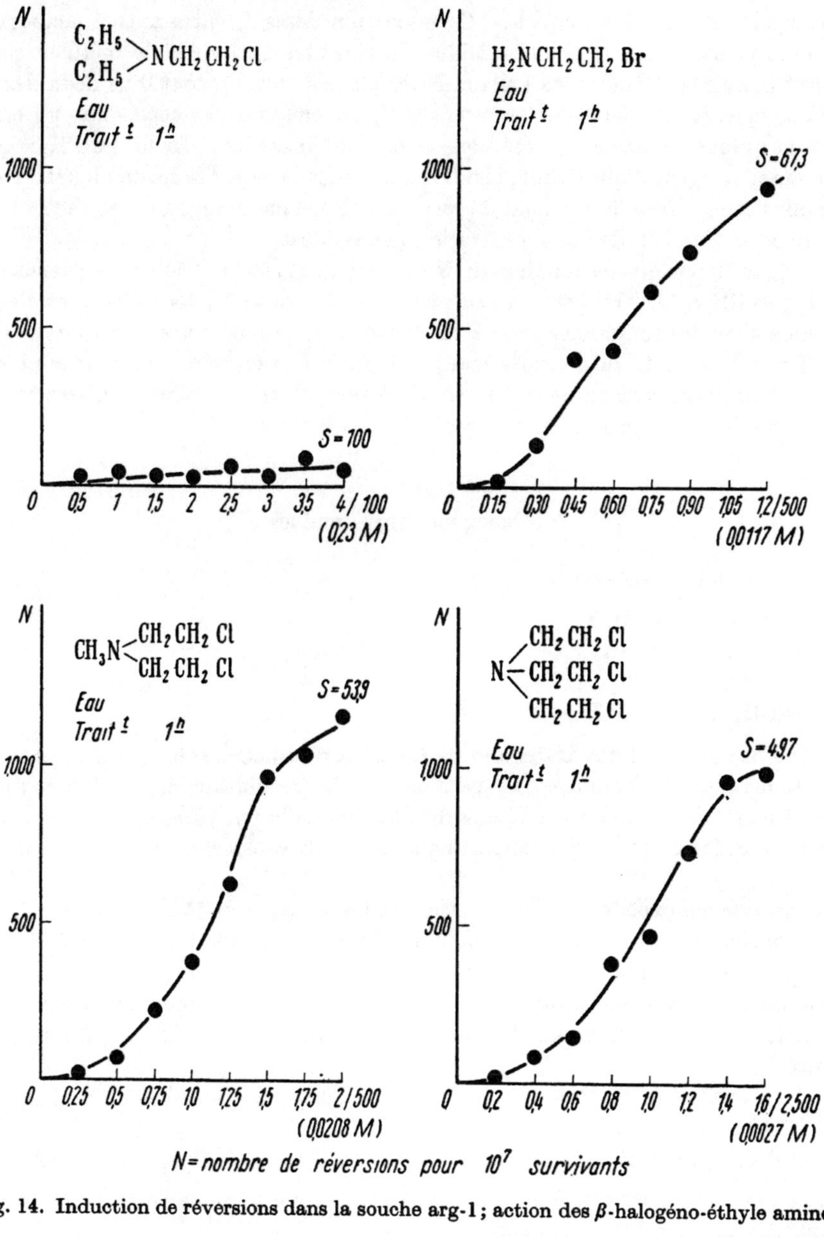

Fig. 14. Induction de réversions dans la souche arg-1; action des β-halogéno-éthyle amines

Pour que ces corps puissent réagir, il est nécessaire que l'ion éthylène-imonium soit d'abord formé, or ceci exige que l'atome d'azote soit trivalent, c'est à dire n'ait pas, en raison de sa basicité, fixé un proton pour donner la forme quaternaire:

$$R_2\underset{+}{\overset{\overset{\displaystyle H}{|}}{N}}CH_2CH_2Cl$$

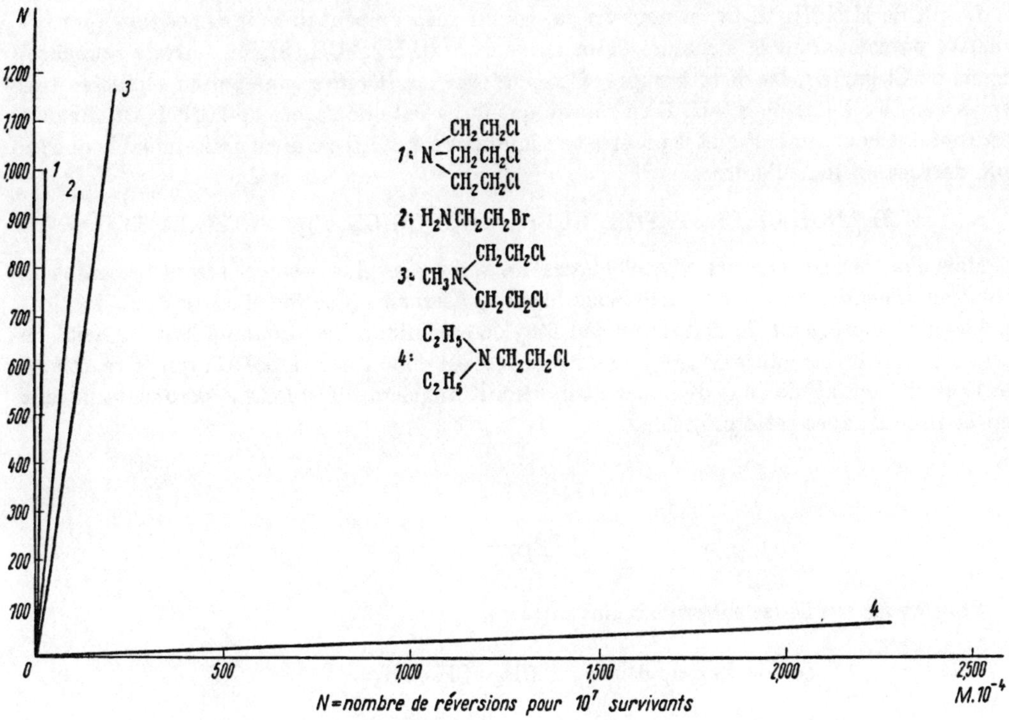

Fig. 15. Induction de réversions dans la souche arg-1; action des β-halogéno-éthyle amines

Or toutes ces substances sont généralement fournies sous forme de chlorhydrate dont la structure est précisément quaternaire:

$$\overset{\displaystyle H}{\underset{\displaystyle \overset{+}{Cl^-}}{R_2\overset{|}{N}}}\,CH_2\,CH_2\,Cl$$

C'est seulement dans la mesure ou une partie au moins des molécules existent sous forme de base libre $R_2NCH_2CH_2Cl$ qu'on peut s'attendre à une réactivité chimique et, partant, à une activité mutagène.

Il résulte de ces faits que les réactions des quatre substances vont dépendre très fortement du pH de la solution. En outre, pour chaque corps, la force basique – mesurée par le pK – est un élément important. D'après les données de la littérature [2], on a:

Composé	pK
$(C_2H_5)_2NCH_2CH_2Cl$	8,82
$CH_3N(CH_2CH_2Cl)_2$	6,45
$N(CH_2CH_2Cl)_3$	4,39

On observe la diminution progressive du pK lorsque le nombre d'atomes de chlore, présents dans la molécule augmente. Ceci est en parfait accord avec les théories électroniques, l'adjonction d'atomes Cl (qui attirent fortement les électrons) tendant à diminuer la densité électronique de l'azote et par conséquent la basicité.

Le pK de $H_2NCH_2CH_2Br$, ne nous est pas connu, mais on peut prévoir sa position approximative en comparant la structure de ce corps avec $(C_2H_5)_2NCH_2CH_2Cl$: outre le remplacement de Cl par Br, les deux groupes $-C_2H_5$ ont sur les électrons une action répulsive très supérieure à celle de deux –H. Il en résulte que la basicité de l'azote de $H_2NCH_2CH_2Br$ doit être notablement diminuée, et le pK être très inférieur à 8,82. Il est donc vraisemblable que les pK décroissent dans l'ordre:

$$(C_2H_5)_2NCH_2CH_2Cl > H_2NCH_2CH_2Br > CH_3N(CH_2CH_2Cl)_2 > N(CH_2CH_2Cl)_3$$

Mais d'autre part, aux pH physiologiques, les amines les plus basiques seront présentes en solution aqueuse en majeure partie sous forme quaternaire (fixation d'un proton), et incapables par conséquent de donner un ion éthylène-imonium. Ceci diminue leur capacité de réaction avec les structures riches en électrons. On est donc amené à prévoir que la réactivité et l'activité mutagène vont diminuer quand le pK augmente. Les faits expérimentaux sont en bon accord avec cette prévision.

d) Epoxides

Nous avons étudié les substances suivantes:

Oxyde de propylène:

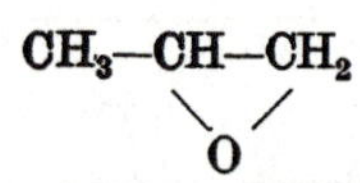

Glycidol:

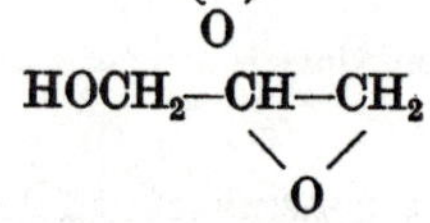

Epichlorhydrine:

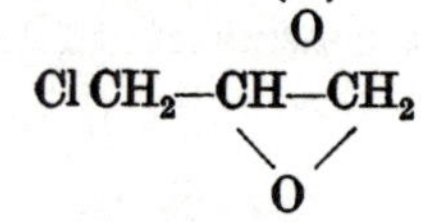

1-2, 3-4-Diépoxybutane:
(3 isomères: D, L, méso)

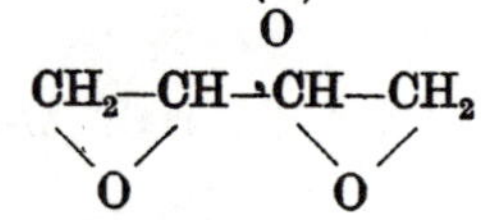

1-2, 5-6-Diépoxyhexane:

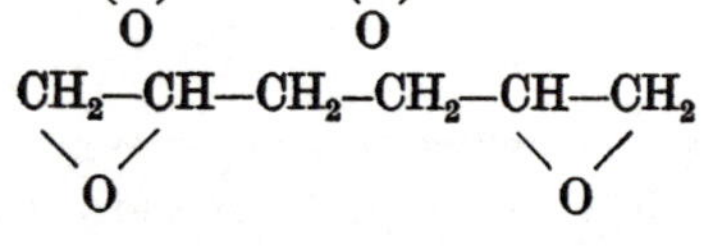

Sur les souches leuc-3 et ur-1, seuls ont été testés l'oxyde de propylène, le glycidol et le dl-diépoxybutane. Les résultats obtenus sont indiqués dans les fig. 16 et 17. Les courbes sont des types parabolique ou sigmoïde déjà rencontrés.

Les études les plus détaillées ont été réalisées sur la souche arg-1. Le temps de traitement choisi a été de 1 heure (ou parfois de 5 heures). Les résultats sont indiqués sur la fig. 16. Pour comparer les activités mutagènes de ces différentes substances, on a calculé les concentrations moléculaires (fig. 17). Il existe des différences dans les activités des trois isomères (D, L et méso) du diépoxybutane et, globalement, par ordre décroissant d'activité (de gauche à droite), on a le classement suivant:

1-2, 3-4 Diépoxybutane $\geq$ Epichlorhydrine > 1-2, 5-6-Diépoxyhexane > Glycidol > Oxyde propylène.

Pour essayer d'interpréter ces résultats, il importe d'examiner quel est le mode de réaction des époxydes en solution aqueuse. Ce problème a été étudié par BRÖNSTEDT et al. [1] et par

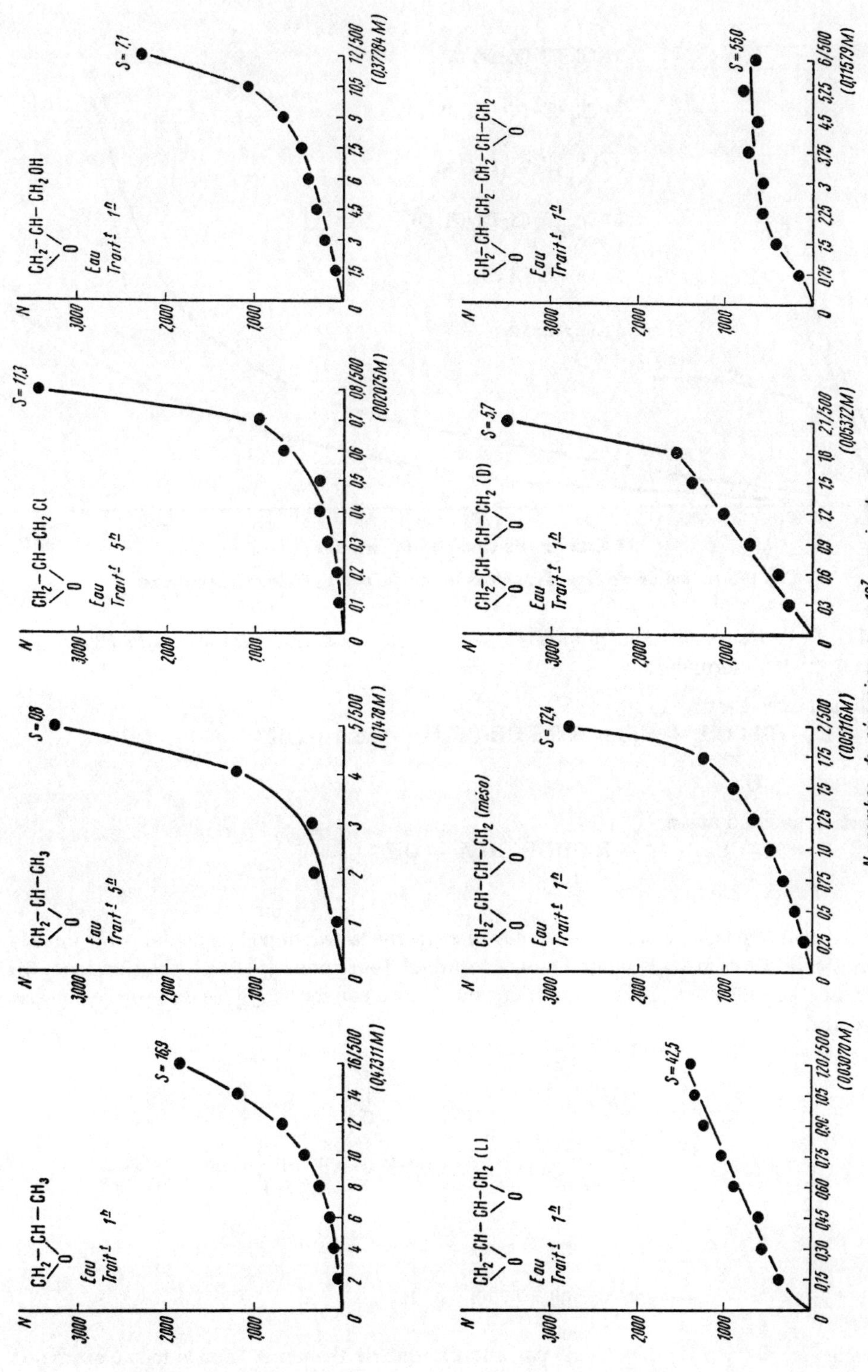

Fig. 16. Induction de réversions dans la souche arg-1; action des époxydes

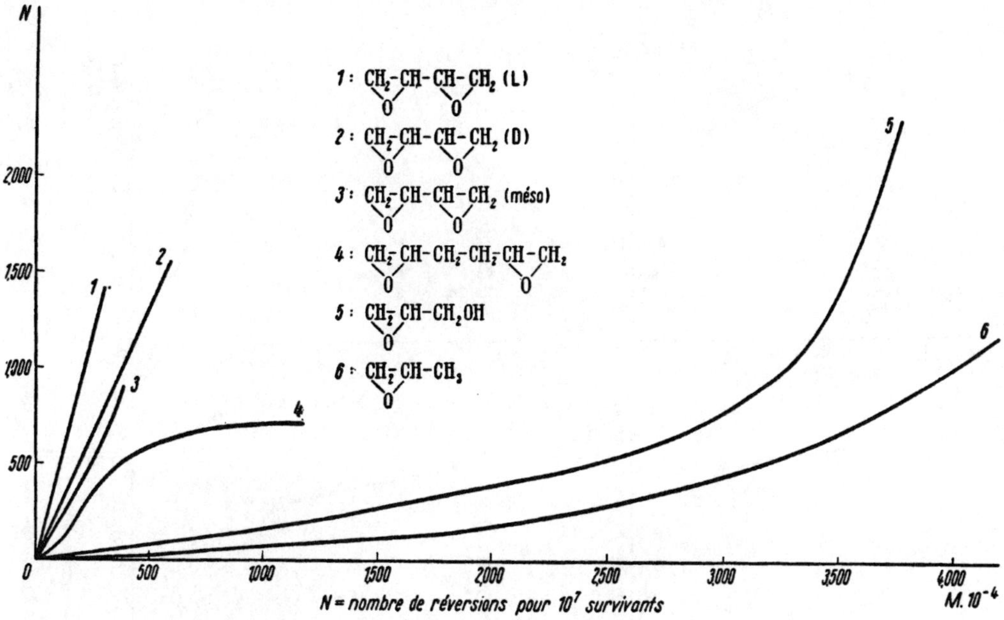

Fig. 17. Induction de réversions dans la souche arg-1; action des époxydes

Ross [31]. Ces auteurs ont montré qu'il s'agissait d'un mécanisme bimoléculaire (S_N 2) qui, à pH = 7, peut se formuler:

(1) Réaction avec l'eau:

$$H_2O + R-\underset{\diagdown O \diagup}{CH-CH_2} + H_2O \rightarrow R\,CH\,OH\,CH_2OH + H^+ + OH^-$$

(2) Réaction avec un anion:

$$H_2O + R-\underset{\diagdown O \diagup}{CH-CH_2} + A^- \rightarrow R\,CHOH\,CH_2A + OH^-$$

Ross a montré de façon concluante, en ce qui concerne la réaction [1], que l'attaque initiale d'une molécule d'eau avait lieu sur l'atome terminal de carbone (polarisé positivement) du cycle oxirane, et était suivie de l'attaque d'une molécule d'eau sur l'oxygène de ce cycle polarisé négativement:

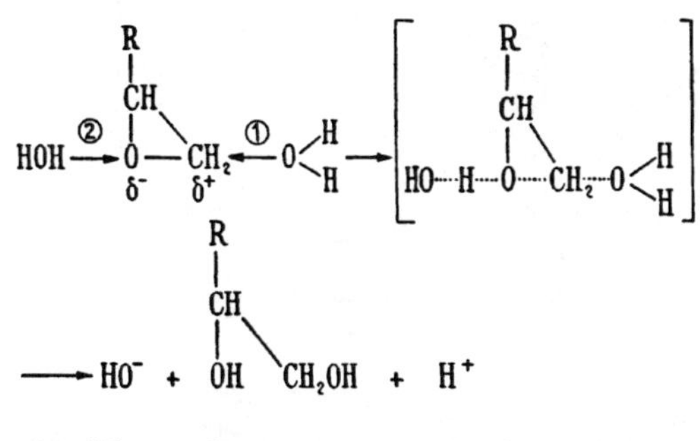

De même, la réaction (2) débuterait par une attaque de l'anion A^- sur le même atome de carbone, porteur d'une charge positive partielle:

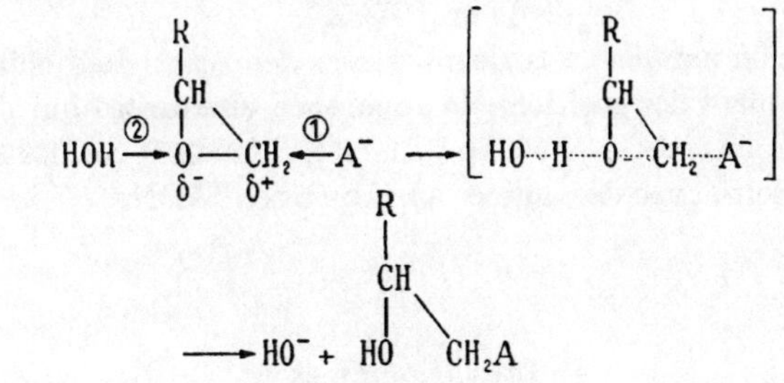

Dès lors, pour un époxyde R—CH—CH$_2$ (avec pont O entre CH et CH$_2$), on peut prévoir que les substituants R qui attirent les électrons vont avoir pour effet d'augmenter la charge positive partielle sur l'atome de carbone terminal et de diminuer la charge négative sur l'oxygène. Ces substituants vont donc favoriser les réactions [1] et [2].

Au contraire, les substituants R qui repoussent les électrons diminuent la charge positive sur C et par conséquent entravent les réactions [1] et [2].

Ross [31] a mesuré la réactivité des époxydes par plusieurs procédés: simple hydrolyse de l'eaŭ, ou réaction avec l'ion thiosulfate (dont le facteur de compétition est très élevé). Il a trouvé les valeurs suivantes des constantes de vitesse pour la réaction avec le thiosulfate:

Composé	K_{thio} [min^{-1}]
Epichlorhydrine	1,30
Glycidol	1,15
1-2, 3-4-Diépoxybutane	0,85
1-2, 5-6-Diépoxyhexane	0,40
Oxyde de propylène	0,36

Ces résultats sont en bon accord avec les prévisions. En effet, pour les monoépoxydes, les substituants R peuvent se classer en:

—CH$_2$Cl
—CH$_2$OH } attirent les électrons et augmentent la réactivité

— CH$_3$ repousse les électrons et diminue la réactivité.

C'est bien ce que l'on observe.

La situation est plus compliquée pour les diépoxydes. Le 1-2, 3-4-diépoxybutane doit sans doute sa haute réactivité au fait que la réaction du premier groupe époxyde donne naissance à un composé de structure voisine de celle du glycidol: CH$_2$ —CH CHOH CH$_2$A (avec pont O entre CH$_2$ et CH). Le deuxième cycle réagira alors avec plus d'énergie. Au total, la vitesse de réaction est plus grande.

En ce qui concerne le 1-2, 5-6-diépoxyhexane, il diffère du précédent par l'adjonction de deux carbones entre les groupes époxydes, ce qui a pour effet de repousser les électrons vers les carbones terminaux en diminuant la réactivité.

Si nous réexaminons maintenant la classification des époxydes par activité mutagène décroissante, à laquelle nous avons abouti (fig. 17), nous constatons qu'effectivement, pour les monoépoxydes nous avons:

Epichlorhydrine > Glycidol > Oxyde de propylène,
c'est à dire une diminution parallèle à la décroissance des réactivités chimiques. Les diépoxydes, par contre, occupent des positions en apparence aberrantes qui demeurent pour le moment inexpliquées. Pour pousser plus loin l'analyse, il faudrait étudier la réaction des époxydes avec d'autres anions, avec des amines ainsi qu'avec l'ADN.

e) Ethylène-imines

Nous avons étudié les dérivés substitués suivants de l'éthylène-imine:

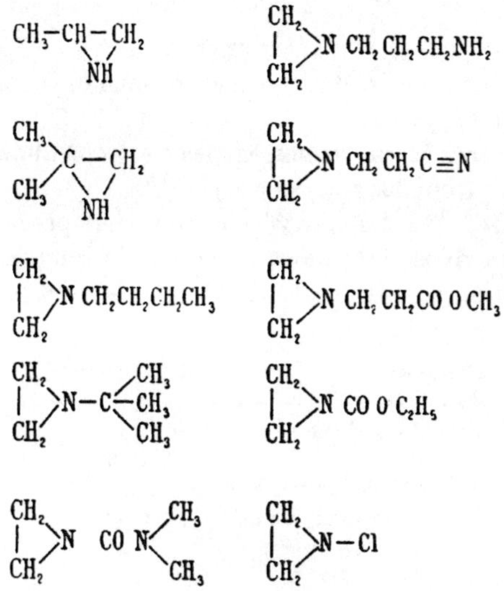

en utilisant la souche arg-1, avec un temps de traitement de 1 heure. Les résultats sont indiqués dans les fig. 18 et 19. Pour comparer les activités mutagènes les concentrations moléculaires ont été calculées (fig. 20) et on a trouvé que les composés se classaient (de gauche à droite) dans l'ordre suivant des activités mutagènes décroissantes:

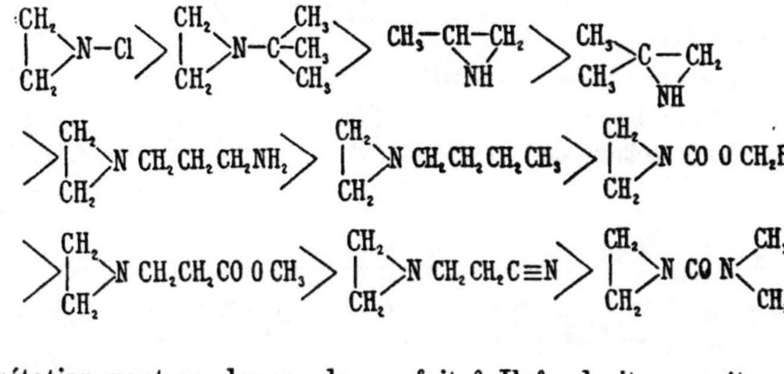

Quelle interprétation peut-on donner de ces faits? Il faudrait connaître la réactivité des précédentes substances à l'égard de l'eau, des anions, des amines. Malheureusement, très peu de travaux ont été jusqu'ici consacrés à l'étude des réactions des éthylène-imines en solution aqueuse.

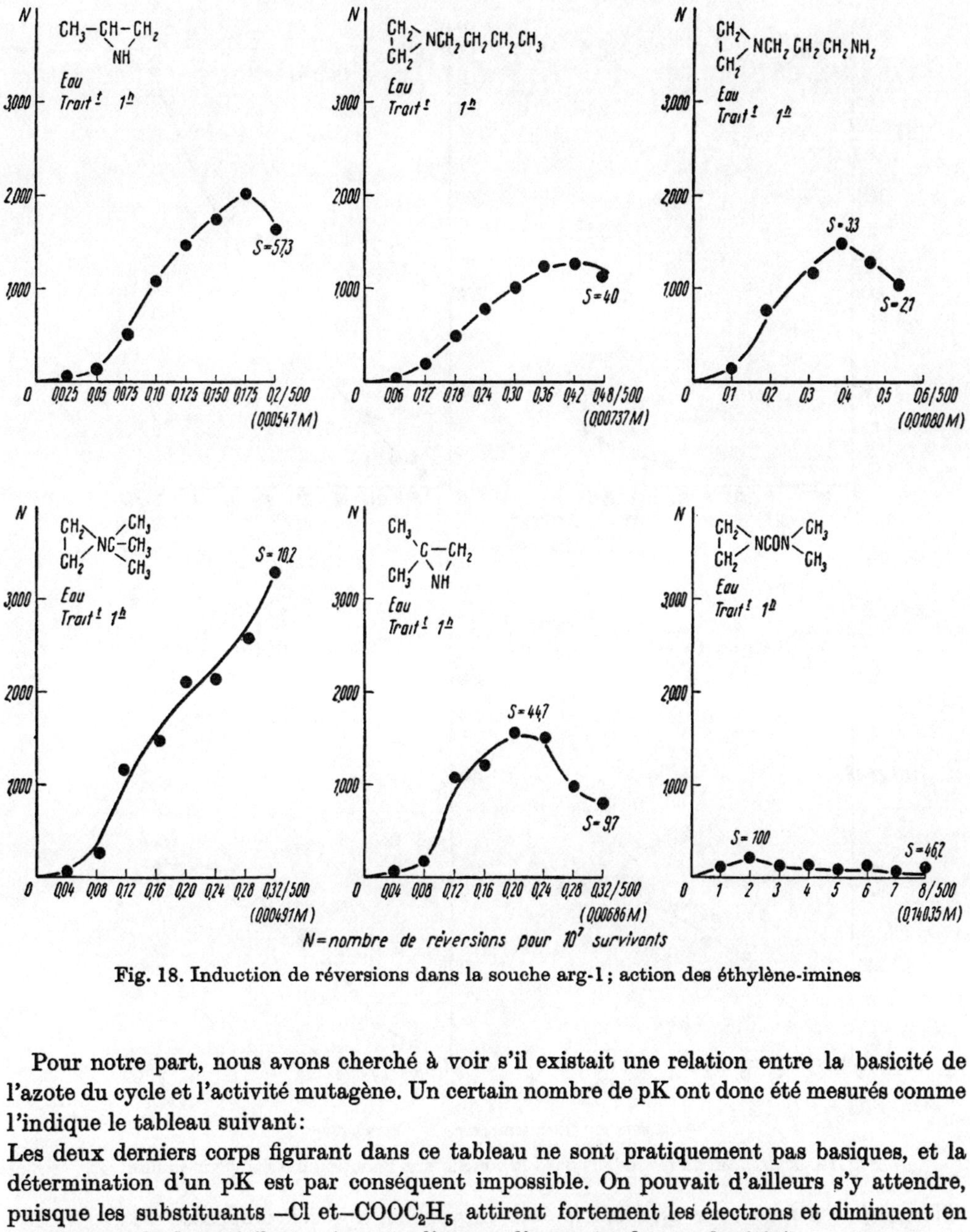

Fig. 18. Induction de réversions dans la souche arg-1 ; action des éthylène-imines

Pour notre part, nous avons cherché à voir s'il existait une relation entre la basicité de l'azote du cycle et l'activité mutagène. Un certain nombre de pK ont donc été mesurés comme l'indique le tableau suivant:
Les deux derniers corps figurant dans ce tableau ne sont pratiquement pas basiques, et la détermination d'un pK est par conséquent impossible. On pouvait d'ailleurs s'y attendre, puisque les substituants –Cl et–$COOC_2H_5$ attirent fortement les électrons et diminuent en conséquence la densité électronique sur l'atome d'azote (et donc sa basicité).

Cependant, la comparaison des pK et de l'activité mutagène ne révèle pas de corrélation évidente. Ainsi, dans la fig. 20 les corps 1 et 7 ont des activités très différentes. Par contre, on observe que les corps 2, 3, 4, 6, dont les substituants repoussent les électrons, ont des activités mutagènes du même ordre.

En résumé, aucune conclusion valable ne peut être proposée tant que des études de cinétique n'auront pas été effectuées, la réactivité chimique pouvant fort bien n'avoir aucun rapport direct avec la basicité de d'azote du cycle.

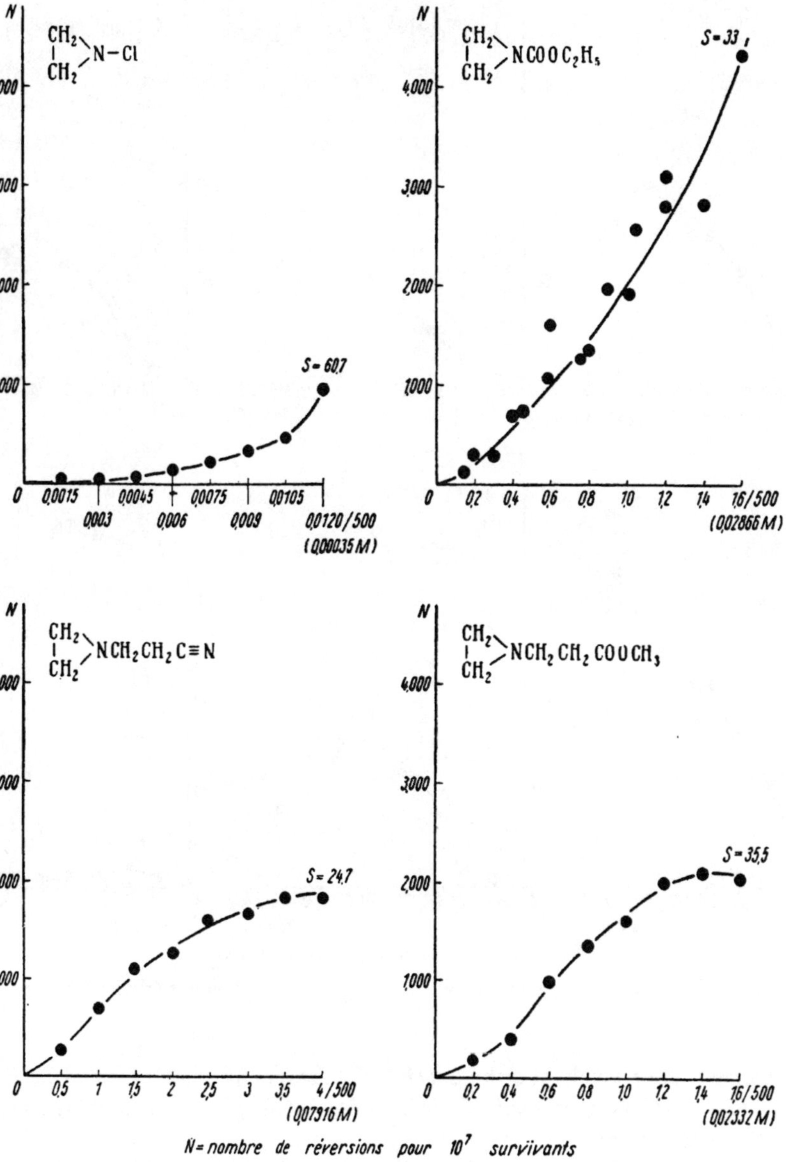

Fig. 19. Induction de réversions dans la souche arg-1; action des éthylène-imines

Néanmoins, il faut noter qu'il existe deux moyens en apparence opposés d'accroître la capacité d'alkylation de l'éthylène-imine:

1. soit augmenter la basicité de l'azote du cycle en plaçant des substituants qui repoussent les électrons (exemple: tert-butyl). Aux pH physiologiques, l'azote fixe alors un proton pour donner un ion éthylène-imonium.

2. soit diminuer la basicité de l'azote du cycle en plaçant des substituants qui attirent fortement les électrons (exemple: chlore).

On ne doit donc pas s'étonner de ce que les corps 1 et 2 (fig. 20) aient des réactivités (et partant des activités mutagènes) du même ordre.

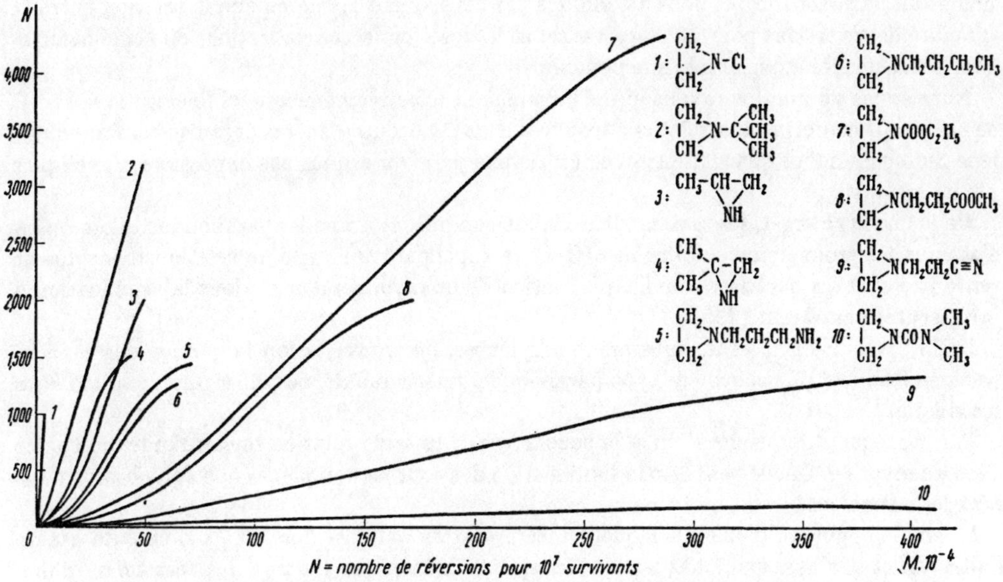

Fig. 20. Induction de réversions dans la souche arg-1; action des éthylène-imines

Composé	pK	pH de la solution aqueuse au départ (0,06 N)
$(CH_3)_2C{-}CH_2$ (cycle NH)	8,8	10,6
$CH_2{-}CH_2$ (cycle NH)	8,1	10,5
$(CH_2)_2$>N $CH_2\,CH_2\,CH_2\,NH_2$	$pK_1 = 10{,}5\ (-NH_2)$; $pK_2 = 6{,}4$ (N cycle)	11,5
$(CH_2)_2$>N $CH_2\,CH_2\,C \equiv N$	5,3	9,3
$(CH_2)_2$>N CO N<$(CH_3)_2$	$pK_1 = 9{,}4\ (-N(CH_3)_2)$; pK_2 (N cycle) non mesurable	11
$(CH_2)_2$>N CO OC_2H_5	non mesurable	8
$(CH_2)_2$>N − Cl	non mesurable	6

6. *Discussion et conclusions*

Schizosaccharomyces pombe paraît être, à première vue, bien adapté à des études quantitatives sur les mutations induites. Il s'agit en effet d'un organisme unicellulaire, uninucléé et haploïde. Nous avons pris, en plus, la précaution de dissocier les amas cellulaires par traitement aux ultra-sons. Nous nous attendions par conséquent [14] à ce que la relation entre N (nombre de réversions pour 10^7 survivants) et la dose (ou la concentration) de l'agent mutagène suive une loi aussi simple que possible.

Nous avons vu que les rayons γ (Co^{60}) donnaient effectivement une loi linéaire avec arg-1, ce qui paraît conforme aux relations trouvées chez beaucoup d'autres organismes. Par contre leuc-3 donne une loi quadratique avec les rayons γ, ce qu'il nous est impossible d'expliquer pour l'instant.

De même avec arg-1, les rayons ultra-violets conduisent à une loi parabolique, alors qu'on s'attendait à trouver une loi linéaire. C'est là cependant un type de relation très souvent rencontré chez les bactéries soumises à l'action de ce rayonnement et dont la signification a été discutée par WITKIN [39].

Enfin, avec $NaNO_2$ et les substances alkylantes, on trouve, selon la gamme des concentrations étudiées, une courbe de type parabolique, une sigmoïde, ou une sigmoïde suivie d'un maximum.

Par conséquent, en se limitant à la souche arg-1, la seule relation linéaire obtenue l'a été avec les rayons γ. Quelle peut être la signification des courbes complexes que l'on observe dans tous les autres cas?

Il semble opportun d'examiner quelle interprétation WITKIN (loc. cit.) a donné du cas de l'ultra-violet. On sait que, chez les bactéries, l'irradiation par l'ultra-violet est suivie d'une période sensible pendant laquelle on peut modifier les conséquences génétiques du traitement. Par des manipulations appropriées au cours de cette période, on peut modifier (en plus ou en moins), dans des proportions considérables, la fréquence des mutations induites.

WITKIN a montré que si l'on irradie une souche de bactéries auxotrophes, la fréquence des mutants prototrophes est proportionnelle à la quantité de protéines synthétisées au cours de la période sensible et, d'autre part, la durée de la période sensible est proportionnelle à la dose d'ultra-violet.

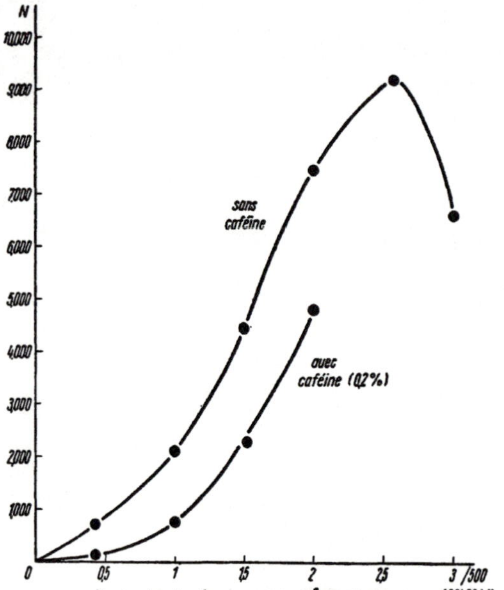

Fig. 21. Action mutagène du sulfate d'éthyle sur arg-1; influence de la caféine

En conséquence, si l'on double la dose, on double le nombre des «prémutants» (accroissement linéaire du nombre de coups conformément à la théorie de la cible), mais on double aussi le temps pendant lequel la synthèse des protéines (peu affectée par les faibles doses d'ultra-violet) se poursuit. Au total donc, un doublement de la dose multiplie par quatre environ la fréquence des prototrophes parmi les bactéries survivantes. Ainsi se trouve expliquée la loi quadratique rencontrée à la place de la relation linéaire à laquelle on s'attendait *a priori*.

WITKIN a aussi montré que si l'on ajoute de la caféine au milieu de culture, il se produit un retard dans la division cellulaire qui est proportionnel à la concentration de caféine. En ensemençant des bactéries irradiées par l'ultra-violet sur un milieu contenant de la caféine (en concentration non toxique aussi élevée que possible, soit 0,1%), la période sensible est accrue de façon importante après de faibles doses d'ultra-violet, ceci ayant pour conséquence d'augmenter la fréquence des mutants à ces doses. On pouvait donc s'attendre à ce que la relation mutation-dose, ordinairement quadratique, devienne linéaire. C'est bien ce que l'expérience, réalisée par WITKIN, a confirmé.

Ces faits peuvent-ils nous aider à interpréter les résultats obtenus avec *Schizosaccharomyces pombe*? Pour ce qui est de l'ultra-violet, nous n'avons pas encore utilisé de milieu contenant de la caféine, mais nous avons recherché si un tel milieu permettait de modifier la relation observée chez arg-1 après traitement par le sulfate neutre d'éthyle. Dans une expérience comparative (fig. 21), nous avons utilisé soit les milieux de culture habituels, soit ces mêmes milieux additionnés de caféine à 0,2%. On voit sur le graphique que la fréquence de mutation est diminuée en présence de caféine mais que la courbe conserve son caractère quadratique.

Nous avons aussi recherché si la composition du milieu minimum avait une influence. Ce milieu comporte, entre autres éléments:

$$\text{glucose} - 1\%,$$
$$SO_4\,(NH_4)_2 - 0{,}5\%.$$

Nous avons utilisé des milieux contenant jusqu'à 20% (x 20) de glucose sans que le taux de mutation et la forme de la courbe soit altérés. Nous avons de même accru la teneur en sulfate d'ammoniaque jusqu'à 2,5% (x 5) et observé, dans ce cas, une augmentation notable du taux de mutation mais, à nouveau, la forme de la loi ne varie pas.

D'autre part, nous avons recherché si le degré de prolifération cellulaire avait une influence. Avec la souche arg-1, nous avons donc traité par le sulfate neutre d'éthyle des cellules ayant poussé sur milieu complet à 30 °C pendant 3, 4, 5, 6, 7, 8, 9 jours. Dans les cellules agées de 9 jours, il est certain qu'il n'y a plus de multiplication. Cependant, nous n'avons observé aucune influence sur la forme de la loi.

Enfin, nous avons obtenu par croisement une souche homothallique arg-1 h^{90} qui sporule spontanément. Nous avons préparé des suspensions d'ascospores libres (qu'on sait être uninucléées) et, après avoir détruit par un traitement à l'alcool les cellules végétatives, nous avons réalisé un traitement par le sulfate neutre d'éthyle. Une fois de plus, la forme de la loi n'a pas été altérée.

Un autre phénomène singulier est le maximum que l'on observe pour N dans certaines expériences: à hautes concentrations (le temps de traitement restant fixe) on obtient moins de mutations qu'à des concentrations plus faibles.

Nous avons cherché à établir s'il s'agissait d'un phénomène de compétition: on peut en effet supposer que les mutants prototrophes induits sont plus sensibles à l'agent mutagène que la souche arg-1 de départ et sont tués sélectivement aux hautes concentrations.

Nous avons donc isolé douze mutants prototrophes induits par le sulfate d'éthyle et nous avons réalisé un mélange comportant 50% de cellules arg-1 et 50% de prototrophes. Ce mélange à été traité par des concentrations croissantes de sulfate d' éthyle (jusqu' à 2,4/500 pendant 1 heure). On observe effectivement, aux concentrations les plus élevées, une faible diminution du pourcentage de prototrophes parmi les cellules survivantes.

Une expérience de même type a été faite avec l'éthylène-imine

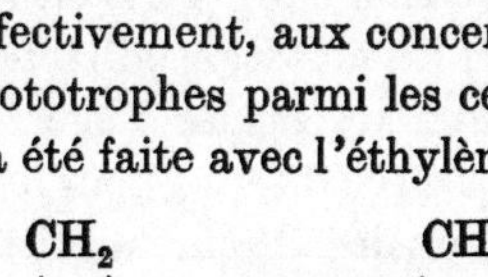

(concentration jusqu'à 8/500 pendant 1 heure). Cette fois, on constate une diminution notable du pourcentage de prototrophes aux hautes concentrations.

Il existe donc une contre-sélection des prototrophes; mais celle-ci ne paraît pas suffisamment marquée pour expliquer la présence d'un maximum dans la courbe de variation de N. Au surplus, au cours du traitement, on a affaire à des «prémutations» et non à des mutations prototrophiques établies. En effet, selon toute vraisemblance, la mutation n'est «fixée» qu'au moment de la première réplication de l'ADN, la présence de bases puriques et pyrimidiques anormales pouvant se traduire alors par des altérations dans leur séquence. Il y aurait donc lieu d'étudier, par une méthode qui reste à préciser, la contre-sélection éventuelle des «prémutations» par les agents mutagènes responsables de leur induction.

Avec le sulfate d'éthyle, qui s'hydrolyse dans l'eau avec production d'acide, le pH de la solution au bout d'une heure est évidemment beaucoup plus bas aux concentrations élevées. Nous nous sommes demandé si ces inégalités de pH expliquaient l'existence d'un maximum pour N. Au lieu de dissoudre le sulfate d'éthyle dans l'eau, nous avons employé un tampon à pH $= 7{,}0$ (Tris + HCl) suffisamment concentré pour que le pH reste fixé à 7 pendant toute la durée de l'expérience et pour toutes les concentrations de l'agent mutagène. Les résultats d'une telle expérience sont indiqués sur la fig. 10. Si les valeurs trouvées pour N sont plus faibles, la courbe conserve néanmoins sa forme générale sigmoïde et tout semble indiquer qu' un maximum serait atteint à des concentrations plus élevées (non réalisables pour des raisons de solubilité).

En conclusion, la signification exacte de la relation entre N (nombre de réversions pour 10^7 cellules survivantes) et la concentration des agents mutagènes demeure pour l'instant inexpliquée. De nouvelles expériences sont nécessaires.

Dans tout ce qui précède, nous avons utilisé le terme réversion pour désigner les cellules mutantes capables de former des colonies sur le milieu minimum. Nous n'avons pas encore précisé s'il s'agissait de mutations réverses vraies ou de la mutation de gènes suppresseurs.

Un certain nombre de réversions ont été isolées dans la souche arg-1, qu'il s'agisse de réversions spontanées ou de réversions induites par divers agents mutagènes. Ces réversions de type h^-, ont été croisées avec la souche sauvage h^+ et la descendance a été analysée en utilisant des suspensions d'ascospores libres (les cellules végétatives étant tuées par l'alcool). Les colonies obtenues sur milieu complet ont fait ensuite l'objet d'une copie sur milieu minimum, ce qui permet de distinguer les prototrophes des auxotrophes.

Dans le cas où il s'agit d'une véritable mutation réverse, la descendance ne doit contenir que des prototrophes. Si au contraire, il s'agit d'un gène suppresseur, on observera une ségrégation de:

25% d'auxotrophes: pour un gène suppresseur non lié à arg-1

$<$ 25% d'auxotrophes: pour un gène lié à arg-1.

L'expérience indique, dans le cas présent, que la mutation réverse vraie de arg-1 – si elle existe – est rare, et que les réversions correspondent à plusieurs gènes suppresseurs, les uns indépendants, les autres plus ou moins étroitement liés à arg-1.

Il en résulte que le système de réversion d'arg-1 est en fait, non un système de mutation réverse, mais un système polygénique de mutation «directe» de suppresseurs.

On comprend dès lors pourquoi, dans nos études d'agents mutagènes, la spécificité paraît être faible ou nulle. Il n'est donc pas surprenant que, pour les esters sulfuriques et sulfoniques, les β-halogéno-éthyle amines et les époxydes, il semble exister une corrélation entre l'activité mutagène et la réactivité chimique mesurée in vitro.

Pour l'exécution de ces recherches, nous avons bénéficié de l'aide de l'Institut National de la Recherche Agronomique et du Commissariat à l'Énergie Atomique (Département de Biologie).

Les agents mutagènes chimiques ont été synthétisés par MM. Lévy et Monard au Service des Poudres (esters sulfoniques et sulfuriques, β-halogéno-éthyle amines) et par M. Tempé (la plupart des éthylène-imines).

Nous devons les sultones à l'obligeance de M. Manecke (Fritz-Haber-Institut, Berlin) et de M. Heyden (Henkel, Düsseldorf).

Les trois isomères de diépoxybutane nous ont été obligeamment fournis par «Löwen Chemische Fabrik» de Copenhague, par l'intermédiaire du Prof. Westergaard.

M. Feunteun, au laboratoire de chimie de l'Institut National Agronomique, a eu l'obligeance d'effectuer la détermination des pK des éthylène-imines et de mesurer la vitesse d'hydrolyse des esters sulfoniques et sulfuriques et des sultones.

Bibliographie

[1] Brönsted, J. N., 1929: Kinetic studies on ethylene-oxides. J. Amer. chem. Soc. **51**, 428–461.

[2] Cohen, B., et al., 1948: Reaction kinetics of aliphatic tertiary β-chloroethylamines in dilute aqueous solutions. I. The cyclization process. J. Amer. chem. Soc. **70**, 281–285.

[3] Davidson, J. N., 1959: The biochemistry of the nucleic acids. Methuen, 288 p.

[4] Dubinin, N. P., B. N. Sidorov et N. N. Sokolov, 1959: The genetic effect of free radicals. C. R. Acad. Sci. URSS 124–129.

[5] Gierer, A., and K. W. Mundry, 1958: Production of mutants of tobacco mosaic virus by chemical alteration of its ribonucleic acid in vitro. Nature (London) 4647, 1457–1458.

[6] Grigg, G. W., 1957: Competitive suppression and the detection of mutations in microbial populations. Austral. J. Biol. Sci. **11**, No. 1, 69–84.

[7] Haber, F., and J. Weiss, 1934: The catalytic decomposition of hydrogen peroxide by iron salts. Proc. Roy. Soc. (London), Ser. A, **147**, 332–351.

[8] Hanby, W. E., et al., 1947: The chemistry of 2-chloroalkyl-amines II. J. chem. Soc. (London) 519–527.

[9] Helberger, J. H., G. Manecke und H. M. Fischer, 1948: Zur Kenntnis organischer Sulfonsäure. II. Mitteilung. Liebigs Ann. Chem. **561**, 23–35.

[10] Helberger, J. H., G. Manecke und R. Heyden, 1949: Zur Kenntnis organischer Sulfonsäure. III. Mitteilung. Liebigs Ann. Chem. **565**, 22–35.

[11] Helberger, J. H., R. Heyden und H. Winter, 1954: Zur Kenntnis organischer Sulfonsäure. IV. Mitteilung. Liebigs Ann. Chem. **586**, 147–157.

[12] Helberger, J. H., und H. Lautermann, 1954: Zur Kenntnis organischer Sulfonsäure. V. Mitteilung. Liebigs Ann. Chem. **586**, 158–164.

[13] Helberger, J. H., 1954: Zur Kenntnis organischer Sulfonsäure. VI. Mitteilung. Liebigs Ann. Chem. **588**, 71–78.

[14] Heslot, H., 1960: *Schizosaccharomyces pombe*: un nouvel organisme pour l'étude de la mutagénèse chimique. Abhandl. Dtsch. Akad. Wiss. Berlin, Klasse für Medizin, **1**, 98–105.

[15] Kaudewitz, F., 1959: Production of bacterial mutants with nitrous acid. Nature (London) 4678, 1829–1830.

[16] Kaudewitz, F., 1959: Inaktivierende und mutagene Wirkung salpetriger Säure auf Zellen von *Escherichia coli*. Z. Naturforsch. **14b**, 528–537.

[17] Kaudewitz, F., 1960: Inaktivierung und Mutagenese von *Escherichia coli* durch Inkubation mit Natriumnitrit. Abhandl. Dtsch. Akad. Wiss. Berlin, Klasse für Medizin, **1**, 86–91.

[18] Kølmark, G., and M. Westergaard, 1949: Induced back-mutation in a specific gene of *Neurospora crassa*. Hereditas **35**, 490–506.

[19] Kølmark, G., and M. Westergaard, 1953: Further studies on chemically induced reversions at the adenine locus of *Neurospora*. Hereditas **39**, 209–224.

[20] Kølmark, G., 1953: Differential response to mutagens as studied by the *Neurospora* reverse mutation test. Hereditas **39**, 270–276.

[21] Kølmark, G., 1955: Comparative studies of monoepoxides as inducers of reverse mutations in *Neurospora*. Genetics **40**, 890–902.

[22] Kølmark, G., 1956: Mutagenic properties of certain esters of inorganic acids investigated by the *Neurospora* back-mutation test. C. R. Laborat. Carlsberg, Sér. Physiol., **26**, 204–220.

[23] Leupold, U., 1950: Die Vererbung von Homothallie und Heterothallie bei *Schizosaccharomyces pombe*. C. R. Laborat. Carlsberg, Sér. Physiol., **24**, 381–480.

[24] Leupold, U., 1955: Methodisches zur Genetik von *Schizosaccharomyces pombe*. Schweiz. Z. allg. Path. **18**, 1141–1146.

[25] Leupold, U., 1955: Versuche zur genetischen Klassifizierung adenin-abhängiger Mutanten von *Schizosaccharomyces pombe*. Arch. Julius-Klaus-Stift. (Zürich) **30**, 506–516.

[26] Leupold, U., 1957: Physiologische Studien an adenin-abhängigen Mutanten von *Schizosaccharomyces pombe*. Ein Beitrag zum Problem der Pseudoallelie. Schweiz. Z. allg. Path. **20**, 535–544.

[27] Leupold, U., 1958: Studies on recombination in *Schizosaccharomyces pombe*. Cold Spring Harbor Symposia on Quant. Biol. **23**, 161–170.

[28] Loveless, A., and S. Howarth, 1959: Mutation of bacteria at high levels of survival by ethyl methane sulfonate. Nature (London) **184**, 1780–1782.

[29] Mundry, K. W., und A. Gierer, 1958: Die Erzeugung von Mutationen des Tabakmosaikvirus durch chemische Behandlung seiner Nucleinsäure in vitro. Z. Vererbungslehre **89**, 614–630.

[30] Rapoport, I. A., 1947: Inheritance changes taking place under the influence of diethyl sulfate and dimethyl sulfate. Doklady Akad. Nauk **12**, No. 10, 12–15.

[31] Ross, W. C. J., 1950: The reactions of certain epoxides in aqueous solutions. J. chem. Soc. (London), 2257–2272.

[32] Ross, W. C. J., 1953: The chemistry of cytotoxic alkylating agents. Advances Cancer Res. **1**, 397–449.

[33] Ross, W. C. J., 1958: In vitro reactions of biological alkylating agents. Ann. N. Y. Acad. Sci. **68**, 669–681.

[34] Schuster, H., und G. Schramm, 1958: Bestimmung der biologisch wirksamen Einheit in der Ribosenucleinsäure des Tabakmosaikvirus auf chemischem Wege. Z. Naturforsch. **13**b, 697–704.

[35] Schuster, H., A. Gierer, und K. W. Mundry, 1960: Inaktivierende und mutagene Wirkung der chemischen Veränderung von Nucleotiden in Virus-Nucleinsäure. Abhandl. Dtsch. Akad. Wiss. Berlin, Klasse für Medizin, **1**, 76–85.

[36] Vielmetter, W., und C. M. Wieder, 1959: Mutagene und inaktivierende Wirkung salpetriger Säure auf freie Partikel des Phagen T2. Z. Naturforsch. **14b**, 312–317.

[37] Vielmetter, W., H. Schuster und C. M. Wieder, 1960: Inaktivierende und mutagene Wirkung salpetriger Säure auf Phagen. Abhandl. Dtsch. Akad. Wiss. Berlin, Klasse für Medizin, **1**, 92–97.

[38] Westergaard, M., 1957: Chemical mutagenesis in relation to the concept of the gene. Experientia **13**, 224–234.

[39] Witkin, E. M., 1959: Post-irradiation metabolism in bacteria. Proc. X^{th} Internat. Congr. Genetics **1**, 280–299.

Radiation-Induced Paramagnetic Centers in Plant Seeds at Different Oxygen Concentrations

A. Ehrenberg, L. Ehrenberg, G. Löfroth

Medicinska Nobelinstitutet, Karolinska institutet;
Institutet för organisk kemi och biokemi, Stockholms Universitet, Stockholm

In radiobiologic experiments with plant seeds such modifying factors as water content and oxygen concentration affect in the same way different types of lesions, e. g., gene mutation, chromosomal aberrations, growth inhibition, and lethal effects. Efforts are being made by means of electron spin resonance (ESR) to study the role of free radicals in the production of the biologic effects (Ehrenberg, A., and Ehrenberg, L. 1958; Sparrman, Ehrenberg, L., and Ehrenberg, A., 1959). Such experiments are partly performed with tissue, partly with simple model systems. The protective action of NO in dry seeds seems to prove that free radicals do play an important role in the production of, e. g., mutation (Ehrenberg, L., 1960; Ehrenberg, A., 1961; Sparrman, Ehrenberg, L., and Ehrenberg, A., 1959). In the following some data will be given to show the possibilities of studying the radiobiologic effects of oxygen by means of ESR spectroscopy.

When seeds of barley or *Agrostis stolonifera* are irradiated with γ-rays, the number of free radicals, measured by means of electron spin resonance, increases linearly with the dose up to about 300 krad. About 40 eV are consumed per radical formed.

When seeds are irradiated in air, a simple narrow absorption line appears at low doses but above about 40 krad a superimposed broader signal becomes more and more evident (see Fig. 1). When measurements are performed at $-80\,°C$ additional lines appear.

When seeds are irradiated at a higher oxygen concentration, obtained for example by pretreatment with oxygen at 80 atm. only signals of the narrow type appear, the amplitude increasing with increasing dose (see Fig. 1).

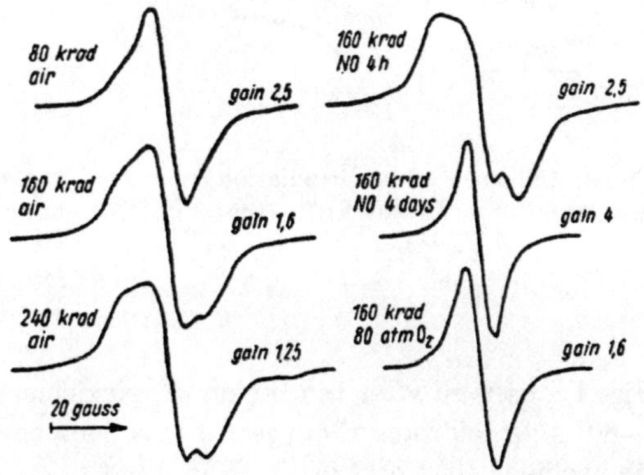

Fig. 1. ESR spectra of seeds irradiated under different conditions. To the left: different doses of γ-radiation on seeds in equilibrium with air. To the right, from the top: "NO 4 h" means pretreatment with NO for 4 hours, NO being removed immediately after irradiation; "NO 4 days" means that NO is not removed until after 4 days storage at room temperature (this strongly decreases the total concentration of free radicals, cf. the gain used and also Fig. 2; 80 atm. of O_2 were applied 12 hours before irradiation; this treatment strongly enhances the radation effect)

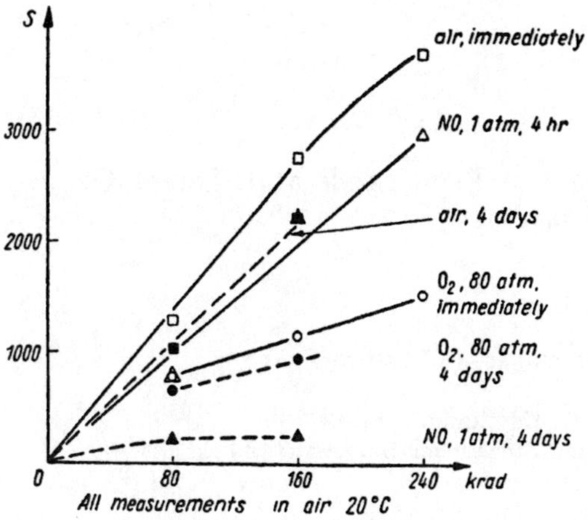

Fig. 2. Dose dependance of integrated signal obtained after irradiation and storage under different conditions (cf. text of Fig. 1)

One possible explanation is that the radicals giving the narrow signal are of a special chemical type, less sensitive to oxygen than the radicals giving the broader signal, which starts to appear when the intracellular oxygen has been consumed. During storage in air the broader signal decays more rapidly than the narrow one. The decay is probably an effect of the diffusion of oxygen into the seeds.

An alternative explanation that we are presently also considering is that the radicals giving the broader signal through reaction with oxygen are converted into the radicals giving the narrow signal.

When the intracellular oxygen concentration is decreased by evacuation or treatment with nitric oxide before irradiation, the ratio between the narrow and broad signals is decreased, as seen in Fig. 1. During storage of irradiated seeds both types of radicals react with nitric oxide, the one giving the broad signal being the more sensitive. The decay of the signals under different conditions and after different doses of radiation is illustrated in Fig. 2.

It is known from the work of Smith and Caldecott (1948), Caldecott (1961) and others that a heat treatment of dry seeds before irradiation leads to a decrease of their radiosensitivity. After a pretreatment with heat, irradiation induces relatively less of the narrow signal radicals, whereas the total number of radicals is not significantly changed (see Fig. 3). Hence, the protective action of the heat treatment might consist, partly at least, in an expulsion of oxygen.

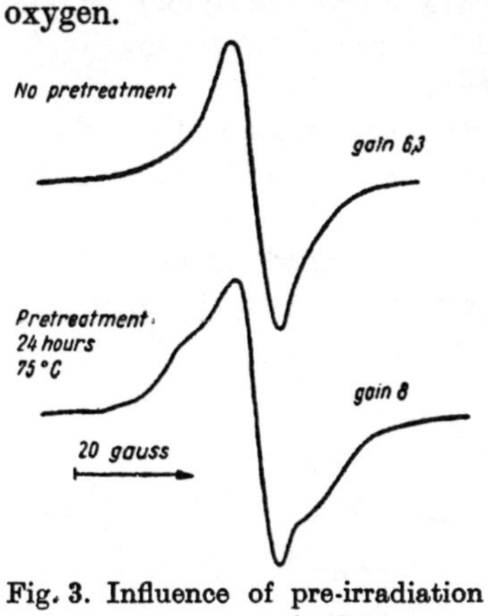

Fig. 3. Influence of pre-irradiation heat treatment on the ESR spectrum

The chemical nature of the radicals is not yet clear. The narrow signal is not of the asymmetric type characteristic of peroxy radicals, shown in

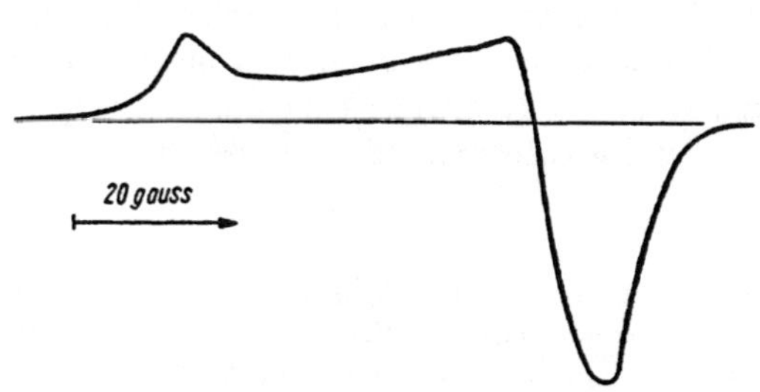

Fig. 4. A typical peroxy radical ESR spectrum, obtained by irradiating (− 80° C, 240 krad), e.g., plexiglas or cholesterol. Compare with earlier publications for the ESR spectra of anoxically irradiated plexiglas (Ingram, 1958) and cholesterol) Ehrenberg, A., Ehrenberg, L., and G. Löfroth, 1960).

Fig. 4, observed after irradiation of plexiglas or cholesterol in the presence of oxygen at −80 °C (or addition of oxygen at this temperature after irradiation). It cannot, however, be excluded that part of the ESR absorption is of the peroxy type.

Integration of the different signals might lead to methods to determine the amount of (free + available) oxygen in the tissue.

Acknowledgement

The investigations have been supported by grants from Knut och Alice Wallenbergs stiftelse, Riksföreningen mot cancer and Statens Råd för atomforskning.

References

CALDECOTT, R. S., 1961: Seedling height, oxygen availability, storage and temperature: their relation to radiation induced genetic and seedling injury in barley. IAEA and FAO Symposium on the Effects of Ionizing Radiations on Seeds, Karlsruhe 1960, JAEA, Vienna, 3—24.

EHRENBERG, A., 1961: Research on free radicals in enzyme chemistry and in radiation biology. In: M. S. BLOIS and al. (Eds.), Free Radicals in Biological Systems. Academic Press, New York, pp. 337—350.

EHRENBERG, A., and L. EHRENBERG, 1958: The decay of X-ray induced free radicals in plant seeds and starch. Arkiv Fysik **14**, 133—141.

EHRENBERG, A., L. EHRENBERG, and G. LÖFROTH, 1960: Nitric oxide titration of free radicals in irradiated materials. Risö Report No. **16**, 21—22.

EHRENBERG, L., 1960: Induced mutation in plants: Mechanisms and principles. Genet. agrar. **12** 364—389.

INGRAM, D. J. E., 1958: Free radicals as studied by electron spin resonance. Butterworths, London, see p. 198 and quoted papers.

SMITH, L., and R. S. CADLECOTT, 1948: Modification of X-ray effects on barley seeds by pre- and posttreatment with heat. J. Heredity **39**, 173—176.

SPARRMAN, B., L. EHRENBERG, and A. EHRENBERG, 1959: Scavenging of free radicals and radiation protection by nitric oxide in plant seeds. Acta Chem. Scand. **13**, 199—200.

HANS STUBBE

Sinn und Bedeutung der Kulturpflanzenforschung

(Vorträge und Schriften der Deutschen Akademie der Wissenschaften zu Berlin, Heft 62)

1957. 21 Seiten – 8° – 1,– DM

In diesem Vortrag auf einer Feier des Leibniz-Tages der Deutschen Akademie der Wissenschaften zu Berlin berichtet der Direktor des Instituts für Kulturpflanzenforschung der Deutschen Akademie der Wissenschaften und Präsident der Deutschen Akademie der Landwirtschaftswissenschaften in allgemein verständlicher Form über die vielseitigen Verflechtungen der Kulturpflanzenforschung. Sie hat entscheidenden Einfluß auf die Menschheitsgeschichte und in der Gegenwart die Aufgabe, daran mitzuarbeiten, daß die ständig wachsende Bevölkerung der Erde sich ernähren und in Glück und Wohlstand leben kann.

Wenn auch die Kulturpflanzenforschung nicht mit so umwälzenden Ergebnissen wie andere Zweige der Naturwissenschaften aufwarten kann, so strahlt sie doch nahezu in alle Bereiche des menschlichen Daseins, selbst in das Gefüge der Weltpolitik aus, wie der Verfasser in eindrucksvoller Weise schildert.

Einige Ergebnisse der Mutationsforschung an Kulturpflanzen

(Sitzungsberichte der Deutschen Akademie der Wissenschaften zu Berlin, Klasse für Medizin, Jahrgang 1959, Nr. 1)

1959. 32 Seiten – 8 Text- und Tafelabbildungen – 8 Tabellen – 8° – 2,20 DM

Bestellungen durch eine Buchhandlung erbeten

THEORIA CUM PRAXI

A K A D E M I E - V E R L A G · B E R L I N

RICHARD B. GOLDSCHMIDT

Theoretische Genetik

Übersetzung aus dem Englischen

In deutscher Sprache bearbeitet und herausgegeben von Franz Schwanitz

Mit einem Geleitwort von Hans Stubbe

1961. XVI, 546 Seiten – 23 Abbildungen – 6 Tabellen – 1 Titelbild – gr. 8° – Ganzleinen DM 32,–

Drei Fragen sind es, deren Beantwortung uns das Wesen der Vererbungserscheinungen verständlich machen kann:

Was sind die stofflichen Grundlagen der Vererbungsvorgänge?

Wie ist ihre Wirkung?

Welche Bedeutung haben ihre Beschaffenheit und ihre Wirksamkeit für die Evolution des Lebens?

Richard Goldschmidt hat in seinem letzten Werk versucht, aus der Fülle der Tatsachen und Erkenntnisse, welche die experimentelle genetische Forschung in einem halben Jahrhundert zusammengetragen hat, die allgemeinen Gesetzmäßigkeiten herauszuarbeiten und so eine Antwort auf diese Fragen zu finden. Das Ergebnis seiner kritischen Bemühungen ist ein umfassendes Bild unseres heutigen Wissens von den Erscheinungen der Vererbung, das auch für die künftige genetische Forschung von größter Bedeutung sein wird.

HERMANN VON GUTTENBERG

Lehrbuch der allgemeinen Botanik

5., durchgesehene Auflage

1956. XVIII, 708 Seiten – 637 Abbildungen – 7 Tafeln – gr. 8° – Ganzleinen DM 25,– (Vergriffen)

6. neubearbeitete Auflage erscheint Ende 1962

Aus den Besprechungen zur 4. Auflage:

„In nicht ganz 5 Jahren die 4. Auflage! Das ist ein schöner Erfolg für das zu den großen deutschsprachigen Lehrbüchern der Botanik zählende Werk und außerdem eine gute Empfehlung. Das Lehrbuch behandelt bewußt nur die allgemeine Botanik; ein Grundriß der Systematik wird deshalb nicht gebracht.

Im morphologischen Teil, der erfreulich umfangreich gehalten und mit zahlreichen Abbildungen (über 500!) versehen ist, wurde der Stoff wie üblich eingeteilt (Zytologie, Histologie, Organographie und auch Fortpflanzung). Insbesondere wurden die Kapitel der Histologie und Organographie (= Morphologie der äußeren Gestalt) eingehender als sonst üblich behandelt und sind bei den Studenten infolge der didaktisch guten Darstellung geschätzt.

Im zweiten Teil, der Physiologie, ist gleichfalls die Stoffanordnung in der bekannten Weise vorgenommen worden: Physiologie des Stoff-, Form- und Ortswechsels. Im einzelnen sind die Teilgebiete klar dargestellt, jedoch mit etwas unterschiedlicher Ausführlichkeit. Auch kurze Einblicke in die pflanzenphysiologischen Untersuchungsmethoden werden geboten. Neue Forschungsergebnisse werden in vielen Fällen soweit als möglich berücksichtigt. So erfährt etwa der ‚Säurecyclus' eine recht eingehende Behandlung. Die Darstellung der Wuchsstoffphysiologie ist erwartungsgemäß modern ...“

(Naturwissenschaftliche Rundschau, Stuttgart)

Bestellungen durch eine Buchhandlung erbeten

AKADEMIE-VERLAG · BERLIN

Zeitfracht Medien GmbH
Ferdinand-Jühlke-Straße 7
99095 Erfurt, Deutschland
produktsicherheit@kolibri360.de